Science Education: A Global Perspective

Ben Akpan
Editor

Science Education: A Global Perspective

Springer

Editor
Ben Akpan
Science Teachers Association of Nigeria
Abuja, Nigeria

ISBN 978-3-319-32350-3 ISBN 978-3-319-32351-0 (eBook)
DOI 10.1007/978-3-319-32351-0

Library of Congress Control Number: 2016947167

Printed on acid-free paper

This Springer imprint is published by Springer Nature
The registered company is Springer International Publishing AG Switzerland

Preface

This book, *Science Education: A Global Perspective*, is 'global' both in content and authorship. It seeks to bring to the fore current developments in science education and their implications. These developments should make us explore how to improve on the quality of delivery of science education by shaping our opinions. Thus, the ideas expressed in this book are expected to promote discourse and thereby advance the cause of science education in all its ramifications.

The book has 17 chapters. Chapter 1 provides an overview of the entire book. The other 16 chapters are divided into four Parts. Part I, which dwells on the nature of science view, demonstrates (in Chap. 2) the high premium currently being placed on the nature of science especially in relation to argumentation as the basis of providing a solid foundation for future growth and deeper understanding of science concepts. Beyond this, and in a natural sequence, is a consideration of equally argumentative subjects as the relationship between science and religion along with the representation of evolution in science education that is explored in Chaps. 3 and 4. The challenge here demonstrates the dilemma that we continue to face, namely, how to impart science education in ways consistent with the nature of science and at the same time being sensitive to the divergent beliefs and traditional views that students hold. Following this is a vivid alternative strategy in Chap. 5 for the teaching of acidity in a marked departure from some time-honoured approaches.

Part II discusses science and national development with particular references to India, Nigeria, and Southeast Asian countries. In India as discussed in Chap. 6, science education has been shown to serve as a catalyst for growth and national development. Similarly Chap. 7, which is on the Southeast Asian Ministers of Education Organisation Regional Centre for Education in Science and Mathematics (SEAMEO-RECSAM – an organisation located in Penang, Malaysia), shows how a regional organisation has provided the platform for the promotion of cooperation in education, science, and culture. In Nigeria, the Science and Technology Education Post-Basic (STEP-B) project highlighted in Chap. 8 has facilitated an increased production of more and better quality medium and higher-level skilled workers through the strengthening of the capacity of federal post-basic science and technology institutions. The first three chapters in this Part provide the required background for the discussion of the issues of endogenous research and innovation in Chap. 9.

Curriculum and pedagogy in science education are the major considerations in Part III with motivational science teaching using a context-based approach anchored on the 'education through science' movement setting the tone for this Part in Chap. 10. This is further elaborated with specific reference to New Zealand in Chap. 11. The clarion call for a paradigm shift towards a socially responsible science education that caters for learners' priorities and needs thus meeting the requirements of multiculturalism is an extension of this discourse in Chap. 12. So also are the examples and illustrations in Chap. 13 from the Portuguese curriculum which is designed and implemented based on the assumption that teachers transform, reinterpreting those guidelines in the light of their knowledge, experiences, and conceptions about the teaching process. A discussion of the in-service teachers' professional programme in Brazil in Chap. 14 concludes this Part.

Providing equitable access is a step towards reducing barriers to high achievement. However, reform programmes need to consider other typical obstacles to Education for All (EFA): teachers' expectations and instructional models. Even when women complete STEM studies, they are less likely than men to work in these sectors. These are some of the ideas coming through in Chap. 15 under Part IV where development and future studies are discussed. In Chap. 16, the book examines future trends for science education research by using selected works to highlight trends, issues, and the implications for research. The book ends on an optimistic note in Chap. 17 by looking at science education in 50 years' time with a recommendation, among others, for stakeholders to take the responsibility of preparing children towards a blossoming science education sector in an anticipated future world.

I am immensely grateful to all the contributing authors to this book and to Springer International Publishing AG for facilitating its publication.

Abuja, Nigeria
21 January, 2016

Ben Akpan

Contents

List of Figures

List of Tables

Chapter 1
Introduction

Ben Akpan

This book, *Science Education: A Global Perspective*, comprises of four separate but interconnected Parts: nature of science, science and national development, curriculum and pedagogical considerations in science education, and development and future studies in science education. The term "science education" is used in its broadest sense in some sections of the book as it encompasses science and technology and science and technology education wherever the context so demands.

In order to achieve scientific literacy, an understanding of the nature of science (NOS) is required. This is more so because when science is viewed as a way of knowledge, NOS portrays and reinforces that perspective. But in order to understand NOS, argumentation comes into play. Indeed, Kuhn (2010) is of the view that argumentation is one path to knowledge alongside inquiry, analysis, and inference, while Gabrielson (2013) contends that science progresses through argumentation. This is just as Erduran et al. (2006) make a strong case for preservice education to promote argumentation in science. The case for argumentation in relation to science education has thus remained very strong as, for example, the Next Generation Science Standards make references to it. To underscore the importance of argumentation in science education, this book opens with a comprehensive review of empirical studies in NOS and argumentation in science education. Still in Part I is a consideration of a major dilemma in science education, namely, how the issues of science and religion are handled. It is generally agreed that scientific knowledge is empirical and tentative, seeking the understanding of nature. On the other hand, religious issues are unchanging and dogmatic as they are based of faith. But the distinction is not as straightforward as it seems. In many societies, belief in religion is very widespread. Indeed, there are situations where science educators (and scientists) also hold religious beliefs. In addition, there are also related issues such as evolution, creationism, and intelligent design. The challenge therefore is on how

B. Akpan (✉)
Science Teachers Association of Nigeria, Abuja, Nigeria
e-mail: ben.b.akpan@gmail.com

B. Akpan (ed.), *Science Education: A Global Perspective*,
DOI 10.1007/978-3-319-32351-0_1

science education is delivered recognizing that people are entitled to whatever beliefs they hold even when these are at variance with well-established principles of science. Frazier (2003) has advised that the matter be handled with great caution by, for example, basing ideas on facts and avoiding the temptation to regard either science or religion as being superior to the other.

The view is commonly being held that nations at the forefront of modern development are those that have invested enormous resources over considerable time in three major areas: first, in the establishment and nurturing of a stable, well-supported science and technology system; second, in the promotion of mission-oriented research in the basic sciences, coupled with long-term strategy for technology development; and third, in the institution of well-articulated programs for the education of a large scientifically and technologically literate work force (Brown and Sarewitz 1991). Thus in spite of the difficulty of precise measurements, there seems to be a general agreement that the real difference between the developed countries of America, Europe, Asia, and the Far East and the developing countries of Africa lies in their scientific and technological capabilities. These capabilities may be considered as the extent to which countries access, utilize, and create science and technology for the solution of socioeconomic problems. This position ascribes much of the phenomenal rate of economic growth shown by China, Japan, and a host of countries in Latin America, to their harnessing the fruits of science and technology. Countries that fail to innovate lose their competitive position (Akpan 2012). It is therefore instructive that in this volume, Part II is devoted to science and national development using several projects and programs in India, Malaysia, and Nigeria. These three countries are in some ways exemplars of modern growth points in their regions.

Part III of the book is devoted to curriculum and pedagogical considerations. We are all ever conscious of the many changes that have continued to characterize the field of education in general and science education in particular. These changes have been propelled in the main by changes in both curricula and pedagogy which in turn are tailored to meet the expectations of the society. Indeed, as schools are microcosms of society, curriculum reinforces the good values that are shared by society while putting in place the good values that are not in existence but are nonetheless desired by it (Abimbola 1993). In this direction, Carr et al. (2000) have explored on the one hand the relationships between curriculum and pedagogy, assessment, social outcomes, and learning outcomes and pedagogy and learning outcomes on the other hand. These very extensive reviews have drawn attention to the ever-increasing role of curriculum and pedagogy in the progress of education. In the area of science and technology, questions are being asked about the relevance of what learning is going on in schools in relation to the phenomenal explosion of knowledge, products, and services in those fields. As knowledge acquisition is dwarfed by the need for knowledge application, science education is focusing more on skills, attitudes, and values that support societal development via context-based teaching approach (Holbrook and Rannikmae 2010), and curricula are being made more open and relevant (Galvoa et al. 2011; Sexton 2011) with calls for socially responsible science education that pays increased attention to issues that have scientific bases and are significantly important to society (Onwu 2012). The central theme for this section, therefore, is the need to assure science education meets the immediate and future needs of students in the context of their societies.

Development and future studies in science education are the main issues discussed in Part IV. The section looks at development from the perspective of equity and gender. At issue is the education of girls in terms of access and quality. Globally, the education of girls has been very topical. There is a school of thought that holds the view that the education of girls, and women generally, is an economic investment. An exponent of this position is Summers (1992):

> An educated mother…has greater value outside the house and thus has an entirely different set of choices than she would have without education. She is married at a later age and is better able to influence family decisions. She has fewer, healthier children and can insist on the development of all of them, ensuring her daughters are given a fair chance. And the education for her daughters makes it much more likely that the next generation of girls, as well as boys, will be healthy and educated as well. The vicious cycle is thus transformed into a virtuous circle. (p. 5)

The other school of thought spearheaded by UNESCO regards education whether of girls or boys as a fundamental human right (UNESCO 2001). The good news is that both positions support the equitable education of girls as a means of assuring future development. Chapter 15 explores this issue further. The second to the last chapter of the book takes a careful look at future trends for science education research especially from the perspective of what promises these trends may impart in a world deeply in search of solutions to a multiplicity of challenges in science education. The final chapter continues from there by exploring the place of science education half a century into the future. With likely positive changes in global economy and attendant boom in science and technology products and services, the corollary will be major changes in curriculum and pedagogy that will best support a blossoming science education sector in an anticipated future world in 2065 and beyond.

References

Abimbola, I. O. (1993). Guiding philosophical perspectives. In U. M. O. Ivowi (Ed.), *Curriculum development in Nigeria* (pp. 4–16). Ibadan: Sam Bookman Educational and Communication Services.

Akpan, B. B. (2008). *Nigeria and the future of science education*. Ibadan: Science Teachers Association of Nigeria.

Akpan, B. B. (2012). Science education in Nigeria. In U. Ivowi & B. Akpan (Eds.), *Education in Nigeria – From the beginning to the future* (pp. 77–104). Lagos: Foremost Educational Services Ltd.

Brown, J. R., & Sarewitz, D. R. (1991). Fiscal alchemy: Transforming debt into research. *Issues in Science and Technology*, pp 70–76.

Carr, M., McGee, C., Jones, A., McKinley, E., Bell, B., Barr, H., & Simpson, T. (2000). *Strategic research: Initiative literature review; the effects of curricula and assessment on pedagogical approaches and on educational outcomes*. Wellington: Ministry of Education.

Erduran, S., Adac, D., & Yakmaci-Guzel, B. (2006). Learning to teach argumentation: case studies of preservice secondary science teachers. *Eurasia Journal of Mathematics, Science and Technology Education, 2*(2). http://www.ejmste.com/022006/d1.pdf. Retrieved 3 Nov 2015.
Frazier, K. (2003). Are science and religion conflicting or complementary? Some thoughts about boundaries. In P. Kurtz (Ed.), *Science and religion: Are they compatible?* (pp. 25–28). New York: Prometheus Books.
Gabrielsen, P. (2013). Teach science through argument, Stanford professor says. *Stanford Report*, April 9.
Galvoa, C., Reis, P., Freire, S., & Almeida, P. (2011). Enhancing the popularity and the relevance of science teaching in Portuguese science classes. *Research in Science Education, 41*(5), 651–666.
Holbrook, J., & Rannikmae, M. (2010). Contextualisation, decontextualisation, recontextualisation – A science teaching approach to enhance meaningful learning for scientific literacy. In I. Eilks & B. Ralle (Eds.), *Contemporary science education* (pp. 69–82). Aachen: Shaker Verlag.
Kuhn, D. (2010). Teaching and learning science as argument. *Science Education, 94,* 5.
Onwu, G. (2012). Towards a culturally relevant and socially responsible science education. *ECTN Association/European Chemistry and Chemical Engineering Education Network (EC2E2N)* 13(05), November Special Issue: Africa, 1–4.
Poole, M. (1990). *A guide to science and belief.* Oxford: Lion Publishing Plc.
Sexton, S. S. (2011). Revelations in the revolution of relevance: Learning in a meaningful context. *The International Journal of Science in Society, 2*(1), 29–40.
Summers, L. H. (1992). *Investing in all the people*. Washington, DC: The World Bank.
UNESCO. (2001). *Medium – Term strategy 2002–2007*. Paris: UNESCO.

Part I
Nature of Science View

Chapter 2
Exploring Nature of Science and Argumentation in Science Education

Christine V. McDonald

2.1 Introduction

Enhancing learners' views of nature of science (NOS) is a central goal of the science education community, and this goal has been the focus of extensive research efforts for over 50 years (Abd-El-Khalick and Lederman 2000a; Lederman 1992). Many reasons have been cited by science education researchers and reform organisations for developing learners' views of NOS, with perhaps the most fundamental reason positing that an understanding of NOS is necessary for achieving scientific literacy (American Association for the Advancement of Science [AAAS] 1990, 1993; Tytler 2007). Cavagnetto (2010, p. 337) states that scientific literacy is supported by '…the integration of science concepts and processes, metacognitive processes, critical reasoning skills, and cultural aspects of science (e.g., the epistemic nature of science)', and collectively these concepts, processes and abilities reflect scientific practice. Within scientific practice, language plays a pivotal role in knowledge construction and interpretation (Ford 2008), and argument is viewed as a critical aspect of the language practices of science. In addition, advances in technological innovations, and increasing globalisation, require students of the twenty-first century to handle vast, and often complex, sets of information from a variety of different sources. Students are expected to be able to evaluate this information, thus requiring them to engage in argumentation to arrive at evidence-based decisions (Jimenez and Erduran 2007). The inclusion of argumentation in the curricula is an important component of contemporary science education in many countries (e.g. AAAS 1993; Australian Curriculum, Assessment and Reporting Authority [ACARA] 2012; National Research Council [NRC] 1996), and many researchers (e.g. Driver et al. 2000; Duschl and Osborne 2002; Erduran et al. 2004; Kuhn 1993;

C.V. McDonald (✉)
School of Education and Professional Studies, Griffith University,
176 Messines Ridge Road, Mt Gravatt 4122, Queensland, Australia
e-mail: c.mcdonald@griffith.edu.au

B. Akpan (ed.), *Science Education: A Global Perspective*,
DOI 10.1007/978-3-319-32351-0_2

Sampson and Clark 2008) have proposed that participating in argumentation aids in the development of scientific literacy. As stated earlier, the achievement of scientific literacy is evidenced in reform documents internationally, and a key requirement for the achievement of scientific literacy is '...to develop an ability to solve problems and make informed, evidence-based decisions about current and future applications of science while taking into account ethical and social implications of decisions' (ACARA 2012, p. 3).

The aim of this chapter is to examine empirical studies which have explored NOS and argumentation in science education. The chapter will explore the nature of this relationship and will examine studies conducted in socioscientific and scientific contexts. To begin, scholarship in NOS will be considered, followed by a discussion of argumentation in science education, with a particular focus on sociocultural perspectives and socioscientific contexts. A brief description of the methods utilised in the review of studies will be provided, followed by a critical review of 23 empirical studies. Emergent themes from this analysis will then be discussed, followed by a consideration of recommendations stemming from the analysis.

2.1.1 Nature of Science

Epistemology as a construct is conceptualised differently in the domains of philosophy and psychology. In philosophy, epistemology focuses on the study of knowledge; hence, philosophers of science are interested in the study of knowledge within the discipline of science. The epistemology of science can be viewed as the logical and philosophical grounds upon which scientific claims are proposed and justified and is considered to be domain specific (Sandoval and Millwood 2007). It encompasses the sources of scientific knowledge, the values used to justify scientific knowledge and the processes utilised by the scientific community to accept scientific knowledge (Sandoval 2005). In psychology, epistemological views are studied from the perspective of the individual. It views personal epistemology as the set of beliefs individuals hold about the nature and justification of knowledge. Hofer and Pintrich (1997) outline four dimensions of epistemology – beliefs about the certainty of knowledge, beliefs about the simplicity of knowledge, beliefs about the source of knowledge and beliefs about the justification of knowledge. Debate exists as to whether individuals' personal epistemologies are considered to be domain specific or domain general, but importantly research has shown that individuals' personal epistemologies influence their learning approaches and reasoning (King and Kitchener 1994; Kuhn 1991; Schommer-Aikins 1993).

Within science education, two overlapping constructs have commonly been utilised to describe learners' epistemological views about science and scientific knowledge – nature of science (NOS) views and scientific epistemological beliefs (SEBs) (Wu and Tsai 2011). Stemming from a philosophical perspective, NOS refers to science as a way of knowing or the values and beliefs inherent to scientific knowledge and its development (Lederman 1992), whereas SEBs draw on the psychological perspective of personal epistemologies (Hofer and Pintrich 1997) and refer to beliefs about the nature of scientific knowledge and the nature of knowing science. Our understanding of students' scientific epistemologies has been further advanced by the scholarship of Hogan (2000) and more recently Sandoval (2005), who make the distinction between the beliefs and ideas that students develop about scientific knowledge through their own engagement and interaction in school-based inquiry (proximal knowledge/practical epistemologies) and the beliefs students' express about knowledge generation in professional science (distal knowledge/formal epistemologies). The primary difference between these two conceptualisations lies in the point of reference – distal knowledge/formal epistemology is generally not experiential and references an external set of knowledge (in this case, scientist's knowledge). In contrast, proximal knowledge/practical epistemology is experiential and references an internal set of knowledge (in this case, the individuals' own science experiences). For the purposes of this review, studies which have utilised all of the epistemological approaches described in this section will be examined and considered under the umbrella term *NOS*.

Despite the extensive amount of research conducted in the field and the prominence of this important component of scientific literacy in the reform documents, the development of informed NOS views has been a difficult goal to achieve, with many studies reporting difficulties in changing learners' NOS views (Duschl 1990; Lederman 1992). Importantly, research has highlighted the effectiveness of explicit NOS instructional approaches in improving learners' views of NOS (e.g. Abd-El-Khalick and Lederman 2000b; Hanuscin et al. 2006; Schwartz et al. 2004). An explicit NOS instructional approach deliberately focuses learners' attention on various aspects of NOS during classroom instruction, discussion and questioning. This type of instructional approach is based on the assumption that NOS instruction should be planned for and implemented in the science classroom as a central component of learning, not as an auxiliary learning outcome. This approach is contrasted with implicit instructional approaches to teaching NOS which are underpinned by the view that an understanding of NOS will result from engaging students in inquiry-based activities, without the addition of deliberately focused (explicit) NOS instruction. Results from studies conducted in this area indicate that implicit NOS instructional approaches are generally not successful in developing learners' NOS views (e.g. Bell et al. 2003; Moss et al. 2001; Sandoval and Morrison 2003).

Although explicit instructional approaches have been shown to be relatively more successful than implicit instructional approaches in developing learners' NOS views, studies continue to show that the implementation of explicit NOS instructional approaches does not result in improved NOS views for all learners (Abd-El-Khalick and Akerson 2004). Scholars working from a sociocultural perspective (e.g. Ford 2008; Sandoval 2005) propose that engaging learners in scientific practices, such as argumentation, may lead to developments in their NOS views. In a similar vein, other scholars have proposed that learners' NOS views may influence their engagement in argumentation (Nussbaum et al. 2008; Sandoval and Millwood 2007). To explore the nature of the relationship between NOS and argumentation further, it is thus important to discuss the role of argumentation in science education.

2.1.2 Argumentation

In its most fundamental form, an argument is an assertion where the premises are stated as a means of justifying a conclusion (Govier 2010). Osborne and Patterson (2011) state that arguments attempt to justify uncertain conclusions with a claim that is supported by data, with warrants acting as links to explain how the data support the claim. They state the goal of argumentation '… is an attempt to persuade the listener of the validity of the conclusions…' (p. 636). Although myriad definitions exist in the science education literature for the terms 'argument', 'argumentation', 'informal reasoning', 'scientific reasoning', 'decision-making' and so on, for the purposes of this review, studies which have utilised all of the above conceptualisations will be examined and considered under the umbrella term *argumentation*.

Despite the worldwide trend to incorporate the teaching of argumentation in science classrooms via recent reform recommendations and curriculum developments (Jimenez and Erduran 2007) and the recommendations stemming from research viewing argumentation as an important instructional strategy and educational goal for science education (Bricker and Bell 2008), both early (e.g. Driver et al. 2000) and more recent (e.g. Berland and Reiser 2009; Simon et al. 2006) empirical research indicate argumentation is rarely effectively incorporated in science classrooms. Previous studies indicate that most classrooms are teacher dominated (Crawford 2005), with students given few opportunities to learn about or engage in argumentation (Lemke 1990), and teachers generally do not possess adequate skills to teach argumentation to their students (Newton et al. 1999; Sampson and Blanchard 2012). Various researchers (e.g. Clark and Sampson 2006; Duschl 1990) have highlighted the limitations of presenting scientific knowledge as a set of facts to be memorised and recited for assessment. This transmissive approach to teaching science not only provides students with an inaccurate image of the nature of science but also fails to encourage an exploration of how scientific ideas have developed and changed over time. Students may not appreciate the purpose of discussing and critiquing these ideas and are less likely to engage in argumentative discourse about how these ideas are developed and validated by the scientific community.

In a similar vein, epistemological ideas about science cannot be transmitted didactically to students, and some researchers have suggested that students must be engaged in the practices of science, including argumentation, to further develop their understandings of epistemological ideas within the discipline of science (Duschl 2008; Sandoval 2005). Other researchers have sought to explore this relationship from the opposite perspective, proposing that students' NOS views may influence their engagement in the practice of argumentation. L. Kuhn and Reiser (2006) assert that if students hold naive views of scientific knowledge as a body of absolute facts, they are unlikely to see the need to engage in debates about scientific issues. Thus, the difficulties students have in engaging in argumentation may be explained by examining their epistemological views, as without developed epistemological views, students may not realise claims are open to challenge and refutation and require the support of empirical evidence (Sampson and Clark 2006).

Early research in the field of argumentation indicated that developmental factors, such as age and previous knowledge, may influence argumentation abilities (Means and Voss 1996), and both children and adults are unable to produce valid arguments (Kuhn 1991, 1993). These findings were the impetus for many studies conducted with the aim of improving learners' argumentation skills via explicit instructional approaches (e.g. Bell and Linn 2000; Zohar and Nemet 2002), with results generally reporting improvements in students' skills and/or quality of argumentation. Explicit instruction in this context refers to the direct teaching of various aspects of argumentation including instruction pertaining to the various definitions, structure, function and application of arguments and the criteria used to assess the validity of arguments. More recent research has cast doubt over earlier assertions regarding children's abilities to engage in effective argumentation; with findings indicating that learners are indeed capable of engaging in scientific argument when they experience the need and possibility of persuading others (Berland and Hammer 2012; Ryu and Sandoval 2012). Thus, research efforts have shifted from viewing argumentation from a predominantly cognitive perspective, focused on teaching skills of argument, to examining argumentation through a sociocultural lens, concerned with examining the social contexts in which students do, and do not, take part in argumentation (Berland and Hammer 2012).

Importantly, a variety of argumentation instructional approaches, learning contexts, interventions and strategies have been utilised by educators over the past 20 years to promote student engagement in argumentation (see Sampson and Clark 2008, for a review). These strategies and interventions have been implemented in a variety of learning contexts, including socioscientific and scientific contexts. In a recent review, Cavagnetto (2010) categorised research conducted in argumentation using three orientations – socioscientific, immersion and structure – and sought to examine the extent to which each orientation promoted scientific literacy. Socioscientific-oriented interventions are designed to use argument as a vehicle to explore social influences on science, including moral, ethical and political issues. Engaging students in these issues provides an authentic context for learning science. Immersion-oriented interventions used argument as a tool for constructing and understanding scientific principles and cultural practices of science. Argument is embedded throughout the inquiry in these studies. These types of interventions are often underpinned by sociocultural perspectives. Structure-oriented interventions are designed to facilitate the transfer of an argument structure to diverse settings. The focus of these interventions is the communication and defence of knowledge claims, and argument is viewed as a product of inquiry, rather than embedded during the inquiry.

For the purposes of this review, studies will be categorised according to the learning context in which they are situated – socioscientific or scientific. Within these learning contexts, researchers may work from cognitive or sociocultural perspectives and may, or may not, implement explicit argumentation instruction. Importantly, there will be areas of overlap within individual studies. The following section will review scholarship in socioscientific contexts, which has yet to be considered. As stated earlier, a shift from viewing argumentation from a predominantly cognitive perspective (typically including explicit instruction) to examining argumentation from a sociocultural perspective has recently occurred. As such, research guided by sociocultural perspectives will also be considered. For a review of research incorporating explicit argumentation approaches, see McDonald and McRobbie (2012).

Socioscientific Contexts The term *socioscientific issues* (SSIs) is utilised in the science education literature to describe issues which are utilised in socioscientific learning contexts. Engagement in SSIs requires students to apply scientific concepts, principles and practices to issues which are also influenced by societal, political, ethical and/or economic considerations (Kolstø 2001; Sadler 2009). SSIs involve ill-structured problems with uncertain solutions and are often complex (Sadler and Fowler 2006) and controversial (Sadler 2009), in nature. Individuals' reasoning processes during their engagement in SSIs have generally been recognised as informal reasoning due to their ill-defined nature and uncertain outcomes. Participating in SSIs requires students to identify premises and make decisions about situations which have opposing sides. As students can take a variety of perspectives or positions on these issues, SSIs entice students to engage in argumentation and decision-making, in an attempt to resolve differences of opinion (Walker and Zeidler 2007). Importantly, although informed argumentation on an SSI requires the application of scientific conceptual knowledge, a consideration of the values a student brings to an SSI is crucial, as the decisions students make during their engagement in SSIs are influenced by the roles played by both scientific evidence and human values (Nielsen 2012).

Sadler (2009) provides a comprehensive review of argumentation studies conducted within the framework of SSIs. Findings from several of these studies reported improvements in students' argumentation during SSI interventions (e.g. Dori et al. 2003; Grace 2009; Pedretti 1999; Tal and Hochberg 2003; Tal and Kedmi 2006; Zohar and Nemet 2002), and Sadler concludes that SSI learning environments provide useful contexts for developing learners' argumentation, although the degree of success in particular interventions is dependent upon the supports provided to students during their engagement in the SSI. Supports such as teacher scaffolding, explicit argumentation instruction and engagement in reflective practices were found to be beneficial in many of the cited studies. Importantly, other studies have reported issues with developing learners' argumentation during SSI interventions (Albe 2008; Harris and Ratcliffe 2005; Kortland 1996) including problems with incorporating scientific evidence in decision-making, making unjustified claims, an inability to consider opposing positions and a lack of competence in evaluating and analysing arguments (Cavagnetto 2010).

Zeidler et al. (2013) state that advancing epistemological beliefs is significant whilst engaged in SSIs as students must coordinate their beliefs about values with both knowledge claims and scientific issues. Engaging students in SSIs provides numerous opportunities for explicit considerations of NOS, and many researchers have proposed relationships between learners' NOS views and their reasoning in SSIs, underpinned by the assumption that learners with more informed NOS understandings should be able to participate more effectively in decision-making in SSIs. These studies will be examined in the later sections of this chapter.

Sociocultural Perspectives As stated earlier, research in argumentation in science education has recently shifted focus to investigating learning contexts which promote engagement in the practices of argumentation (Berland and Hammer 2012, Kuhn 2010). Studies conducted in this area are concerned with examining the social contexts in which students do, and do not, take part in argumentation (Kelly et al. 1998; McNeill and Pimentel 2010), with results indicating that students may only engage in argumentation when they experience the need and possibility of persuading others (e.g. Ryu and Sandoval 2012). As such, L. Kuhn and Reiser (2006) suggest that educators need to design learning environments to 'create a need' for students to engage in argumentation.

In a recent publication, Sandoval and colleagues criticised cognitive approaches to argumentation, which view learning to argue as the acquisition and development of decontextualised skills needed before an individual can effectively engage in argumentation (Ryu and Sandoval 2012). They propose an alternative view which is guided by the assumption that a learner needs to engage in argumentation in order to learn how to argue effectively. Guided by a sociocultural perspective, argumentation is defined as a form of discourse to be appropriated and internalised by learners. Thus, this perspective does not discount the development of cognitive skills of argument, but instead attributes the emergence of these skills to the particular social practice a learner operates within. In this view, learning to argue scientifically cannot be achieved simply by acquiring a set of cognitive skills, but instead requires immersion in the activities of authentic science, and is therefore situated in disciplinary practice. Importantly, appropriating the norms of a classroom culture, guided by a sociocultural approach, requires an extended period of time to enact, and the authors caution that changes in practices of argumentation are unlikely to occur if interventions are implemented over short durations.

Importantly, sociocultural perspectives can guide research conducted in both scientific and socioscientific contexts. Thus, although socioscientific contexts were discussed separately in the previous subsection, it is necessary to remember that studies conducted in these learning contexts can also be underpinned by sociocultural theoretical perspectives (e.g. studies conducted by Sadler and Zeidler's research group are informed by these perspectives). Studies which have incorporated sociocultural perspectives in socioscientific and scientific contexts, as a framework for examining learners' NOS views and engagement in argumentation, will be examined in the later sections of this chapter.

2.2 Review of Studies

2.2.1 Methods

The purpose of this review is to examine empirical studies which have explored NOS and argumentation in science education. The review seeks to explore the nature of this relationship and will examine studies conducted in socioscientific and scientific contexts. This review is an extension of a previous examination of empirical research examining nature of science and argumentation (see McDonald and McRobbie 2012) which evaluated studies reported prior to 2009. Since this time, research in the field has intensified, and the present review includes all relevant, empirical studies published in high-quality, peer-reviewed, science education journals (e.g. *International Journal of Science Education*, *Journal of Research in Science Teaching*, *Research in Science Education*, *Science Education*) up until the end of 2012. Keywords utilised to facilitate the search included *nature of science*, *epistemology*, *scientific epistemological views/beliefs*, *personal epistemology*, *argument*, *argumentation*, *informal reasoning*, *scientific reasoning*, *decision-making*, *socioscientific issues (SSI)* and *socioscientific reasoning*. Results of the literature search yielded 23 empirical papers which addressed the relationship between nature of science and argumentation. The previous review (McDonald and McRobbie 2012) categorised studies according to two learning contexts – socioscientific and scientific. After searching for common themes in the 23 papers which are the focus of the current review, preliminary analysis indicated that the previous categorisation, of socioscientific and scientific contexts, was still relevant for the present review. Thirteen research reports explored NOS and argumentation in socioscientific contexts, nine research reports explored NOS and argumentation in scientific contexts, and one research report explored NOS and argumentation in both contexts.

Thus, studies in this review were categorised according to the dominant learning context in which they are situated – socioscientific contexts or scientific contexts. Scientific contexts for argumentation are concerned with the application of scientific reasoning to enable an understanding of the justification for hypotheses, the validity and limitations of scientific evidence and the evaluation of competing models and theories (Giere 1979). Scientific arguments utilise theoretical and empirical evidence to evaluate knowledge claims (Jimenez-Alexiandre and Erduran 2007). Socioscientific contexts for argumentation are concerned with the application of scientific ideas and reasoning to an issue and also invoke a consideration of moral and ethical values and personal and social concerns (Evagorou 2011; Osborne et al. 2004). As stated earlier, the term *socioscientific issues* (SSIs) is utilised in the science education literature to describe issues which are used in these contexts. SSIs do not have straightforward solutions and cannot be fully resolved by solely applying theoretical and empirical evidence (Sadler 2011, cited in Evagorou and Osborne 2013).

A further classification utilised in this review considered the nature of the study (assessment or intervention). Assessment studies are studies which did not include an intervention and simply sought to assess the relationship between participants' NOS views and their argumentation. Many of these studies utilised open-ended surveys and interviews, and participants were asked to respond to provided scenarios or problem-solving tasks. Intervention studies are studies which engaged participants in a planned intervention designed to engage them in argumentation and/or NOS instruction. Examples of these interventions included immersion in SSIs, immersion in inquiry-based learning environments and engagement in computer-scaffolded learning environments. Intervention studies also assessed participants' NOS views and argumentation. Table 2.1 provides a summary of the reviewed research reports.

2.2.2 *Socioscientific and Scientific Contexts*

This section will review the sole study identified in the literature which sought to explore the relationship between NOS and argumentation in both socioscientific and scientific contexts. McDonald (2010) explored the influence of a science content course incorporating explicit NOS instruction and explicit argumentation instruction on five Australian preservice primary teachers' views of NOS. The course utilised both scientific and socioscientific contexts for argumentation, to provide opportunities for participants to apply their NOS understandings to their arguments. Data sources included questionnaires and surveys (*VNOS-C*, Abd-El-Khalick 1998; *global warming survey*, Sadler et al. 2004; *superconductors survey*, Leach et al. 2000), interviews, audio- and videotaped class sessions and written artefacts. Results indicated that the science content course was effective in enabling four of the five participants' views of NOS to be improved. Engaging in argumentation in scientific contexts was found to be more difficult than engaging in argumentation in socioscientific contexts, and participants did not recognise a need to explain their data in some scientific contexts. The inclusion of argumentation scaffolds facilitated participants' engagement in argumentation in some socioscientific contexts, and conversely a lack of consideration of alternative data and explanations hindered participants' engagement in argumentation in some scientific contexts. The inclusion of epistemological probes was found to aid in the development of participants' views of NOS and used in conjunction with explicit NOS instruction; these written or verbal prompts were successful in orienting participants' attention to relevant NOS aspects highlighted in a task. Similarly, a lack of epistemological probes in some tasks hindered participants' abilities to apply their views of NOS to their reasoning during argumentation. The author proposes that simply engaging learners in argumentation may not be sufficient to ensure their NOS views are developed and recommends explicit attention to specific NOS aspects during argumentation-based interventions to provide cognitive anchor points allowing learners to access and engage in epistemological discourse during argumentation. The following section will examine studies conducted in socioscientific contexts.

Table 2.1 Summary of reviewed studies

Study	Aim	Context	Nature of study	Participants	Outcomes
Bell and Lederman (2003)	Investigated the role of NOS in decision-making about socioscientific issues	Socioscientific	Assessment – no intervention	21 university professors and research scientists in the United States	NOS views did not influence reasoning on the issues. Recommend explicit instruction on applying their NOS views to decision-making
Bell and Linn (2000)	Explored the relationship between students' views of NOS and their argument construction	Scientific	Intervention – engagement in computer-scaffolded learning environment (6-week project). No explicit NOS instruction, supported argumentation instruction via software tool	172 middle school students in the United States	Students with more informed NOS views created more complex arguments. Claim that engaging students in argumentation improves their understanding of NOS
Bell et al. (2011)	Investigated the relative impacts of NOS instructional approach and context of NOS instruction, on preservice teachers' NOS views	Socioscientific	Intervention – immersion in SSI learning environment (one semester). Explicit and implicit NOS instruction, contextualised and decontextualised NOS instruction	75 preservice elementary teachers in the United States	Preservice teachers experiencing explicit NOS instruction exhibited improved NOS understandings. Teaching NOS explicitly and within an SSI produced significant gains in applying NOS views to decision-making. Claim evidence of effectiveness of SSIs for facilitating application of NOS to reasoning

Eastwood et al. (2012)	Investigated the influence of an SSI context versus traditional content context on students' NOS views	Socioscientific	Intervention – immersion in SSI learning environment for treatment group (full year). Explicit NOS instruction	4 classes of 11th and 12th grade students in the United States	NOS improved for all students, but SSI group showed better articulation of the social and cultural NOS. Claim engaging students in argumentation and debate promoted more informed views of social and cultural NOS
Hogan and Maglienti (2001)	Investigated scientific reasoning processes by examining the epistemological criteria used to rate the validity of conclusions	Scientific	Assessment – no intervention	45 middle school students, non-scientists, research scientists and technicians in the United States	Differences in the use of epistemological criteria between scientists and non-scientists were evident. Claim that these differences are due to contextual frameworks from differing spheres of cultural practice
Khishfe (2012a)	Investigated the relationship between students' NOS views and argumentation skills in two SSIs	Socioscientific	Assessment – no intervention	219 11th grade students in Lebanon	Students who constructed stronger arguments also showed more informed NOS views. Claim evidence of relationship between NOS and argumentation skills
Khishfe (2012b)	Investigated the influence of NOS on students' decision-making in an SSI	Socioscientific	Intervention – immersion in SSI learning environment (4-week unit). All students received explicit argumentation instruction, and treatment group students also received explicit NOS instruction	4 sections of 9th grade students in the United States	NOS instruction did not influence decision-making, although the reasons students provided to explain their decision-making were informed by NOS in the treatment group

(continued)

Table 2.1 (continued)

Study	Aim	Context	Nature of study	Participants	Outcomes
Liu et al. (2011)	Investigated the relationship between scientific epistemological views (SEVs) and reasoning processes during decision-making on an SSI	Socioscientific	Assessment – no intervention	177 college students in Taiwan	SEVs were related to students' decision-making about the SSI. Recommend students participate in programmes which provide opportunities for multiple reasoning and interdisciplinary thinking
Matkins and Bell (2007)	Investigated the influence of preservice teachers' NOS views on their decision-making in an SSI	Socioscientific	Intervention – immersion in SSI learning environment (one semester). Explicit NOS instruction	15 preservice elementary teachers in the United States	Preservice teachers' views of NOS improved, and they were able to apply their NOS views to their decision-making in the SSI. Claim explicit, contextualised NOS instruction informed their decision-making
McDonald (2010)	Investigated the effectiveness of an argumentation-based, science content course on preservice teachers' NOS views	Scientific and socioscientific	Intervention – immersion in a science content course including inquiry-based activities and engagement in SSIs (one semester). Explicit NOS instruction and explicit argumentation instruction	5 preservice elementary teachers in Australia	Preservice teachers' NOS views improved, and these views were reflected in their reasoning in socioscientific contexts, but not in scientific contexts. Claim evidence of effectiveness of argumentation-based course incorporating explicit NOS instruction, on NOS views

Nussbaum et al. (2008)	Investigated the influence of students' epistemic beliefs and exposure to argumentation instruction, on the quality of argumentation	Scientific	Intervention – engagement in computer-scaffolded learning environment (within a single semester). No explicit NOS instruction, supported argumentation instruction via software tool for treatment group only	88 undergraduate students in the United States	Student's epistemic orientations influenced their engagement in argumentation
Ryu and Sandoval (2012)	Investigated the influence of a sustained focus on argumentation on students' understanding of epistemic criteria for scientific argument	Scientific	Intervention – immersion in inquiry-based context emphasising a sustained instructional focus on argumentation (full year). No explicit NOS	21 elementary students in the United States	Students learned how to apply epistemic criteria whilst engaging in argumentation. Claim evidence that elementary students are capable of engaging in argumentation and, in turn, develop understanding of the core epistemic criteria of scientific argument
Sadler et al. (2004)	Investigated the influence of students' NOS views on their decision-making in an SSI	Socioscientific	Assessment – no intervention	84 high school students in the United States	Views of the social NOS influenced their reasoning in the SSI. Claim evidence of SSIs as effective contexts for addressing NOS in the classroom

(continued)

Table 2.1 (continued)

Study	Aim	Context	Nature of study	Participants	Outcomes
Sandoval and Cam (2011)	Investigated student's judgements of the epistemic status of justifications for causal claims	Scientific	Assessment – no intervention	26 elementary students in the United States	Two epistemological resources were activated by engaging in decision-making in the study – the idea that claims require empirical evidence and the idea that causal mechanisms must be plausible
Sandoval and Millwood (2005)	Examined the influence of epistemological views about argumentation on inquiry practices	Scientific	Intervention – engagement in computer-scaffolded learning environment (4-week unit). No explicit NOS instruction, supported argumentation instruction via software tool for treatment group only	87 high school students in the United States	Students constructed logical arguments, but had difficulty citing sufficient data to support claims and providing warrants. Implications from this study suggest that naive epistemological views may constraint engagement in argumentation
Sandoval and Millwood (2007)	Examined students' epistemological views about warranting claims	Scientific	Intervention – immersion in inquiry-based learning environment (3-week unit). No explicit NOS instruction, supported argumentation instruction via software tool	33 7th grade students in the United States	The majority of students were able to articulate claims, but most did not provide warrants for their claims. Claim evidence that students' epistemological views influence their engagement in argumentation

Schalk (2012)	Investigated the influence of an inquiry-based SSI intervention on undergraduate students' NOS conceptualisations with respect to their ability to reason	Socioscientific	Intervention – immersion in inquiry-based SSI learning environment (15-week course). Explicit NOS instruction	26 undergraduate students in the United States	Undergraduates' conceptions of NOS developed, and their reasoning ability improved. They were able to link their NOS views to their reasoning. Claim improved reasoning associated with more developed conceptions of NOS
Walker and Zeidler (2007)	Examined the role of NOS in decision-making in an SSI	Socioscientific	Intervention – immersion in SSI learning environment (7-week unit). Explicit NOS instruction	36 high school students in the United States	Students' NOS views improved, although NOS views did not influence their reasoning. Recommend explicit argumentation instruction and guidance in applying their NOS views to their reasoning
Wu and Tsai (2011)	Investigated the relationship between students' scientific epistemological views (SEBs), cognitive structures and informal reasoning in an SSI	Socioscientific	Assessment – no intervention	68 10th grade students in Taiwan	Some aspects of students' SEBs significantly correlated with high-quality reasoning. Claim evidence of relationship between NOS and quality of informal reasoning
Yerrick (2000)	Investigated changes in students' argumentation in an open-inquiry setting	Scientific	Intervention – immersion in an open-inquiry learning environment (full year). Explicit argumentation instruction, no explicit NOS instruction	5 high school students in the United States	Students' views of NOS improved, and some improvements in their skills of argument were evident. Students' views of NOS were also reflected in their arguments

(continued)

Table 2.1 (continued)

Study	Aim	Context	Nature of study	Participants	Outcomes
Zeidler et al. (2002)	Investigated the relationship between NOS views and evidence that challenged beliefs in an SSI	Socioscientific	Assessment – no intervention	82 high school students/ preservice elementary teachers in the United States	Views of some aspects of NOS reflected in arguments in the SSI. Recommend explicit NOS and argumentation instruction needed to facilitate development
Zeidler et al. (2009)	Investigated the influence of an SSI context versus traditional content context on students' reflective judgement	Socioscientific	Intervention – immersion in SSI learning environment for treatment group (full year). Explicit NOS instruction	4 classes of 11th and 12th grade students in the United States	Personal epistemological views improved for SSI group, no improvement for content group. Claim engaging students in argumentation aided in developing reflective judgement, which is comparable to the development of NOS understandings
Zeineddin and Abd-El-Khalick (2010)	Examined the relationship between epistemological commitments and scientific reasoning	Scientific	Assessment – no intervention	139 college students in the United States	Prior knowledge and epistemological commitment were found to influence scientific reasoning, with results indicating that the higher the epistemological commitment, the higher the quality of reasoning, for comparable levels of prior knowledge

2.2.3 Socioscientific Contexts

The following subsections will critically review the 13 research reports which have explored NOS and argumentation in socioscientific contexts. The first subsection will review early studies which sought to assess the hypothesised relationship between NOS and argumentation and will be followed by a consideration of more recent studies which have examined the nature of the relationship between these two constructs. Intervention studies which have investigated the influence of NOS on argumentation will then be examined, as well as a consideration of intervention studies which have investigated the influence of argumentation on NOS.

Early Assessments of the Relationship Between NOS and Argumentation The first empirical study to emerge in a socioscientific context which investigated NOS and argumentation was conducted by Zeidler et al. (2002). The study was designed to investigate the relationship between students' views of NOS and their reactions to evidence that challenged their beliefs about a socioscientific issue. Participants consisted of 82 students ranging from junior (years 9 and 10) high school science students to preservice elementary teachers. Data were collected from students' responses to questionnaires, written responses to a socioscientific scenario on animal rights and interviews. Data analysis indicated that, in a few cases, students' views of NOS were reflected in the arguments they presented on a moral and ethical issue. Results also indicated that many participants' responses were based on personal opinions and failed to integrate relevant scientific evidence. The authors recommend that teacher preparation programmes need to expose their students to both explicit instruction about NOS and argumentation.

Sadler et al. (2004) examined 84 high school biology students' views of NOS and their interpretation and evaluation of evidence regarding a socioscientific issue. These researchers were interested in how students interpret and evaluate contradictory evidence when engaged in a global warming scenario. Students read two contradictory reports on global warming and responded to questions eliciting understanding of targeted NOS aspects and factors influencing their decision-making. Results indicated that students displayed an understanding of both the tentative and social NOS, although just under half of the students displayed naive views of the empirical NOS. Their views of the social NOS were found to considerably influence their reasoning and argumentation in the socioscientific context, and the authors recommend that explicit NOS instruction is necessary to ensure students are provided with the opportunity to form developed views of NOS. They also highlighted that SSI could provide many opportunities for addressing NOS in the science classroom.

Conversely, a study which challenged the findings of the previous studies was conducted by Bell and Lederman (2003) who also investigated the role of NOS in decision-making about socioscientific issues. The underlying rationale for the study was based on the premise that if there is a relationship between NOS and decision-making, then participants with diverse views of NOS should exhibit different reasoning about socioscientific issues. Twenty-one university professors and research scientists were purposively selected to provide divergent views of NOS and were placed in two groups representing disparate views of NOS. Data sources included an open-ended NOS questionnaire (VNOS-B), an open-ended questionnaire designed to obtain information about their decision-making in a variety of socioscientific contexts and individual interviews. Results indicated that participants' NOS views were not a significant contributing factor in the decisions reached by the participants in either group, with reasoning patterns tending to focus on personal, social and political aspects of the issue. There was little reference to scientific evidence as a contributing factor in their reasoning. The authors recommend that learners need to be explicitly instructed on how to utilise and apply their NOS views when engaged in decision-making on issues.

In summary, mixed results were reported with respect to the influence of learners' NOS views on their reasoning. Importantly, both studies conducted with high school students (Sadler et al. 2004; Zeidler et al. 2002) reported favourable findings, although implications from all of the studies highlight the importance of providing explicit NOS instruction to aid in the development of learners' NOS views, with Bell and Lederman (2003) also highlighting the necessity of providing explicit instruction on applying NOS views to decision-making.

Recent Assessments of the Relationship Between NOS and Argumentation In contrast to the three early studies which sought to assess the relationship between NOS and argumentation, all of which were conducted in the United States, the three recent studies reported in this section all emerged from countries outside of the United States. In Taiwan, Liu et al. (2011) sought to examine the relationship between scientific epistemological views (SEVs) and reasoning processes of 177 college students during decision-making on an SSI. Two instruments were utilised in this study, an SEV instrument (Tsai and Liu 2005) administered in the first week of the semester and a decision-making (DM) questionnaire on a local environmental problem administered in the sixth week of the semester. The DM questionnaire contained an expository text which provided different perspectives of the issue and a set of related, open-ended questions to probe participants' decision-making on the issue. Participants engaged in their regular course content throughout the study. Using mixed methods, data analysis indicated that SEVs were related to students' decision-making about the SSI. Tentativeness and creativity were found to be the significant constructs directly manifested in the socioscientific decision-making process. Other findings indicated that over half the participants (particularly science majors) made decisions based on a single disciplinary perspective, and the authors suggest that due to the complex nature of SSIs, students should be encouraged to participate in educational programmes which provide opportunities to utilise multiple reasoning modes and interdisciplinary thinking.

In another study conducted in Taiwan, Wu and Tsai (2011) examined the relationship between high school students' scientific epistemological beliefs (SEBs), cognitive structures and informal reasoning about nuclear power use. The authors were interested in ascertaining whether students' conceptual understandings, or their scientific epistemological views, were the more dominant factor contributing to their reasoning on an SSI. Participants were 68 tenth grade students in two classes. A correlational research design was used, and data collection utilised mixed methods and included questionnaires and interviews. Students' SEBs were assessed using a questionnaire, and data pertaining to cognitive structures was collected via interviews. An open-ended questionnaire (Wu and Tsai 2007) was used to collect data about students' informal reasoning. Results indicated that students with more extended and integrated cognitive structures did not make more evidence-based decisions and did not apply relevant knowledge to their reasoning. Other results indicated that scores on two of the scales of the SEB instrument, 'development' (related to beliefs on the nature of scientific knowledge) and 'justification' (related to the beliefs on the nature of knowing science), were significantly correlated with high-quality reasoning (specifically, rebuttal construction). The authors conclude that cognitive structures, particularly the information processing mode 'comparing', were the most significant factor for predicting the quality of their reasoning on the chosen SSI, although beliefs about justification of scientific knowledge were also an important predictor of reasoning quality. The authors conclude that the findings of this study contribute to the growing body of research supporting a relationship between students' NOS views and the quality of their informal reasoning.

More recently, Khishfe (2012a) examined the relationship between high school students' views of NOS and their argumentation skills as they engaged in two controversial socioscientific issues focused on genetically modified (GM) foods and water fluoridation. Using a mixed methods approach, 219 grade 11 students from five schools in Lebanon were administered a survey consisting of two open-ended scenarios and follow-up questions requiring them to generate arguments, counterarguments and rebuttals and present their views about various NOS aspects. In addition, 34 students took part in semi-structured interviews to further elaborate and clarify their responses. Data analysis showed strong correlations between the development of counterarguments and the three examined NOS aspects, with results indicating that students who constructed stronger arguments showed more informed NOS views. Other results indicated that the majority of students held naive views of NOS and were not able to construct sound arguments with valid justifications. Significant differences in students' argumentation skills and NOS views across the two examined scenarios were also found. Contextual factors such as prior content knowledge, issue exposure and familiarity and personal relevance were highlighted as possible factors affecting students' engagement in the socioscientific scenarios in this study. The author concluded that the findings from this study lend support to the assertion that a relationship exists between NOS and argumentation skills in socioscientific contexts.

In summary, results from the research conducted by Liu et al. (2011) support earlier findings indicating that students' NOS views influence their argumentation and extend these findings to college-level students. The other two studies (Khishfe 2012a; Wu and Tsai 2011) provide evidence of a positive relationship between NOS and argumentation for high school students, with findings showing a link between informed NOS views and higher-quality argumentation.

Interventions Exploring the Influence of NOS on Argumentation Five intervention studies, all of which were conducted in the United States, have sought to examine the influence of learners' NOS views on their reasoning. Two studies were conducted with high school students (Khishfe 2012b; Walker and Zeidler 2007), and three studies were conducted with college-level students (Bell et al. 2011; Matkins and Bell 2007; Schalk 2012). Walker and Zeidler (2007) examined the role of NOS in decision-making about a socioscientific issue. The purpose of the study was to assess how a 7-week, web-based instructional unit on genetically modified foods (GMFs) might elicit, reveal and develop 36 high school students' understanding of NOS and inform their decision-making. The study was designed to incorporate specific science content knowledge about GMFs, explicit NOS instruction and supported argumentation instruction in the form of guidance in evidence selection. Prior to the intervention, students completed a NOS questionnaire to assess their views of some aspects of NOS. At the conclusion of the intervention, student pairs took part in semi-structured interviews utilising questions from an open-ended NOS questionnaire (VNOS) to assess their views of NOS. Findings indicated that students' views of NOS developed over the duration of the study and were aligned with dynamic views of NOS at the conclusion of the study. Other results indicated that NOS was not explicitly referred to in their arguments, although the issue-based activity did enable their views to be elicited and revealed. In general, the students were not able to develop sound, evidence-based arguments, and the authors propose that more time and explicit instruction in argumentation is necessary to develop students' abilities to construct sound arguments. They also recommend that teachers need to develop the necessary pedagogical skills to guide their students in effectively applying their NOS understandings to SSIs.

In a more recent study, Khishfe (2012b) investigated the relationship between NOS and decision-making in a controversial SSI. This study was guided by the assumption that decision-making within SSI is not a spontaneous or intuitive process, and students need to be provided with opportunities to develop their abilities to formulate arguments, as a prerequisite for informed decision-making. The author sought to test the assumption of whether learning about NOS would inform students' decision-making as they engaged in an SSI. Four sections of ninth grade students were randomly allocated to comparison and treatment groups, and all students participated in explicit argumentation instruction, genetic engineering instruction and decision-making, during the 4-week unit. In addition, students in the treatment groups participated in explicit NOS instruction and instruction in applying their NOS understandings to their arguments and decisions during the SSI. Students' NOS views were assessed pre- and post-intervention using the open-ended VNOS-C, and their decision-making was assessed with an open-ended questionnaire which provided a scenario about genetically modified foods and follow-up questions. A subset of students were also individually interviewed. Results showed significant differences between the comparison and the treatment groups in relation to their understandings of the examined NOS aspects. No significant differences were found in the decisions of treatment group participants after engaging in explicit NOS instruction in the SSI over the duration of the study, although almost half of these participants referred to NOS as a factor which influenced their decision-making at the end of the study. Students in the comparison groups showed no change in their factors which affected their decision-making at the end of the study, with the majority of their decisions still dominated by factors other than NOS. The author concluded that although NOS instruction did not change students' decisions, the reasons students provided to explain their decision-making were informed by NOS when explicit NOS instruction, and instruction in how to apply their views of NOS to their decision-making, was provided.

Thus, results from interventions conducted with high school students show limited success. Conversely, studies conducted with college-level student's report more favourable findings. Matkins and Bell (2007) investigated 15 female preservice elementary teachers' views of NOS and global climate change in an SSI and whether they were able to apply their NOS conceptions to their decision-making whilst engaged in the SSI. Explicit NOS instruction was embedded in a single-semester, SSI-based, science methods course. The course was designed not only to aid students in developing their views of NOS but also to facilitate the application of their NOS views to the global climate change controversy. Thus, links between global climate change content and relevant NOS aspects were regularly made, in addition to providing opportunities for students to apply their NOS views to the controversy. Participants' views of NOS and their views of global climate change were assessed with open-ended questionnaires and interviews, at the commencement and conclusion of the study. Results indicated that participants' views of NOS and global climate change improved over the duration of the course, and participants were able to successfully apply these views to their decision-making in the SSI.

In a more recent study, Bell et al. (2011) compared the relative impacts of NOS instructional approach (explicit versus implicit) and context of NOS instruction (contextualised 'in an SSI' versus decontextualised 'as an isolated topic'), on 75 preservice elementary teachers' NOS understandings. The authors were also interested in whether the NOS instructional approach utilised influenced participants' ability to apply their NOS understandings in novel situations. Participants were enrolled in four sections of a single-semester science methods course, and data collection included open-ended questionnaires (VNOS-B) administered pre- and post-intervention, semi-structured interviews and classroom artefacts. Results indicated that participants who took part in explicit NOS instruction made substantial gains in their NOS understandings and were able to apply their NOS understandings appropriately to novel contexts, in both the contextualised and the decontextualised groups. Neither of the groups experiencing implicit NOS instruction (contextualised or decontextualised) exhibited gains in their NOS understandings. Importantly, other findings indicated that teaching NOS both explicitly and contextually (within the SSI) produced significant gains in participants' abilities to apply their NOS understandings to their decision-making. Shifts in decision-making from predominantly personal, compartmental reasoning at the beginning of the study to more scientific reasoning at the end of the study were observed. The authors propose that findings from this study provide some evidence to suggest that teaching NOS within an SSI is effective for facilitating the application of learners' NOS views to their decision-making.

Positive findings were also reported by Schalk (2012) who examined whether an inquiry-based SSI intervention effectively developed 26 undergraduate students' NOS conceptualisations, with respect to their ability to reason. This study utilised Hogan's (2000) conceptualisations of NOS as distal and/or proximal and examined undergraduates' distal and proximal NOS conceptualisations in a single-semester, microbiology course. During the SSI-based course, undergraduates took part in explicit-reflective NOS instruction and also participated in inquiry-based activities, including designing their own experiments and researching SSIs, which were supported by discussion and reflection on underlying epistemological conceptions. Data collection instruments included student journals, laboratory reports, research projects, an end-of-semester laboratory quiz, an end-of-semester evaluation and a start- and end-of-semester evidence evaluation activity (Sadler et al. 2004). Qualitative data from these multiple instruments were inductively analysed, and results indicated that undergraduates' distal knowledge of NOS developed, in addition to an improvement in reasoning skills. Undergraduates reasoning was shown to change as they recognised that knowledge was not absolute, and they began to reason with an increased awareness of relevant epistemological considerations. The author states that their improved reasoning was associated with more developed conceptions of NOS, and students were successful in linking their NOS views to their reasoning. Implications of this research highlight the importance of providing explicit discussions and reflective opportunities about NOS during inquiry-based activities to ensure learners make connections between their epistemological views and their reasoning during engagement in SSIs.

In summary, findings from these studies indicate that interventions conducted with college-level students were relatively more successful than interventions conducted with high school students, in terms of enabling students to apply their NOS views to their arguments within the context of SSIs. Although high school students' NOS views developed in the interventions conducted by Khishfe (2012b) and Walker and Zeidler (2007), findings indicated that, in general, students' NOS views did not influence decision-making during the SSIs. Conversely, college-level students were able to successfully apply their developed NOS views to their reasoning (Bell et al. 2011; Matkins and Bell 2007; Schalk 2012).

Interventions Exploring the Influence of Argumentation on NOS Only two interventions have sought to examine the nature of the relationship between NOS and argumentation from the opposite perspective. These two related studies were guided by a sociocultural perspective where learners are enculturated into a classroom community to enable them to develop socioscientific discourses. Eastwood et al. (2012) explored the influence of socioscientific issue (SSI)-driven, and content-driven, learning contexts on four classes of 11th and 12th grade students' NOS views over a full school year. All students experienced explicit-reflective NOS instruction throughout the intervention, with two classes participating in a content-based curriculum and two classes participating in an SSI-based curriculum. The SSI-based curriculum embedded course content within a series of SSIs underpinned by Kolstø's (2001) 'content transcending' themes and utilised pedagogical strategies for decision-making including a consideration of moral consequences, establishing criteria for evidence and examining scenarios with uncertain outcomes. Students spent the majority of class time engaged in argumentation, small group work, discussion and role plays. In contrast, students in the content curriculum were engaged in traditional teaching and learning activities and spent the majority of class time engaged in lectures, laboratory tasks, discussion of core concepts and completion of worksheets. All students were administered an open-ended NOS questionnaire (VNOS-C) at the commencement and conclusion of the school year, and responses were analysed to generate student profiles. Quantitative analyses showed significant gains in the majority of examined NOS aspects for all participants. No significant differences in gains in NOS understandings were found between the SSI and content group, although qualitative analysis of students' post-intervention VNOS responses indicated that students in the SSI group were more likely to use socioscientific examples to illustrate their views of the social and cultural aspect of NOS. The authors conclude that embedding NOS in SSI was found to be effective in promoting more informed views of the social and cultural NOS, and the authors infer that this was a likely consequence of engaging in classroom instruction where social factors were debated and reflected upon, in addition to engaging students in argumentation and debate.

In a study utilising the same participant group, Zeidler et al. (2009) explored relationships between SSI instruction and the development of reflective judgement of 11th and 12th grade students over a full school year. The authors were interested in exploring the reasoning patterns students utilised whilst engaged in solving ill-structured problems. Students were provided with opportunities to reflect on issues to enable them to analyse evidence, assess claims and evaluate differing views on ethical issues in science, via social interaction and discourse. The authors proposed that by engaging in these practices, aspects of NOS would be evoked. Participants took part in the same explicit NOS instruction as the previously reviewed study (Eastwood et al. 2012), in either the assigned treatment groups (SSI groups) or the comparison groups (content groups). The *Reflective Judgment Model* (RJM) was utilised to evaluate the efficacy of the SSI instruction, and the *Prototypic Reflective Judgment Interview* (PRJI) (King and Kitchener 1994, 2004) was the instrument used to assess students' reflective judgement. The instrument utilises interview questions designed to elicit personal epistemological views including the adequacy of alternative explanations and views about the certainty of knowledge. A mixed methods analysis strategy was used, and results indicated that students in the SSI group showed evidence of epistemological development, whereas students in the content group showed no evidence of development of their epistemological views. Other results indicated that students in the SSI group learned more science content (anatomy and physiology concepts) than students in the content group. The authors conclude that their results provide evidence that utilising an SSI approach can promote reflective judgement, which is comparable to the development of understandings of NOS. They state that by engaging in SSI, students are able to reflect the normative activity of scientific argumentation, by engaging in the evaluation of claims by various stakeholders. Results of this study highlight the importance of incorporating argumentation and evidence-based reasoning in contexts that aim to develop learners' reflective judgement.

In summary, findings from these two studies (Eastwood et al. 2012; Zeidler et al. 2009) suggest that engaging high school students in SSI learning environments underpinned by sociocultural perspectives, and incorporating argumentation and evidence-based reasoning, leads to favourable developments in their personal epistemological views, in addition to their views of NOS. The following section will examine studies conducted in scientific contexts.

2.2.4 Scientific Contexts

The following subsections will critically review the nine research reports which have explored NOS and argumentation in scientific contexts. The first subsection will review studies which sought to assess the influence of NOS on argumentation and will be followed by an analysis of intervention studies which have examined the influence of NOS on argumentation. A consideration of intervention studies which have examined the influence of argumentation on NOS will conclude the section.

Assessments of the Influence of NOS on Argumentation The first study identified in the literature which sought to examine the influence of NOS on argumentation in scientific context was conducted by Hogan and Maglienti (2001) who investigated science professionals' and non-scientists' scientific reasoning, by examining the epistemological criteria they used to rate the validity of conclusions drawn from a body of evidence. The participant group consisted of 45 volunteers, including 24 middle school students and five non-scientist adults (comprising the 'non-scientist' group) and six research scientists and ten science technicians (comprising the 'science professionals' group). Data was collected via one-to-one interviews, where participants were asked to judge ten conclusions and explain their reasoning for rating the conclusions as valid or invalid. Results indicated that the science professionals' predominant epistemological criterion for judging conclusions was the coherence of conclusions with the range of available evidence, thus prioritising empirical consistency. The non-scientists were also found to apply that criterion some of the time, but also applied alternative criteria more frequently than the science professionals, which were based on the plausibility of conclusions and often included personal views. The authors argue that the differences in the two group's use of epistemological criteria did not appear to be related to age, developmental level or prior knowledge and postulate that the differences may be due to the contextual frameworks that individuals brought to the task from their differing spheres of cultural practice. They conclude that efforts to develop scientific reasoning must broaden their focus to not only consider cognitive perspectives on scientific reasoning but also encompass sociocultural perspectives on the origin of epistemological criteria in the cultural activities individuals engage in. They suggest that simply engaging students in activity-based science is not sufficient to provide them with insights into the epistemological underpinnings of the scientific enterprise.

More recent research has also been conducted with younger students in this area. Sandoval and Cam (2011) examined 26 elementary students' judgements of the epistemic status of justifications for causal claims. The authors were interested in whether children would show preferences for evidence, plausible mechanisms or appeals to authority. The research was guided by an approach to epistemological development which viewed cognition as situated (Brown et al. 1989) and operated under the assumption that neither explicit instruction about NOS nor inquiry-oriented science instruction would lead to changes in students' epistemological beliefs, unless the classroom instruction triggers these beliefs. Students took part in individual interviews where they were asked to help story characters choose the best reason for a claim. Results indicated that the students showed a broad preference for empirical evidence, with data viewed as the best justification. Appeals to plausible mechanisms were also shown when the data appeared inconclusive, and appeals to authority were least preferable, with preferences linked to the credibility of the justification. The authors state that at least two epistemological resources were activated during decision-making in this study – the idea that claims require evidence and the idea that causal mechanisms must be plausible. They recommend the instruction geared towards highlighting the attributes of experimentation that make data credible be utilised to aid in the development of students' epistemological views.

Building on Hogan and Maglienti's (2001) earlier work, Zeineddin and Abd-El-Khalick (2010) sought to examine the relationship between 139 college students' epistemological commitments and their scientific reasoning. Epistemological commitments in this study refer to beliefs about the nature of knowledge and the processes of knowing. An integrative approach was utilised to examine students' reasoning by contextualising reasoning tasks in a specific science domain 'hydrostatics', thus also enabling a consideration of students' prior knowledge. A quasi-experimental, factorial design was used to examine the impact of a specific epistemological commitment 'consistency of theory with evidence' on the quality of students' reasoning processes. Participants were enrolled in two physics courses and completed an online questionnaire assessing their prior knowledge of the topic and epistemological commitment to the consistency of theory with evidence. After an analysis of the responses to the online questionnaire, four groups of ten students were selected to represent varying levels of prior knowledge (PK) and epistemological commitment (EC): high PK-high EC, high PK-low EC, low PK-high EC and low PK-low EC. Students in each of the groups took part in individual interviews to assess the quality of their reasoning by engaging them in a set of four problem-solving tasks based on hydrostatics. Results indicated significant differences in reasoning quality among the four groups. Prior knowledge and epistemological commitment were found to be important determinants of reasoning quality, with results showing that the higher the ECs were, the higher the quality of reasoning was, for comparable levels of PK. Other findings indicated that PK impacted reasoning more strongly when ECs were weaker. The authors propose that advanced scientific reasoning is more likely to occur when prior knowledge is accurate and epistemological commitments are sophisticated, although, importantly, results of this study indicated that high-quality reasoning can also be present if either prior knowledge or epistemological commitment is strong. The authors recommend the inclusion of instructional activities that explicitly aid in developing students' epistemological beliefs to help facilitate high-quality reasoning.

In summary, findings from these studies provide evidence of the influence of epistemological views on reasoning in scientific contexts. Hogan and Maglienti (2001) and Sandoval and Cam (2011) highlight the importance of situating classroom contexts as knowledge-building communities to enable students to participate in instruction which focuses on constructing evidence-based explanations, thus triggering their epistemological views. Other findings provide evidence of a positive relationship between epistemological views and argumentation, with findings showing a link between higher levels of epistemological commitment and higher-quality reasoning (Zeineddin and Abd-El-Khalick 2010).

Interventions Exploring the Influence of NOS on Argumentation The first intervention study identified in the literature which sought to examine the influence of students' NOS views about scientific argumentation on their inquiry practices was conducted by Sandoval and Millwood (2005). They investigated the quality of 87 high school biology students' written explanations about natural selection over a 4-week period, utilising a software tool designed to support scientific inquiry and guide students in constructing theory-based scientific explanations. Their research was guided by the assumption that implicit epistemological ideas are reflected in students' selection and use of data in their scientific explanations. Results indicated that the software tool successfully provided supported argumentation instruction via scaffolds that allowed students to construct logical arguments. Nevertheless, students had difficulties citing sufficient data to support claims and also had difficulties providing warrants for some claims. Other results indicated that many students viewed data as self-evident and did not provide an explanation of the data in their scientific explanations. The authors proposed that students may not distinguish claims from data and may believe that data are an objective representation of scientific knowledge. Implications from this research suggest that students who display naive views of NOS may not provide explanations or warrants for their claims, thus influencing their ability to engage in scientific argumentation effectively.

A more recent study conducted by the same authors examined 33 grade 7 students' ideas about how to warrant claims (Sandoval and Millwood 2007). They explored how students' argumentation practices developed during a 3-week unit exploring plant adaptation and the possible influence of the inquiry on their ideas about NOS. Students completed the POSE (Perspectives on Scientific Epistemology; Abd-El-Khalick 2002) to assess their scientific epistemological views at the beginning of the study, which were found to be naive. Supported argumentation instruction was provided during an online investigation where students were instructed to present data-based arguments. Data analysis indicated that the majority of students were able to articulate claims, but most students did not warrant their claims. During individual interviews, over half of the students cited that warrants were the reason they believed their claims, even though less than 25 % of the students explicitly provided warrants in their written essays. The authors propose that the students may not have been motivated to provide explicit evidence (in the form of warrants) as the audience for the students' written arguments was their teacher and that their primary role was in providing the correct answer. Thus, the results of this study provide further empirical support for the assertion that learners' epistemological views influence their engagement in argumentation.

Nussbaum et al. (2008) examined the influence of college students' epistemic beliefs and exposure to argumentation instruction, on the quality of their arguments. Participants were 88 undergraduates (94 % were seeking a teaching credential) randomly assigned to the treatment or control group, where only the treatment group received supported argumentation instruction. A web-based learning environment provided the context for the investigation over a single semester, where both the control group and treatment group participants engaged in pair-based discussions of several physics problems. All participants completed a number of online surveys, including a survey assessing participants' tendency to approach or avoid arguments and an epistemic belief survey. Results indicated that treatment group participants developed better-quality arguments than control group participants, and the authors propose that more direct instruction (i.e. explicit argumentation instruction) would result in greater gains in argumentation. Another finding was that a significant proportion of treatment group participants expressed conceptually correct responses to one of the physics tasks. Other results indicated that participants' epistemic orientations influenced their argumentation. For example, participants with the most developed epistemological orientations brought up different ideas than their partners, rarely displayed inconsistent reasoning and tended to engage in more critical argumentation. Thus, findings from this study support the assertion that participants' epistemological views influence their engagement in argumentation.

In summary, findings from these three studies provide some evidence to support the assertion that learners' views of NOS influence their engagement in scientific argumentation in middle, high school and college-level school contexts. In general, naive epistemological views appear to constraint learners' engagement in scientific argumentation (Sandoval and Millwood 2005, 2007), and more informed epistemological views appear to promote engagement in scientific argumentation (Nussbaum et al. 2008).

Interventions Exploring the Influence of Argumentation on NOS Similarly to studies conducted in socioscientific contexts, some researchers have sought to explore the relationship between NOS and argumentation in scientific contexts from the opposite perspective. Three interventions have examined the influence of argumentation on students' NOS views in elementary, middle and high school contexts. Yerrick (2000) investigated five low-achieving high school students' participation in a general science unit which focused on argument construction, question generation and experimental design. The research was guided by a sociocultural perspective, and the researcher was interested in assessing changes in students' abilities to construct arguments in an open-inquiry setting. He explicitly taught argumentation to the students over one school year, and no explicit NOS instruction was implemented during the intervention. Results indicated that students' views of the tentative nature of scientific knowledge, the use of scientific evidence and the source of scientific authority developed over the duration of the study to be closely aligned with informed views of these NOS aspects. Students' views of the above aspects of NOS were also reflected in their arguments, and some improvements in their skills of argument were also evident. The author states that aspects of scientists' interactions as a community were able to be acquired by students as they participated in the social practices of the developed community.

Similar findings were reported by Bell and Linn (2000), who assessed 172 middle school students' argument constructions during a 6-week Knowledge Integration Environment (KIE) debate project. The classroom was structured as knowledge-building community where students learned to construct and apply evidence-based explanations, to provide them with insights into the epistemological underpinnings of the scientific enterprise. The study was guided by the assumption that arguments formulated by students will reflect aspects of their views about NOS. Supported argumentation instruction was implemented in the study via a software tool designed to make the structure of an argument visible to students. No explicit NOS instruction was implemented during the intervention, although students completed a multiple choice survey about their views of NOS at the commencement and conclusion of the study. Results indicated that students with more informed views of NOS created more complex arguments which integrated unique warrants, an increased frequency of warrant usage and more conceptual frames. Results also indicated that students' knowledge integration and skills of argumentation improved over the duration of the study. The authors state that their study provides evidence for the claim that engaging students in the process of argumentation improves their understanding of NOS, as post-test results of participants' NOS views indicated an improvement in NOS understandings.

More recently, Ryu and Sandoval (2012) investigated the influence of a sustained instructional focus on argumentation, on 21 elementary students' understanding and application of key epistemic criteria for scientific arguments. Guided by a sociocultural perspective, these authors view argument as a collective, cultural practice, which contrasts with cognitive views of argument as skill acquisition. They state that cognitive views of argument prioritise argumentation as a task that students 'should' engage in and may lead to the development of isolated 'argument lessons'. Instead, they propose that argumentation should be a core discursive feature of the classroom. Data sources included argument construction and evaluation tasks, videotaped observations and field notes of science lessons. No explicit NOS instruction was implemented during the intervention. The study was conducted over the course of the school year, and results indicated students learned how to apply evidentiary criteria whilst constructing and evaluating arguments and showed dramatic improvements in their use of explicit justifications. The authors attribute these developments to student's sustained engagement in the collective practices of argumentation enacted in the classroom, which were underpinned by the goals of persuasion and consensus. Implications from this study show that elementary children are capable of engaging in the practices of argumentation and, through engagement in these practices, are capable of developing their understandings of core epistemic criteria of scientific argument. The authors suggest that future research should focus less on teaching argumentation as a task and focus more on integrating a sustained approach to argumentation as a discourse that runs through the culture of the classroom.

In summary, findings from these studies conducted with elementary, middle and high school students provide evidence to support the assertion that engaging students in scientific argumentation leads to developments in their NOS views (Bell and Linn 2000; Ryu and Sandoval 2012; Yerrick 2000). The following section will discuss the themes emerging from the critical analysis and provide recommendations for future research.

2.3 Discussion and Recommendations

The achievement of scientific literacy is considered the primary goal of school science education. Previous research suggests that improving students' views of nature of science (NOS) is pivotal in helping to achieve this aim. More recent research proposes that engaging students in argumentation also aids in the development of scientific literacy, and emerging research has sought to examine the relationship between these two constructs. Some researchers have proposed that engaging learners in argumentation may lead to developments in their NOS views (e.g. Ford 2008; Sandoval 2005), whilst others have suggested that learners' NOS views may influence their engagement in argumentation (e.g. Nussbaum et al. 2008; Sandoval and Millwood 2007). This review critically evaluated 23 research reports from high-quality science education journals, across socioscientific and scientific learning contexts, to explore the nature of this relationship. The single study conducted in both socioscientific and scientific contexts (McDonald 2010) found differences in college-level students' abilities to engage in argumentation in the differing contexts. Although the majority of students developed informed NOS views over the duration of the single-semester intervention, students engaged in argumentation more effectively in socioscientific contexts and were able to apply their NOS views to their decision-making in these contexts. Importantly, a lack of argumentation scaffolds and epistemological probes may have constrained students' engagement in scientific contexts; thus, a direct comparison of the relative influences of socioscientific versus scientific contexts on students' NOS views is not possible to determine.

Research conducted in socioscientific contexts indicates that, in general, students' NOS views were found to influence their decision-making in assessment studies designed to investigate this relationship (Liu et al. 2011; Sadler et al. 2004; Zeidler et al. 2002). In addition, findings from recent assessment studies conducted with high school students (Khishfe 2012a; Wu and Tsai 2011) provide evidence of a positive relationship between NOS and argumentation, with results showing a link between informed NOS views and higher-quality argumentation. Importantly, a degree of epistemological consistency was evident in the results obtained for high school students from different countries in the examined assessment studies, with comparable findings reported in studies conducted in Taiwan, Lebanon and the United States. Similar findings were reported by Zeidler et al. (2013) who examined high school students' epistemological reasoning patterns about SSIs from a cross-cultural perspective in five countries (Jamaica, South Africa, Sweden, Taiwan and the United States) and found a degree of epistemological similarity among the student samples from the selected countries. Related patterns were found with the framing of reasoning on the SSI and for the justifications offered to support their epistemological beliefs. The authors state that these findings lend support to the claim that common elements of epistemological beliefs exist which may transcend culture in terms of socioscientific reasoning patterns on SSIs. This claim is further supported by the findings emerging from this review.

Regarding intervention studies conducted in socioscientific contexts, results indicated that interventions conducted with college-level students were relatively more successful than interventions conducted with high school students, in terms of enabling students to apply their NOS views to their arguments, within the context of SSIs. Although high school students' NOS views developed in the interventions conducted by Khishfe (2012b) and Walker and Zeidler (2007), findings indicated that, in general, students' NOS views did not influence their decision-making during the SSIs. Conversely, college-level students were able to successfully apply their developed NOS views to their reasoning during the examined SSIs (Bell et al. 2011; Matkins and Bell 2007; Schalk 2012). Possible reasons for this difference may be attributed to age or developmental factors, although previous research conducted with high school students has shown they are developmentally capable of engaging in decision-making informed by epistemological considerations (e.g. Sadler et al. 2004; Zeidler et al. 2002). A closer analysis of the intervention studies highlighted a difference in the length of the interventions in the studies conducted with high school students, compared to the studies conducted with college-level students. The interventions conducted with high school students ranged from 4 to 7 weeks in duration, whereas the interventions conducted with college-level students were all one semester (approximately 15 weeks) in length. This suggests that the length of the intervention may be a pivotal factor in determining whether learners are able to apply their NOS understandings to the reasoning. Findings reported in scientific contexts (Ryu and Sandoval 2012; Yerrick 2000) lend support to the necessity of providing an extended time to immerse learners in argumentation in the learning context. This is an important area of future research in socioscientific contexts.

Findings from the two intervention studies examining the influence of argumentation on NOS views in socioscientific contexts (Eastwood et al. 2012; Zeidler et al. 2009) indicated that engaging high school students in these learning environments led to favourable developments in their personal epistemological views, in addition to their views of NOS. Importantly, these studies were conducted over an extended time period (1 full year) and were underpinned by sociocultural perspectives. As such, these findings lend further support to the assertion that learners need to be immersed in a classroom culture emphasising argumentation and evidence-based reasoning to enable them to appropriate the norms of the classroom, over an extended period of time (Ryu and Sandoval 2012; Yerrick 2000).

Finally, all of the socioscientific interventions which reported improvements in learners' NOS views (for treatment group participants) implemented explicit NOS instruction. Thus, it could be concluded that the inclusion of explicit NOS instruction is a necessary prerequisite to aid in the development of learners' NOS views, and this assertion is supported by a large body of empirical evidence (e.g. Abd-El-Khalick and Lederman 2000b; Hanuscin et al. 2006; Schwartz et al. 2004). Importantly, one must caution the reader not to presuppose that explicit NOS instruction alone is sufficient to ensure learners will apply their NOS views to their reasoning, as results from some of the examined studies (e.g. Bell and Lederman 2003; Walker and Zeidler 2007) indicated that students were not always successful in achieving this goal. Nielsen (2012) reminds us of the complexity of decision-making whilst engaged in SSIs and the crucial role of student values in reasoning on SSIs. Reasoning during engagement in SSIs is not a straightforward process and is heavily influenced by personal opinion and human values, in addition to scientific evidence and views of NOS. Thus, further research investigating strategies to facilitate the application of learners' NOS views to their reasoning in these contexts is needed, in addition to exploring possible reasons why learners may, or may not, incorporate epistemological considerations when engaged in socioscientific decision-making.

Research conducted in scientific contexts provides evidence of the influence of epistemological views on reasoning in assessment studies conducted with elementary students, middle school students, college-level students, non-scientist adults and science professionals (Hogan and Maglienti 2001; Sandoval and Cam 2011; Zeineddin and Abd-El-Khalick 2010). Research conducted by Zeineddin and Abd-El-Khalick (2010) showed evidence of a positive relationship between epistemological views and argumentation, with findings showing a link between higher levels of epistemological commitment and higher-quality reasoning. These results are supported by similar research findings in socioscientific contexts (Khishfe 2012a; Wu and Tsai 2011) and further strengthen the claim that informed NOS views are correlated with higher-quality argumentation.

Intervention studies conducted in scientific contexts show a degree of consistency in the reporting of their findings, with studies conducted with middle school students (Sandoval and Millwood 2007), high school students (Sandoval and Millwood 2005) and college-level students (Nussbaum et al. 2008), providing evidence to support the assertion that students' views of NOS influence their engagement in scientific argumentation. In general, naive epistemological views appear to constraint learners' engagement in scientific argumentation (Sandoval and Millwood 2005, 2007), and more informed epistemological views appear to promote engagement in scientific argumentation (Nussbaum et al. 2008). Interestingly, the studies conducted by Sandoval and Millwood (2005, 2007) were of a short duration (3–4 weeks), with results generally indicating that students did not develop quality arguments during this period. Conversely, the study by Nussbaum et al. (2008) was conducted over a full semester (approximately 15 weeks), and students in the treatment group developed quality arguments. This finding provides further evidence to support the assertion that the length of the intervention may be a pivotal factor in studies which explore NOS and argumentation and is supported by findings in socioscientific contexts which showed that longer interventions (e.g. Bell et al. 2011; Eastwood et al. 2012; Matkins and Bell 2007; Schalk 2012; Zeidler et al. 2009) were generally more successful than shorter interventions (e.g. Khishfe 2012b; Walker and Zeidler 2007).

Similar to the results reported in socioscientific contexts (Eastwood et al. 2012; Zeidler et al. 2009), findings from intervention studies examining the influence of argumentation on NOS views in scientific contexts provide evidence to support the assertion that engaging students in scientific argumentation leads to developments in their NOS views, in elementary (Ryu and Sandoval 2012), middle (Bell and Linn 2000) and high school (Yerrick 2000) contexts. An interesting theme emerging from a closer examination of these studies indicated students' NOS views developed *without* the incorporation of explicit NOS instruction. This finding runs contrary to the results reported in socioscientific contexts and the significant body of empirical research advocating the necessity of explicit NOS instruction in interventions which aim to improve learners' NOS. All three examined studies were underpinned by sociocultural perspectives, and two of the studies (Ryu and Sandoval 2012; Yerrick 2000) were conducted over an extended time period (1 full year). As stated earlier, some researchers have suggested that epistemological ideas about science cannot be transmitted didactically to students and propose that students must be engaged in the practices of science, including argumentation, to further develop their NOS understandings (Duschl 2008; Sandoval 2005). All of these examined studies immersed learners in a classroom culture emphasising argumentation and evidence-based reasoning, to create a need for learners to engage in argumentation and apply their epistemological views to their arguments. These types of learning environments are synonymous with Cavagnetto's (2010) 'immersion-oriented interventions' which use argument as a tool for constructing and understanding scientific principles and cultural practices of science. As a result of his review, Cavagnetto (2010) concluded the immersion orientation held the greatest potential for promoting scientific literacy as it was the only orientation to fully encapsulate epistemic aspects of science embedded in the practices of science. Thus, further studies underpinned by sociocultural perspectives are needed to add to the growing literature base supporting these types of interventions.

To conclude, many similarities were found regarding the nature of the relationship between NOS and argumentation in socioscientific and scientific contexts. Assessment studies in both contexts found evidence of the influence of NOS views on reasoning and positive relationships linking informed views of NOS to high-quality argumentation. Intervention studies in both contexts indicate that NOS views influence argumentation, although in different ways. In socioscientific contexts, the focus of studies is on the ability of students to apply their NOS views to their reasoning, whereas in scientific contexts, the focus of studies lies in whether

NOS views constraint or facilitate engagement in argumentation. Intervention studies conducted in both contexts also explored the relationship between NOS and argumentation from the opposite perspective, with results indicating that argumentation influences NOS views.

The duration of the intervention was found to be an important factor in the success of the examined studies, with longer interventions producing more favourable results. In addition, studies underpinned by sociocultural perspectives were generally successful in achieving their desired aims. A key point of difference between socioscientific and scientific contexts relates to explicit NOS instruction. All of the socioscientific interventions which reported improvements in learners' NOS views implemented explicit NOS instruction, whereas all of the scientific interventions which reported improvements in learners' NOS views did not implement explicit NOS instruction. Thus, further research is needed to examine the role of explicit NOS instruction in interventions which explore the relationship between NOS and argumentation.

References

(AAAS) American Association for the Advancement of Science. (1990). *Science for all Americans*. New York: Oxford University Press.

(AAAS) American Association for the Advancement of Science. (1993). *Benchmarks for science literacy: A project 2061 report*. New York: Oxford University Press.

(ACARA) Australian Curriculum, Assessment and Reporting Authority. (2012). *Australian curriculum: Science F-10 version 3.0*. Sydney: Commonwealth of Australia.

(NRC) National Research Council, (1996). *National science education standards*. Washington, DC: National Academic Press.

Abd-El-Khalick, F. S. (1998). *The influence of history of science courses on students' conceptions of the nature of science*. Unpublished doctoral dissertation, Corvallis: Oregon State University.

Abd-El-Khalick, F. (2002). *The development of conceptions of the nature of scientific knowledge and knowing in the middle and high school years: A cross-sectional study*. Paper presented at the annual meeting of the National Association for Research in Science Teaching (NARST), New Orleans, LA.

Abd-El-Khalick, F., & Akerson, V. L. (2004). Learning as conceptual change: Factors mediating the development of preservice elementary teachers' views of nature of science. *Science Education, 88*, 785–810.

Abd-El-Khalick, F., & Lederman, N. G. (2000a). Improving science teachers' conceptions of nature of science: A critical review of the literature. *International Journal of Science Education, 22*(7), 665–701.

Abd-El-Khalick, F., & Lederman, N. G. (2000b). The influence of history of science courses on students' views of nature of science. *Journal of Research in Science Teaching, 37*(10), 1057–1095.

Albe, V. (2008). When scientific knowledge, daily life experience, epistemological and social considerations intersect: Students' argumentation in group discussion on a socio-scientific issue. *Research in Science Education, 38*, 67–90.

Bell, R. L., & Lederman, N. G. (2003). Understandings of the nature of science and decision making on science and technology based issues. *Science Education, 87*, 352–377.

Bell, P., & Linn, M. C. (2000). Scientific arguments as learning artifacts: Designing for learning from the web with KIE. *International Journal of Science Education, 22*(8), 797–817.

Bell, R. L., Blair, L. M., Crawford, B. A., & Lederman, N. G. (2003). Just do it? The impact of a science apprenticeship program on high school students' understandings of the nature of science and scientific inquiry. *Journal of Research in Science Teaching, 40*, 487–509.

Bell, R. L., Matkins, J. J., & Gansneder, B. M. (2011). Impacts of contextual and explicit instruction on preservice elementary teachers' understandings of the nature of science. *Journal of Research in Science Teaching, 48*(4), 414–436.

Berland, L. K., & Hammer, D. (2012). Framing for scientific argumentation. *Journal of Research in Science Teaching, 49*(1), 68–94.

Berland, L. K., & Reiser, B. J. (2009). Making sense of argumentation and explanation. *Science Education, 93*(1), 26–55.

Bricker, L. A., & Bell, P. (2008). Conceptualisations of argumentation from science studies and the learning sciences and their implications for the practices of science education. *Science Education, 92*, 473–498.

Brown, J. S., Collins, A., & Duguid, P. (1989). Situated cognition and the culture of learning. *Educational Researcher, 18*(1), 34–41.

Cavagnetto, A. R. (2010). Argument to foster scientific literacy: A review of argument interventions in K-12 science contexts. *Review of Educational Research, 80*(3), 336–371.

Clark, D., & Sampson, V. (2006, April). *Characteristics of students' argumentation practices when supported by online personally-seeded discussions.* Paper presented at the annual meeting of the National Association for Research in Science Teaching (NARST), San Francisco, CA.

Crawford, T. (2005). What counts as knowing: Constructing a communicative repertoire for student demonstration of knowledge in science. *Journal of Research in Science Teaching, 42*(2), 139–165.

Dori, Y. J., Tal, R., & Tsaushu, M. (2003). Teaching biotechnology through case studies: Can we improve higher-order thinking skills of non-science majors? *Science Education, 87*, 767–793.

Driver, R., Newton, P., & Osborne, J. (2000). Establishing the norms of scientific argumentation in classrooms. *Science Education, 84*, 287–312.

Duschl, R. A. (1990). *Restructuring science education: The importance of theories and their development.* New York: Teachers College Press.

Duschl, R. A. (2008). Science Education in 3-part harmony: Balancing conceptual, epistemic and social goals. *Review of Research in Education, 32*, 268–291.

Duschl, R. A., & Osborne, J. (2002). Supporting and promoting argumentation discourse in science education. *Studies in Science Education, 38*, 39–72.

Eastwood, J. L., Sadler, T. D., Zeidler, D. L., Lewis, A., Amiri, L., & Applebaum, S. (2012). Contextualizing nature of science instruction in socioscientific issues. *International Journal of Science Education, 34*(15), 2289–2315.

Erduran, S., Simon, S., & Osborne, J. (2004). Tapping into argumentation: Developments in the application of Toulmin's argument pattern for studying science discourse. *Science Education, 88*(6), 915–933.

Evagorou, M. (2011). Discussing a socioscientific issue in a primary school classroom. The case of using a technology-supported environment in formal and nonformal settings. In T. Sadler (Ed.), *Socioscientific issues in the classroom* (pp. 133–160). New York: Springer.

Evagorou, M., & Osborne, J. (2013). Exploring young students' collaborative argumentation within a socioscientific issue. *Journal of Research in Science Teaching, 50*(2), 209–237.

Ford, M. (2008). Disciplinary authority and accountability in scientific practice and learning. *Science Education, 92*(3), 404–423.

Giere, R. N. (1979). *Understanding scientific reasoning*. New York: Holt, Rinehart, & Winston.

Govier, T. (2010). *A practical study of argument* (7th ed.). Belmont: Wadsworth Cengage Learning.

Grace, M. (2009). Developing high quality decision-making discussions about biological conservation in a normal classroom setting. *International Journal of Science Education, 31*(4), 551–570.

Hanuscin, D. L., Akerson, V. L., & Phillipson-Mower, T. (2006). Integrating nature of science instruction into a physical science content course for preservice elementary teachers: NOS views of teaching assistants. *Science Education, 90*, 912–935.

Harris, R., & Ratcliffe, M. (2005). Socio-scientific issues and the quality of exploratory talk: What can be learned from schools involved in a 'collapsed day' project? *Curriculum Journal, 16*, 439–453.

Hofer, B. K., & Pintrich, P. R. (1997). The development of epistemological theories: Beliefs about knowledge and knowing and their relation to learning. *Review of Educational Research, 67*, 88–140.

Hogan, K. (2000). Exploring a process view of students' knowledge about the nature of science. *Science Education, 84*(1), 51–70.

Hogan, K., & Maglienti, M. (2001). Comparing the epistemological underpinnings of students' and scientists' reasoning about conclusions. *Journal of Research in Science Teaching, 38*, 663–687.

Jimenez-Aleixandre, M.-P., & Erduran, S. (2007). Argumentation in science education: An overview. In S. Erduran & M.-P. Jimenez-Aleixandre (Eds.), *Argumentation in science education: Perspectives from classroom-based research* (pp. 3–27). Dordrecht: Springer.

Kelly, G. J., Druker, S., & Chen, K. (1998). Students' reasoning about electricity: Combining performance assessment with argumentation analysis. *International Journal of Science Education, 20*(7), 849–871.

Khishfe, R. (2012a). Relationship between nature of science understandings and argumentation skills: A role for counterargument and contextual factors. *Journal of Research in Science Teaching, 49*(4), 489–514.

Khishfe, R. (2012b). Nature of science and decision-making. *International Journal of Science Education, 34*(1), 67–100.

King, P. M., & Kitchener, K. S. (1994). *Developing reflective judgment: Understanding and promoting intellectual growth and critical thinking in adolescents and adults*. San Francisco: Jossey-Bass.

King, P. M., & Kitchener, K. S. (2004). Reflective judgment: Theory and research on the development of epistemic assumptions through adulthood. *Educational Psychologist, 39*, 5–18.

Kolstø, S. D. (2001). Scientific literacy for citizenship: Tools for dealing with the science dimension of controversial socioscientific issues. *Science Education, 85*, 291–310.

Kortland, K. (1996). An STS case study about students' decision making on the waste issue. *Science Education, 80*, 673–689.

Kuhn, D. (1991). *The skills of argument*. Cambridge: Cambridge University Press.

Kuhn, D. (1993). Science as argument: Implications for teaching and learning scientific thinking. *Science Education, 77*(3), 319–337.

Kuhn, D. (2010). Teaching and learning science as argument. *Science Education, 94*, 810–824.

Kuhn, L. & Reiser, B. J. (2006, April). *Structuring activities to foster argumentative discourse.* Paper presented at the annual meeting of the American Educational Research Association (AERA), San Francisco, CA.

Leach, J., Millar, R., Ryder, J., & Sere, M.-G. (2000). Epistemological understanding in science learning: The consistency of representations across contexts. *Learning & Instruction, 10*, 497–527.

Lederman, N. G. (1992). Students' and teachers' conceptions of the nature of science: A review of the research. *Journal of Research in Science Teaching, 29*(4), 331–359.

Lemke, J. (1990). *Talking science: Language, learning, and values*. Norwood: Ablex.

Liu, S.-Y., Lin, C.-S., & Tsai, C.-C. (2011). College students' scientific epistemological views and thinking patterns in socioscientific decision-making. *Science Education, 95*, 497–517.

Matkins, J. J., & Bell, R. L. (2007). Awakening the scientist inside: Global climate change and the nature of science in an elementary science methods course. *Journal of Science Teacher Education, 18*, 137–163.

McDonald, C. V. (2010). The influence of explicit nature of science and argumentation instruction on preservice primary teachers' views of nature of science. *Journal of Research in Science Teaching, 47*(9), 1137–1164.
McDonald, C. V., & McRobbie, C. J. (2012). Utilising argumentation to teach nature of science. In B. Fraser, K. Tobin, & C. McRobbie (Eds.), *Second international handbook of science education* (pp. 969–986). Dordrecht: Springer.
McNeill, K. L., & Pimentel, D. S. (2010). Scientific discourse in three urban classrooms: The role of the teacher in engaging high school students in argumentation. *Science Education, 94*(2), 203–229.
Means, M. L., & Voss, J. F. (1996). Who reasons well? Two studies of informal reasoning among children of different grade, ability, and knowledge levels. *Cognition & Instruction, 14*, 139–178.
Moss, D. M., Abrams, E. D., & Robb, J. (2001). Examining student conceptions of the nature of science. *International Journal of Science Education, 23*(8), 771–790.
Newton, P., Driver, R., & Osborne, J. (1999). The place of argumentation in the pedagogy of school science. *International Journal of Science Education, 21*(5), 553–576.
Nielsen, J. A. (2012). Co-opting science: A preliminary study of how students invoke science in value-laden discussions. *International Journal of Science Education, 34*(2), 275–299.
Nussbaum, E. M., Sinatra, G. M., & Poliquin, A. (2008). Role of epistemic beliefs and scientific argumentation in science learning. *International Journal of Science Education, 30*(15), 1977–1999.
Osborne, J. F., & Patterson, A. (2011). Scientific argument and explanation: A necessary distinction? *Science Education, 95*, 627–638.
Osborne, J., Erduran, S., & Simon, S. (2004). Enhancing the quality of argumentation in school science. *Journal of Research in Science Teaching, 41*(10), 994–1020.
Pedretti, E. (1999). Decision making and STS education: Exploring scientific knowledge and social responsibility in schools and science centers through an issues-based approach. *School Science and Mathematics, 99*, 174–181.
Ryu, S., & Sandoval, W. A. (2012). Improvements to elementary children's understanding from sustained argumentation. *Science Education, 96*, 488–526.
Sadler, T. D. (2009). Situated learning in science education: Socio-scientific issues as contexts for practice. *Studies in Science Education, 45*(1), 1–42.
Sadler, T. D., & Fowler, S. (2006). A threshold model of content knowledge transfer for socioscientific argumentation. *Science Education, 90*, 986–1004.
Sadler, T. D., Chambers, F. W., & Zeidler, D. L. (2004). Student conceptualisations of the nature of science in response to a socioscientific issue. *International Journal of Science Education, 26*(4), 387–409.
Sampson, V., & Blanchard, M. R. (2012). Science teachers and scientific argumentation: Trends in views and practice. *Journal of Research in Science Teaching, 49*(9), 1122–1148.
Sampson, V. D. & Clark, D. B. (2006, April). *The development and validation of the nature of science as argument questionnaire (NSAAQ).* Paper presented at the annual meeting of the National Association for Research in Science Teaching (NARST), San Francisco, CA.
Sampson, V., & Clark, D. B. (2008). Assessment of the ways students generate arguments in science education: Current perspectives and recommendations for future directions. *Science Education, 92*(3), 447–472.
Sandoval, W. A. (2005). Understanding students' practical epistemologies and their influence on learning through inquiry. *Science Education, 89*, 634–656.
Sandoval, W. A., & Cam, A. (2011). Elementary children's judgments of the epistemic status of sources of justification. *Science Education, 95*, 383–408.
Sandoval, W. A., & Millwood, K. A. (2005). The quality of students' use of evidence in written scientific explanations. *Cognition and Instruction, 23*(1), 23–55.

Sandoval, W. A., & Millwood, K. A. (2007). What can argumentation tell us about epistemology? In S. Erduran & M.-P. Jimenez-Aleixandre (Eds.), *Argumentation in science education: Perspectives from classroom-based research* (pp. 71–88). Dordrecht: Springer.

Sandoval, W. A., & Morrison, K. (2003). High school students' ideas about theories and theory change after a biological inquiry unit. *Journal of Research in Science Teaching, 40*(4), 369–392.

Schalk, K. A. (2012). A socioscientific curriculum facilitating the development of distal and proximal NOS conceptualizations. *International Journal of Science Education, 34*(1), 1–24.

Schommer-Aitkins, M. (1993). Epistemological development and academic performance among secondary students. *Journal of Educational Psychology, 85*, 406–411.

Schwartz, R. S., Lederman, N. G., & Crawford, B. (2004). Developing views of nature of science in an authentic context: An explicit approach to bridging the gap between nature of science and scientific inquiry. *Science Education, 88*, 610–645.

Simon, S., Erduran, S., & Osborne, J. (2006). Learning to teach argumentation: Research and development in the science classroom. *International Journal of Science Education, 28*, 235–260.

Tal, R., & Hochberg, N. (2003). Assessing high order thinking of students participating in the 'WISE' project in Israel. *Studies in Educational Evaluation, 29*, 69–89.

Tal, T., & Kedmi, Y. (2006). Teaching socio-scientific issues: Classroom culture and students' performances. *Cultural Studies of Science Education, 1*(4), 615–644.

Tsai, C.-C., & Liu, S.-Y. (2005). Developing a multi-dimensional instrument for assessing students' epistemological views toward science. *International Journal of Science Education, 27*, 1621–1638.

Tytler, R. (2007). *Re-imagining science education: Engaging students in science for Australia's future*. Camberwell: Australian Council for Educational Research (ACER) Press.

Walker, K., & Zeidler, D. L. (2007). Promoting discourse about socioscientific issues through scaffolded inquiry. *International Journal of Science Education, 29*(11), 1387–1410.

Wu, Y.-T., & Tsai, C.-C. (2007). High school students' informal reasoning on a socio-scientific issue: Qualitative and quantitative analyses. *International Journal of Science Education, 29*(2), 1163–1187.

Wu, Y.-T., & Tsai, C.-C. (2011). High school students' informal reasoning regarding a socio-scientific issue, with relation to scientific epistemological beliefs and cognitive structures. *International Journal of Science Education, 33*(3), 371–400.

Yerrick, R. K. (2000). Lower track science students' argumentation and open inquiry instruction. *Journal of Research in Science Teaching, 37*(8), 807–838.

Zeidler, D. L., Walker, K. A., Ackett, W. A., & Simmons, M. L. (2002). Tangled up in views: Beliefs in the nature of science and responses to socioscientific dilemmas. *Science Education, 86*, 343–367.

Zeidler, D. L., Sadler, T. D., Applebaum, S., & Callahan, B. E. (2009). Advancing reflective judgment through socio-scientific issues. *Journal of Research in Science Teaching, 46*, 74–101.

Zeidler, D. L., Herman, B. C., Ruzek, M., Linder, A., & Lin, S.-S. (2013). Cross-cultural epistemological orientations to socioscientific issues. *Journal of Research in Science Teaching, 50*(3), 251–283.

Zeineddin, A., & Abd-El-Khalick, F. (2010). Scientific reasoning and epistemological commitments: Coordination of theory and evidence among college science students. *Journal of Research in Science Teaching, 47*(9), 1064–1093.

Zohar, A., & Nemet, F. (2002). Fostering students' knowledge and argumentation skills through dilemmas in human genetics. *Journal of Research in Science Teaching, 39*(1), 35–62.

Chapter 3
The Relationship Between Science and Religion: A Contentious and Complex Issue Facing Science Education

Keith S. Taber

3.1 Introduction

Traditionally, at least in many national contexts, science and religion have been seen as quite distinct and largely unrelated school and college subjects and indeed as generally nonoverlapping activities in society in general. However, this is both an oversimplification that ignores the complex and sometimes nuanced interactions between science and religion and a misleadingly comfortable position that is being challenged in many parts of the world.

Consequently, there has recently been increased attention to the relationship between science and religion and how this might impact on teaching and learning science. A particular area of contention is the teaching of scientific theories and models related to origins, including ideas about the origins of the universe and evolutionary ideas about human origins. (The particular issues surrounding the challenges of teaching evolution are discussed in more detail in Chap. 4). However, there is no room for complacency for those not teaching such potentially controversial topics, as the difficulties being faced by those charged with teaching natural selection in some national contexts are a symptom of a deeper underlying tension between understandings of the nature of science and the worldview commitments of many of those holding religious faiths. This chapter offers an introduction to key features of this issue and explains why this is an important topic that all working in science education need to take seriously.

K.S. Taber (✉)
Faculty of Education, University of Cambridge, Cambridge, UK
e-mail: kst24@cam.ac.uk

B. Akpan (ed.), *Science Education: A Global Perspective*,
DOI 10.1007/978-3-319-32351-0_3

3.2 Symptoms of a Problem in Science Education

It is clear that in some national contexts, there is an active focus on the 'science and religion issue', which is causing concern for science educators. This has perhaps had the highest profile in the USA where there have been a series of legal battles focused on the teaching of evolutionary theory in school and where a new perspective ('intelligent design', ID), claiming scientific credibility, has been presented as an 'alternative' to widely accepted evolutionary theory (see Chap. 4). At one level science and science education have held firm in the USA, in the sense that the judiciary has tended to support the teaching of evolutionary theory in US schools, whilst denying requests to also impose teaching (in science lessons) of alternative 'theories' of human origins. However, although the science/science education community in the USA may be winning what we might think of as the 'battle for minds', there is much evidence that the 'battle for hearts' often proceeds rather differently, with individual teachers and students often ignoring legal decisions to choose for themselves which topics they will teach, or study, according to conscience (Long 2011).

This is not an issue for the USA alone. Controversy around teaching of evolution has arisen in various other national contexts where previously science and religion have long coexisted largely independently within mainstream education (Billingsley et al. 2014). As in the USA, the focus of debate is often on the teaching of evolution, and/or more specifically natural selection, as evolution has become something of an intellectual battleground for those adopting different worldview commitments. Moreover, as this chapter will illustrate, it would be a mistake to see the conflict as purely the result of some people of particular religious persuasions objecting to the teaching of some scientific content. That is certainly an important aspect, but increasingly the controversy has also been encouraged by some scientists who are in principle opposed to religion and its influence in society.

Symptomatic of the tensions in science education concerning the relationship between science and religion was an incident in 2008 when a well-respected science educator, and a strong advocate of evolution, Professor Michael Reiss, was pressured to step down from his role as education director in a learned scientific society. In the UK, the Royal Society (RS) – one of the most esteemed scientific societies in the world – was embarrassed by the advice Reiss (also a professor of science education at a prestigious university) offered on how teachers should deal with student responses to the teaching of evolution in classrooms and required his resignation. In effect, Reiss recommended that teachers should engage with students' questions and objections when teaching about evolution. That of course is the approach generally taken when teaching challenging science topics and can be seen as a key strategy in constructivist approaches to science education (Taber 2009) – if not indeed just good sense.

Reiss's comments were reported (and indeed misreported) in the media. From an educational perspective, Reiss was only making a common-sense point, that learners' 'worldview' commitments (such as religious convictions) need to be considered when we teach them, but *the impression given* by the RS's response was that any suggestion that teachers should enter into dialogue about learners' grounds for rejecting or questioning evolution was unacceptable.

At first sight, this seems a very strange position for the RS to adopt, when its very motto ('*nullius in verba*') emphasises the importance of questioning authority rather than simply relying on the word of an authority figure (such as a science teacher). Not only did the decision to dismiss Reiss look illogical to many outsiders, but it provided propaganda fodder for those who campaign against the teaching of evolution. Antievolutionists were able to interpret and present the RS's actions as evidence that scientists hold evolution as above question and wish to avoid debate about the issue. So, for example, one website set up by a group rejecting evolution reported:

> This week, in Britain, we have had the highest profile proof that even a hint that your views on evolution might differ from those of the scientific establishment is enough to force you out. Prof. Michael Reiss, an evolutionist and the Royal Society's director of education, resigned under pressure (given the push) within a couple of days of merely suggesting that creationism and ID could be discussed in classrooms—even if it was in order to explain why they were, in his view, wrong.
>
> (Halloway 2008)

The action of the RS not only gave the impression that Fellows of the organisation were ignorant of fundamental pedagogic principles but inadvertently lent support to those like Halloway who wished to suggest that evolution is an irrational commitment of the scientific community that needs to be protected by a kind of 'thought-police' mentality. For those working in science education, the RS's action seemed doubly counterproductive: it was unhelpful to those who (like Reiss) are strongly committed to scholarly debate that can contribute to effective teaching of evolution, and it undermined the public understanding of science by appearing to present science as an authoritarian endeavour that requires scientists to follow an official party line, rather than being open to freethinkers and diverse views. However, the RS's decision can be understood better when recognised as informed by discourse among scientists about the very nature of science itself. In considering the nature of the relationship between science and religion, it is important to realise that just as there are many religions, and diverse traditions within major world religions such as Christianity and Islam, there are also different ways of understanding the nature of science. The particular conceptualisations of religion and science being adopted *both* need to be taken into account when considering how science and religion can be related.

This chapter will explain why there can be no consensus on an issue such as 'the' relationship between science and religion. Two important themes that will be drawn upon in this account are those of *worldview* and of the *metaphysical commitments* of science. Before considering these notions, it is useful to consider some of the common stances that can be adopted regarding the relationship between science and religion.

3.3 'The' Relationship(s) Between Science and Religion

It is inappropriate to think of 'the' relationship between science and religion. For one thing there are many different religions, which hold different things to be true. Even within what might seem to be a single religion such as Christianity, there are quite diverse traditions. This explains why the Anglican and Roman Catholic churches, for example, do not teach against scientific theories such as cosmological and evolutionary ideas about origins, and yet some other Christian churches (such as some Baptist traditions originating in parts of the USA) consider their Christian tenets to completely exclude the possibility that evolution (in the sense of 'macro' evolution – that leads to completely new forms of living thing) occurs, leading some Christians to reject key scientific ideas from biology, geology, astronomy and cosmology.

Moreover, just as there is not one form of religion, neither is there a single understanding of the nature of science, and different views of what science is, and what its limits might be, can lead to very different views about how science might be understood to relate to religion. Given these complications – that there is more than one religion, and there is more than one way of understanding the nature of science – it is not surprising that very different notions of how science stands in relation to religion have been suggested.

3.3.1 Different General Stances to the Relationship Between Science and Religion

One suggestion is that there are four broad approaches that can be taken to understanding the relationship between science and religion (Barbour 2002), along the following lines:

Independence: that science and religion are actually quite independent of each other and therefore being religious has no bearing on scientific work, and being a scientist has no bearing on religious commitment. This perspective may be adopted by those who see religion as about morality and the purpose and meaning of human lives (areas outside science) but having no authority in accounts of the material world. The evolutionist and essayist Stephen Jay Gould (2001) adopted a version of this perspective suggesting that science and religion were nonoverlapping magisteria.

Integration: that it is quite possible to see science and religion as totally consistent with each other and in agreement about any areas where they might overlap (Moreland and Reynolds 1999).

Dialogue: that science and religion do sometimes have the same concerns but are not obviously in agreement. However, in these cases it is possible to explore how they could be understood as consistent by engaging in a dialogue (Fulljames and Stolberg 2000) between the two starting points (religion and science).

Conflict: that science and religion do sometimes have overlapping concerns, and in at least some of these areas of overlap, they offer totally inconsistent and opposing conclusions (Dawkins 1995/2001). In effect, in these matters, accepting science requires rejecting religion, and accepting religious teaching requires rejecting aspects of science.

Moreover, there is increasingly a version of the 'conflict' stance which is not linked to particular issues (such as evolution by natural selection) but is understood to suggest that science and religion are by their very natures exclusive ways of understanding the world: from this perspective one cannot be both devoutly religious and a fundamentally sound scientist. To adapt a saying from Christianity, no person can have two intellectual masters. This position is often based on a view that religious conviction requires the adoption of metaphysical commitments that are inconsistent with a scientific perspective.

3.4 The Significance of Metaphysical Commitments

Metaphysics is a branch of philosophy, and when scientists class a question as metaphysical, they tend to imply it is a philosophical question that is not open to empirical investigation and so not able to be answered by scientific enquiry. The prefix 'meta' is a little misleading, because metaphysics does not *go beyond* physics but rather if anything precedes physics or any other empirical science. Metaphysics concerns that which is a priori, that which is determined independent of experience, whereas scientific enquiry is concerned with developing knowledge a posteriori, that is, empirically, from experience. The term metaphysics is thought to derive from the way an early editor of Aristotle's works decided that the set of Aristotle's books, now called the 'Metaphysics' concerned with more philosophical and theological questions, were best tackled after reading Aristotle's writings on physics.

Metaphysical commitments are those things a person believes firmly but which cannot be tested empirically. This can be illustrated with an example of two contrary assumptions one might make:

(a) The experienced world is a material world that has existence independent of the observer.
(b) The experienced world is an illusion that is no more than a dream in the mind of the person who thinks they are observing the world.

At first sight many readers may think that option (a) is clearly more sensible and perhaps even that option (b) is a rather fanciful, if not silly, suggestion. Yet most of us have experienced dreams that *at the time* we considered to be real experiences, and many well-respected thinkers have given serious consideration to alternatives to (a) such as (b). Plato considered this world not to represent ultimate reality but to be an imperfect image of a more perfect reality beyond. That is perhaps not so different from the position taken in those religions that consider this world as a temporary staging post ahead of potential admission to a more perfect and less corruptible reality.

A belief that this physical world is all that exists is a metaphysical commitment, as is the belief that we human beings are spirits that have existence prior to and beyond this material life. Neither idea can be proven or falsified in any strict sense – although at one time scientists seriously tested the notion that the body lost weight at the moment of death as the person's soul left the body (Fisher 2004). Similarly, belief that the world has some kind of purpose, being part of the operation of some cosmic plan, is a metaphysical commitment, as is a commitment to the alternative idea that there is no meaning or purpose to the world or to life, beyond the meanings and purposes that we as human beings create for ourselves.

As such metaphysical commitments are 'a priori', and seem to those that are strongly committed to them to be obviously the way things just are, they provide interpretive frameworks through which experience is filtered. Someone who considers that the physical world is all there is will interpret their experiences differently from someone who thinks that clearly this world must have been created by a superior intelligence with an important purpose in mind. In both cases the interpretations will not only inevitably be consistent with their prior assumptions but will also tend to reinforce those commitment by seeming to provide further supportive evidence of the sense of that position (Taber 2013a).

That is not to say that all people have strong commitments on such issues, as it is certainly possible to remain open minded about such questions. Yet where people do hold strong commitments, they can be highly influential in their thinking. A genuine belief that a person who dies in a holy war will be well rewarded in another life, or that the destitute or disabled are simply reaping the punishment due to them because of their wrongdoing in a previous life, makes certain behaviours rational that can seem bizarre, foolhardy or cruel to those who do not share those beliefs.

3.4.1 The Adoption of a Worldview

An increasingly important idea in science education is the notion of a worldview, a 'way people have of looking at reality, the basic assumptions and images that provide a more or less coherent way of thinking about the world, the cognitive structure into which an individual fits new information' (Allen and Crawley 1998, p. 113). Cobern (1994, p. 6) describes a worldview as 'the set of fundamental non rational presuppositions on which … conceptions of reality are grounded', that is, a worldview comprises a coherent set of metaphysical commitments that a person holds. Worldview is often at least in part derived from, and reinforced by, the beliefs of the wider community. This is particularly so in traditional or relatively closed societies where youngsters are surrounded by others who all take certain metaphysical commitments for granted and where those commitments infuse all aspects of behaviour and language.

Worldview has become an important consideration in science education because it has increasingly been recognised that some learners' worldviews may include commitments at odds with what is to be taught in science (Hewitt 2000). If science teachers present ideas in lessons that seem obviously false, or simply nonsense, to learners (because they are inconsistent with their worldviews), then such ideas are likely to be rejected rather than taken as deserving serious consideration by the students. Worldview often acts tacitly, so learners will not necessarily identify a clear contradiction between teaching and some personal commitment: that can happen, but it is also possible that teaching may be misinterpreted or just not seem to make any sense from within the learner's worldview.

Worldview conflicts with science education may occur at the level of ontology (the kinds of things that exist) or epistemology (how we come to obtain reliable knowledge). There may also seem to be clashes in terms of axiology or the values that people take for granted in living their lives. If a student is brought up in a society that includes evil spirits as a core part of its worldview, and considers them responsible for disease or misfortune (Morris 2006), then scientific models of disease may seem both incoherent and irrelevant (Thagard 2008). Similarly, in many tribal societies, traditional ecological knowledge (TEK) is understood and communicated in a holistic manner that does not separate out the analytical reductive approach of ('Western') science from other – metaphorical, spiritual, poetic, etc. – ways of knowing (Berkes 1993). Such societies have often developed effective TEK that allows them to manage environmental resources in sustainable ways (Freeman 1992), but formal science education with its abstract concepts and deductive logic-dominated forms of argument may make little sense in terms of those traditional ways of knowing. For these societies, TEK is not a discrete aspect of culture (as science may appear in many 'developed' nations) but is closely integrated with a worldview that posits a particular relationship between the people, the biota and the environment (Inglis 1993). So, for example, students from some cultures may object to the use of animals for display or dissection in school science (Allen and Crawley 1998).

Religious beliefs are often key components of worldviews. So in many religions, there is a belief in an active non-human, immaterial, being of great intelligence able to act directly in the world through supernatural means. For those who strongly believe that God created the world; and did so for a purpose; and that people are an integral part of that purpose, such commitments may colour all aspects of their thinking about the world. Many of those who believe (e.g. as in Christianity) that God has a personal relationship with them, is interested in them as people, offers them eternal life and guides (and judges) their behaviour, are likely to consider that such a God is relevant to *all* aspects of their lives.

3.5 Metaphysical Commitments in Religious Worldviews

It is an oversimplification to equate religion with worldview, as central tenets of particular religions often blend with other cultural traditions in particular locations to inform the worldview of a society. In particular national contexts, religious beliefs and practice may be syncretic, that is, they may reflect previously distinct traditions that have become blended. There are many examples of this where people have converted to major world religions in ways which adopt and reflect pre-existing local beliefs and rituals (Morris 2006).

So, for example, in Sudan, adherents to Zar cults recognise the existence of three types of jinn, designated as white (generally benign), black (malevolent) and red jinn – the latter being associated with various forms of minor illness or distress. Now the existence of jinn is an accepted part of Islam, where it is considered that God created men from clay and jinn from fire. Jinn, like men, may do either good or bad deeds. So Islam includes an ontological commitment to the existence of a class of beings (jinn) that although invisible to humans are real components of the world and as integral to God's creation as humanity. However, the red jinn are not a conventional feature of Islamic belief and represent aspects of more traditional African beliefs that have been grafted onto Islam in that part of Africa.

Another example would be Vodou (or Voodoo), the traditional folk religion in Haiti. Vodou is a tradition that is often misunderstood (perhaps due to rather misleading sensationalist accounts) but is actually a local adaption of Roman Catholicism incorporating rituals and beliefs not considered part of Catholicism elsewhere. Similarly, there is a religious cult in Brazil, Candomblé, which encompasses a whole range of spirit beings – including some of African derivation, some from Native American traditions and Catholic Saints.

Religion can be understood to draw upon, as well as inform, a group's worldview, as for a new religion to be successfully established, it must seem sensible from the existing worldviews of potential adherents. For some scholars this is quite relevant to important aspects of the science and religion debate and how scriptures and traditions need to be interpreted. So, for example, it has been suggested that the accounts of the creation in the Jewish and Christian scriptures were intended to convey an important spiritual truth (and in particular the notion of a single God, who was the creator) in a form that the tribe with their existing worldview could understand and relate to and learn to accommodate within their incumbent model of the world (Habgood 2002).

3.5.1 *Examples of Ontological Commitments in Religious Worldviews*

There are many examples of beliefs about the nature of what exists and the nature of existence that are incorporated into religious worldviews. The aim within this chapter is not to survey all the potential commitments of different religious traditions that might be seen relevant from a scientific perspective (which would be a massive task) but simply to offer sufficient examples for the reader to appreciate the range of potential points of contact between science and religion.

A key feature considered common to religions is the belief in a spiritual world that in some sense goes beyond what is recognised in the natural world reported in scientific ontologies. This has been referred to as 'intuitive certainty of another world' (Morris 2006, p. 18). The existence of God (of gods) is common to most religions, and there may be beliefs in various other types of entity that are not recorded in scientific accounts of what exists. Some examples have already been met in the jinn and spirits associated with various natural phenomena in cults such as Candomblé. Other examples would be angels, the Devil, another place beyond the physical universe where spirits might reside (Heaven), etc. The difficulty with making such generalisations is that many of these entities may be understood in ontologically different ways in different religions or even within distinct traditions within a religion.

So, for example, the Devil (also known as Satan, Lucifer, Beelzebub, etc.) is important in Christian traditions but may be seen from different traditions within the religion as personifying evil in a metaphorical sense or actually being a person-like creature, either corporeal, or at least able to take the appearance of human form. Clearly belief in the devil as a notion reflecting temptation to do evil things is rather different from believing in an actual individual being who is engaged in a real battle with the (equally real) forces of good. This kind of distinction is very important for teachers to bear in mind, as knowing that a student in a class believes, for example, that *God sends angels to look after or communicate with people* could refer to a specific type of non-human being, or could alternatively be a reference to how God is considered to work through people (who become his angels in particular circumstances by virtue of their actions).

Interpreting just what people mean by their beliefs is important as some particular religious ontologies are more likely to be seen as contrary to scientific perspectives or principles than others. Shamanistic religions, for example, hold to a two-world system, where an alternative reality (experienced in a trance state) is just as real as the everyday reality experienced in the normal waking state (Morris 2006). This contrasts with common Western views (drawing on the naturalistic explanations typical of science) that would see the alternative reality as an illusionary experience, for example, triggered by psychoactive drugs, fatigue and immersion in repetitive rhythmic rituals. The shaman would recognise the same triggers but see them as facilitating entry into another reality, not offering imaginary delusions. To the naturalist, however, these are real experiences, but they are not experiences of something real.

The Buddhist worldview is built around metaphysical commitments to suffering being an intrinsic feature of existence, impermanence and decay as inherent qualities of the world, and the experience of self as an illusion (Morris 2006). A modern scientific perspective might well consider the self as an emergent property, or even an epiphenomenon, if not actually an illusion, and the notion of a fixed self that passes through life is certainly open to critical consideration. (In this matter the Buddhist view might have more in common with scientific accounts than everyday Western 'common-sense' notions!) Some interpretations of evolutionary theory would also see suffering as necessary in the living world, but as a contingent feature of any individual's experiences rather than an inherent feature of existence (Dawkins 1995/2001). Modern theories of matter would consider the world observed in flux to be constructed from a more stable and permanent substrate (at least on timescales that are short compared with the history of the universe). Arguably, each of these three themes is open to dialogue between the Buddhist commitment and scientific understandings. Indeed, there is something of a tradition of looking for parallels between Eastern philosophies and modern interpretations of physics (Capra 1975).

Among those that believe in a creator God, there can be different views about the nature of God and of his actions. A deist would consider that a creator God would set up the universe ready to play out and then observe without taking any further action. Isaac Newton's theological views seemed to approximate to this position, although he thought God needed to occasionally fine-tune His creation (Cooper 1984). Theists may see the nature of creation rather differently. For many theists, the universe was not only initially created by God at some point in the past, but its continued existence depends upon God. From this perspective 'the creation' *was* not an event, but *is* an ongoing process (Alexander 2009).

Some theists may distinguish two kinds of action in the world by God – those things which are a kind of playing out of an original plan or programme and those which involve God acting directly in the world to intervene in the usual cause of events, such as during various events in the Old Testament or Jewish Torah that are considered miraculous. One example is an event during the Exodus story where the Israelites having escaped from slavery in Egypt needed to cross the River Jordan to finally enter their promised land. The waters stopped flowing from upstream as soon as the priests walked into the river.

Traditionally such miracles have been seen as God intervening in the normal natural course of events to interrupt nature (Moreland and Reynolds 1999). From this perspective a supernatural force stopped the waters for long enough for the Israelites to cross. However, other theistic interpretations consider that these events are not supernatural, but rather due to an unusual confluence of particular natural circumstance (e.g. recent rainfall levels, unusual wind direction, mudslides upstream diverting flow, etc.). This interpretation is not intended to make the event a mere coincidence or 'lucky break', but rather represents *a miracle of timing* where God (having omniscience) has set up the natural world in such a way that the improbable circumstances occurred at just the right moment in history (Humphreys 2004).

This raises complications for a common notion that science enquires into nature but has no jurisdiction over a supernatural world should one exist. That position would suggest that the existence of supernatural beings and powers is outside the range of application of science, which only concerns itself with the natural. Yet for many modern-day theists, a simple distinction between the natural and the supernatural is not meaningful as God may be seen as immanent in nature and sustaining it, not through supernatural means, but by being the very basis of nature itself (Alexander 2009).

3.5.2 *Examples of Epistemological Commitments in Religious Worldviews*

As suggested earlier, science is based on empirical enquiry, and as the RS suggests, a premise of the scientific revolution was the notion that scientists should seek to find out for themselves and 'take no one's word for it'. That said, it is now widely recognised that even science is based upon some a priori commitments (as discussed below) that must be adopted prior to empirical enquiry. Also, in practice, the modern scientist inevitably has to assume a good deal of what is already accepted, as science is iterative and progress is only possible because each generation of scientists comes to take for granted what are considered secure findings produced by their predecessors. However, it is still assumed that any scientist has a right to question any scientific principle or result and – indeed – a responsibility to make reasonable efforts to bring significant counter-evidence apparently falsifying accepted ideas to the notice of the scientific community. *In principle*, then, scepticism is encouraged, and interpretation of empirical evidence is considered the basis of scientific authority.

Religious traditions, however, are often quite different. In Islam, for instance, the Qur'an (or Koran) is considered to record the actual word of God as dictated directly to the Prophet (Muhammad). The Qur'an therefore represents absolute authority and is not open to revision or modification (although of course – as with any text, sacred or otherwise – it must be interpreted). In a similar way, the Ten Commandments recorded in the Torah are considered to have been given directly to Moses by God (inscribed on tablets of stone, giving little scope for misreporting of them in scripture). Scripture is therefore seen in some religions as a direct source of knowledge.

Another major source of authority in some religious traditions is the priest, or similar person, who is considered to be able to act as an intermediary between God and other people. Within Christianity, for example, some traditions consider that the interpretation of scripture, and so pronouncements on what is right or wrong, is a responsibility of the church hierarchy – an issue that Galileo Galilei faced when teaching about the heliocentric system (that he was allowed to discuss as a hypothesis or heuristic, but something he was forbidden from presenting as factual).

The Roman Catholic church has changed its view on that particular point, but today has official positions on matters such as abortion and the use of contraceptive technology, that are supposed to be binding on all members of the Church. In some circumstances the Pope as supreme pontiff of the Universal Church (considered to have authority handed down through each Pope since the Apostle Peter) with his advisors is considered to be infallible on matters of religious teaching. By contrast, in some protestant traditions (where there are no formal priests), much more emphasis is given to individual Christians following their own conscience in matters of morality (following advice offered by the Apostle Paul).

Another epistemological commitment in some religions (e.g. Islam and commonly in Christianity) is that humans are able to understand nature in a meaningful way – as it is God's intention that his people should appreciate his creation (Moreland

and Reynolds 1999). Yet in some Eastern religions (Hinduism, Buddhism, Sikhism), there is a strong emphasis on Māyā, that is, on aspects of the perceived world being illusionary.

3.5.3 *Examples of Axiological Commitments in Religious Worldviews*

Religions often offer moral guidance on the right way to live and what comprises good rather than bad or evil. Such guidance may relate to social relations (respect, compassion, forgiveness) or particular practices and rites, for example, certain foods that are prescribed or excluded either on particular occasions or generally. So, in religions such as Judaism and Islam, there are strict dietary codes of this type, both relating to what can be eaten and how the food is sourced and prepared. This means that, for instance, advances in slaughterhouse techniques that might be considered to cause less suffering to the animals being killed would not be acceptable if they were seen as inconsistent with the traditional codes for preparing food.

Religious traditions also often consider that there is a purpose in the universe (e.g. God's purpose), that individual people may have a place in an overall plan as part of that purpose, and that therefore a person should make choices in keeping with God's plan for them. Sometimes people who adopt such views find comfort in considering that misfortune or apparent tragedy (like the death of a child) is for good reason and part of God's overall plan, and this may be associated with a form of fatalism – that whatever will be, will be.

The rejection of contraceptive technology (such as condoms) by the Roman Catholic church, referred to above, is part of the Church's moral teaching, based on axiological commitments to the sanctity of life and there being a special nature to the relationship represented by marriage between a man and a woman. Science has no value positions on such issues as whether the use of contraception or sexual relations outside of marriage is in principle right or wrong, but science does offer strong evidence of the potential consequences of the use or rejection of contraceptives in those countries where HIV/AIDS is rife and/or where much of the population is already at or near subsistence level. So-called sociocultural or socio-scientific issues are increasingly being explored in science classes (Sadler 2011). These are those issues where science offers evidence and theoretical perspectives but where decision-making necessarily also draws on values that are external to science itself. Student worldviews can be highly significant in such lessons.

3.6 Metaphysical Commitments of the Scientific Perspective

Science itself is also based on a set of metaphysical commitments (Taber 2013a). Some of these may seem obvious, but then that is how metaphysical commitments often work – as what we take for granted. So, for example, science not only assumes the existence of the material world independent of our minds but also that that world is stable and ordered enough to make it worth our while enquiring into it (e.g. so it is worthwhile looking for 'universal' laws) and that it is knowable to a sufficient extent to make scientific knowledge possible. The existence and relative stability of a world beyond our minds is an ontological commitment, and the assumption that humans are able to learn about its true nature is an epistemological commitment. There remains scope for debate among those operating with such commitments: for example, regarding which features can be considered to be fundamental and universal and *to what extent* human knowledge is able to comprehend nature (Taber 2013a).

These assumptions are a priori as they cannot be demonstrated without first assuming them! Furthermore, once we make these assumptions, we go on to interpret our experiences accordingly. A further epistemological assumption adopted in science is that the way to learn about the world is to undertake observation and experiment (supported by reflection and analysis) – rather than, for example, by seeking visions during meditation.

If these assumptions seem fairly unproblematic – as is likely to be the case for many of those working in science education – then it is worth noting that they are not always universally shared beyond scientific circles. For example, in Plato's philosophy where the directly experienced world was a kind of distorted image of a more ultimate reality, it made more sense to focus one's efforts on seeking knowledge of that ideal world, rather than enquiring into the illusion of the directly experienced world (e.g. Plato's student Aristotle, however, adopted a somewhat different metaphysics and put a good deal of effort into empirical studies in biology.)

Science does not offer a moral system like those of many religions, but nonetheless there are certain values – axiological commitments – that are adopted. So science values open-mindedness and indeed a willingness to refute one's own ideas (Popper 1934/1959), values that are not always aligned with those adopted in religious traditions.

3.6.1 Worldviews Consistent with Scientific Metaphysics: Theism

The view of scientific metaphysics offered here suggests that the foundational assumptions of science have nothing to say about God or the supernatural. Of course science is a socially constructed enterprise, and its tenets have changed over time and could change again (Taber 2013a). Scientists themselves, historically and today, have taken very different views about religious matters and indeed on whether science has anything to say about belief in a God (Coll et al. 2008). There is no consensus about such matters, so it is wrong to suggest that scientists in general (and so therefore science itself) either judge science as compatible with religions or inherently inconsistent with religious faith. There are many professional scientists today who would adopt each of these positions (as well as many who would simply see religion as entirely irrelevant to science).

3.6.2 Worldviews Consistent with Scientific Metaphysics: Natural Theology

Many famous scientists of the 'scientific revolution' (Westfall 1971) such as Isaac Newton, Robert Hooke and Michael Faraday believed the world was God's creation. Many of these scientists adopted an approach sometimes called natural theology (Grumett 2009). In this approach it was assumed that because God had created the world for people to live in, He would have created it so that people could understand its workings. Studying God's work (his creation) was considered complementary to studying His Word (scripture), and an appropriate activity for a devout believer (McCalla 2006). This perspective is also very common in the Islamic world today (Dagher 2009).

There is a widely adopted tradition in Christianity (at least since Augustine who lived in the fourth and fifth centuries CE) that where investigation of the natural world offered strong evidence that was clearly inconsistent with scripture, then the *apparent* contradiction indicated that the interpretation of scripture should be re-examined. In this tradition scripture was never considered *wrong*, but was capable of being misinterpreted by fallible human readers. This was a tradition that Galileo Galilei called upon when his astronomical observations seemed inconsistent with his Church's interpretation of scripture.

3.6.3 Worldviews Consistent with Scientific Metaphysics: Agnosticism

An alternative tradition of agnosticism, more sceptical of religious beliefs, developed in the nineteenth century. The word agnosticism was suggested by T H Huxley (Gilley and Loades 1981), who was known as 'Darwin's bulldog' because of his strong advocacy for natural selection ('Darwinism'). The agnostic view was that the existence or otherwise of supernatural beings, such as a creator God, was a metaphysical issue that was not open to scientific investigation. Therefore, a scientist should not commit to a belief in God, nor claim that God did not exist, but rather should focus on matters that could be settled by scientific work. (Sometimes people use the term agnostic to refer to someone who is undecided about the existence of God, but strictly it refers to an epistemological commitment that we cannot be sure whether God exists.)

This approach rather assumes that being a scientist is an approach to life, rather than just an occupation. Many scientists who are religious would accept that the question of the existence of God is not open to scientific enquiry, but however claim to have had personal experience of God that provided them with proof of His existence. Their experience is subjective, and not presented as scientific evidence, but acts as sufficient grounds for their *personal* belief, completely independently of their scientific work. As natural science is not able to investigate supernatural phenomena, such beliefs should not impact on their scientific work (but may give them moral guidance about such matters as the desirability of different applications of science and so preferred areas of science to engage with).

3.6.4 Worldviews Consistent with Scientific Metaphysics: Methodological Naturalism

Many scientists who are theists, and consider that God acts in the world, tend to seek to keep such notions separate from their scientific work. Often such scientists adopt 'methodological naturalism' (Moreland and Reynolds 1999), a perspective that assumes that science is the appropriate means by which to investigate the natural world and accepts that supernatural explanations should not be admitted into scientific work.

Historically there has been a tendency for some commentators to recommend acceptance of scientific accounts as far as they go, but to reserve God as the cause to explain anything science cannot explain. This approach, known as a 'God of the gaps', has never been a respectable philosophical position though and, if adopted, is likely to result in regular ceding of more ground 'from' theology 'to' science, as science expands the range of its viable explanations for the natural world (Fergusson 2009; Sagan 1985/2006). Methodological naturalists rather assume that the material world can generally *in principle* be explained by science but may accept that there is another level of explanation and cause underlying the existence of the world that complements (but is not threatened by) scientific explanation (Moreland and Reynolds 1999).

So, in this perspective, God has created the material world and set up its physical laws such that its material form and the laws are themselves open to scientific enquiry. From this view science has precedence as long as it remains in its range of application, but has no apparatus to examine questions relating to ultimate causes. Different types of causes are not seen as being in competition, but working at different levels of explanation (Habgood 2002). By analogy, a question about how the reader comes to be reading this chapter could be answered in terms of the physiology and neurology of reading or in terms of the motivations of the reader. The existence of an explanation in terms of synapses and nerve impulses neither excludes nor is excluded by an explanation in terms of a desire to read about the 'science and religion' issue. In the same sense, a mechanistic explanation of the physics of the big bang can be seen as complementary to an ultimate explanation that the universe exists because God brought it into being (perhaps through the big bang as physics suggests).

As noted above when considering miracles, some theists who adopt this kind of perspective will consider that God *only* acts through natural forces and events – so in principle miracles can be shown to have natural immediate causes, whilst their ultimate supernatural cause remains undetected by scientific methods. Others consider that God can and does intervene in the usual natural state of affairs, but that because these actions are supernatural, they are not open to scientific investigation.

In effect agnosticism (see above) is also consistent with methodological naturalism as strictly agnostics are not those who have *not decided whether* to believe in God, but those who consider that the existence or otherwise of the supernatural is inherently beyond human knowledge. That is, where theism and atheism are ontological commitments (to the existence or nonexistence of God), agnosticism is more an epistemological commitment relating to the limits of human knowledge.

3.6.5 Worldviews Consistent with Scientific Metaphysics: Atheism

Just as many scientists are religious, and many are agnostic, there are also many scientists who are atheists, who do not believe in a God. Many (but not all) of these would see their atheism as having little to do with their scientific work, just as many theist scientists would see these two aspects of their life as largely separate.

3.6.6 Worldviews Consistent with Scientific Metaphysics: Philosophical Materialism and Scientism

In recent decades, however, a group of scientists who claim that atheism should be inherent in science has been increasingly vocal. These scientists adopt an approach sometimes called 'metaphysical naturalism' (Moreland and Reynolds 1999) or materialism which not only considers that science is limited to studying the material world, but that indeed that is *all there is*. For these scientists, God is an illusion. Human spirit or soul is at best an emergent property of the complexity of the nervous system and perhaps even an epiphenomenon – simply a by-product of effective cognition (Taber 2013b) – and that the only forces that operate in the world are natural forces open to scientific investigation.

Often these scientists – of whom Richard Dawkins is the most high profile (Cray et al. 2006) – take the stance that only science can offer reliable knowledge (Moreland and Reynolds 1999) and sometimes also suggest that in principle everything can be explained by science and so that scientific laws should be seen as prescriptive (how the universe must behave) rather than descriptive (how the universe appears to behave). In effect this view is one of scientism – that all there is can be explained by science – so in effect that science is the one true epistemology.

This group of atheist scientists is especially significant for the public understanding of science because they wish to expand the basic metaphysical commitments of science to include those of their own materialist worldview. Whereas the scientific perspective has previously been that science is limited to studying the material

Table 3.1 Scientists may differ in both their belief in God and their views on whether religious convictions have any direct relevance to their scientific work

	Religious belief is seen as irrelevant to scientific work	Religious belief is seen as relevant to scientific work
Theist	Considers belief in the supernatural to be largely irrelevant to scientific enquiry into nature. Spiritual and professional life kept separate except in terms of ethical decision-making	Sees God at work in all things and underpinning nature and its laws. Science seeks to understand God's creation
Atheist	Does not believe in the supernatural, but does not consider religious beliefs an impediment to science, providing the religious scientist does not let them take priority over empirical evidence	Sees atheism as a proper commitment for scientific work and so considers religious scientists to be compromising the scientific attitude by believing in beings/forces for which there is no objective evidence

world and has nothing to say about the supernatural, these materialistic naturalists suggest that a true scientific perspective is that all that exists is the material world, and science excludes the possibility of the supernatural. Dawkins, for example, claims that 'any belief in miracles is flat contradictory not just to the facts of science but to the spirit of science' (Cray et al. 2006).

It would seem then that scientists vary not only in terms of their personal religious convictions, but also in whether they feel such beliefs potentially interact with how science can be understood (see Table 3.1).

3.7 Implications of Worldview on Understanding Scientific Knowledge

People with different worldviews adopt different stances to scientific knowledge and religious beliefs. The brief analysis presented above suggests not only that metaphysical commitments of a religious nature can impact upon how people (e.g. school or college students) understand and respond to scientific teaching, but also that some scientists who conceptualise the nature of science itself within an atheistic materialist worldview perceive religious beliefs as *necessarily* counter to science.

Consider, for example, how prayer may be understood. In some religious traditions, prayer is seen as communication with God that has the potential to lead to changes in future events (e.g. recovery of a loved one from illness) should God choose to act in the world in response to prayers. Some scientists who adopt theistic worldviews would find this acceptable on the grounds that as God is the all-powerful creator of the world, so He is free to intervene to influence it at any time, and as such interventions are supernatural – they cannot be explained by science.

Other scientists who hold theistic worldviews will consider that although God acts in the world, His extreme intelligence allows this to occur within the natural processes He has set up, and so there will be nothing outside of scientific explanation (and so no evidence of miraculous intervention). However, the atheistic materialists will exclude any possibility of there being a God and suggest that if prayer is shown to have an effect then this must be due to influences open to scientific investigation and totally depending on natural processes, even if possibly processes that science has yet to discover. Both of these latter groups will in principle consider that any measurable effects of prayer will potentially have a scientific explanation – but the atheists consider such an explanation excludes a complementary theological explanation that the theists would hold.

It seems then that both religion and science offer accommodating and exclusive positions where they might be considered to potentially overlap (see Table 3.1). Religious perspectives may either lead us to expect to see theological explanations that take precedence over scientific ones or alternatively consider that theological and scientific accounts are parallel and complementary.

In a similar way, core scientific commitments do not exclude a supernatural realm or religious accounts, but lead us to seek physical explanations for the phenomena of the material world. Yet within the broad 'scientific church' are those who would expect there are some special events in the material world beyond scientific explanation; others who expect science to be able to address all material phenomena, but limited to immediate rather than ultimate causes; and yet others who adopt a worldview that excludes the possibility of anything supernatural and considers everything to fall wholly within the realm of science. There are also variations and graduations within and between these gross positions.

3.7.1 The Prominence of Debates Around Evolution

Based on this analysis, the rather high-profile arguments about how science educators should respond to the rejection of evolution by people from some religious communities can be understood at two very different levels. This debate has largely been played out in the context of objections to evolution in some Christian, and increasingly Muslim, communities.

There is certainly an issue in terms of meeting curricular aims that set out evolution as a teaching topic, as some learners may reject evolution on principle (see Chap. 4). However, debate does not just depend upon people with religious convictions rejecting scientific accounts, as there is also a distinct argument being made by some scientists drawing on their particular (atheistic, materialist) understanding of the nature of science rejecting the religious beliefs of theists who themselves accept the scientific account (such as many religious scientists).

Interpretation of religious accounts of the origins of life, and interpretation of the nature of science itself, interact. Therefore the question of whether religion and evolutionary theory are compatible depends upon both (i) which religious, and (ii) which scientific, position is adopted. Reaching one of these positions in relation to a particular topic will clearly depend upon both one's religious beliefs and one's understanding of the nature of science. This is illustrated in Table 3.2 using the examples of the case of evolution and Christianity, where the teaching of some Christian churches absolutely excludes the possibility of humans evolving from other species by natural selection, yet other Christian churches have no problem with this issue.

The columns of Table 3.2 represent two possible Christian understandings of teaching about human origins. In both cases a core tenet of Christianity is represented: that there is a creator God who is responsible for the creation of human beings. In some Christian traditions (represented by religious perspective 2), the accounts of how God created the world (the two accounts in the early chapters of Genesis) are understood to be technical accounts: God created the first man from dust and then formed a woman from one of the man's ribs so that he could have a companion. In this account all the rest of humanity subsequently descended from this original couple. However, there are other Christian traditions (represented by religious perspective 1) that see these accounts in scripture as presenting narratives offering a theological truth about humanity (that humankind is part of God's creation and that people are considered to have a special relationship with their God) in terms that are poetic and would have been accessible to the original historical audience several thousand years ago.

Table 3.2 How two different scientific and religious perspectives can interact in relation to evolution

	Religious perspective 1: religion teaches that God created humankind but does not specify the mechanisms he used	*Religious perspective 2*: religion teaches that God created the first human directly from dust by an act of special creation
Scientific perspective 1: science teaches that there is extensive evidence that humans evolved over a long period from other earlier species by a natural process called 'natural selection' – but has no view on whether there was a purpose or intelligence behind this process	It is perfectly consistent to believe in a creator God and to accept natural selection as the means by which God created human beings	Notions that man evolved from other non-human species are directly contradicted by scripture that reveals how God created humans: religion excludes the scientific account
Scientific perspective 2: science teaches that events in the world occur by natural processes that occur without any supernatural input and that there is extensive evidence that humans evolved over a long period from other earlier species by a natural process called 'natural selection'	Seeing evolution as anything other than a series of events following natural laws inherent in the universe (and needing no supernatural explanation) is contrary to scientific principles: science excludes religion	The scientific and religious perspectives exclude each other both because religion rejects scientific accounts and because science rejects the foundational tenets of religion

The rows in Table 3.2 represent two possible positions that scientists may take about the scientific theory of human origins, that is, evolution by natural selection. One of these views (represented by scientific perspective 1) sees science as offering a mechanism for how humans came to be on this planet, but takes no position on whether this mechanism reflects the will of a God or other supernatural entities. However, the other perspective (scientific perspective 2) considers it inherent in the natural mechanism that it has to be a sufficient explanation without being driven by any supernatural force or agent (such as God).

Public debates about evolution, and the teaching of evolution, are therefore often being played out at two very different levels, one relating to empirical issues and the other to metaphysical ones. The first issue relates to how evolution is a central idea in biology, and worldview commitments that lead to learners misunderstanding or rejecting evolution are therefore problematic for science learning. That issue is considered in more detail in Chap. 4. For science educators *this* issue is quite straightforward – evolution is the currently accepted scientific theory, and it should be taught as just that: a well-accepted idea supported by a wide evidence base.

The other aspect of the evolution debate revolves about scientists using the topic as a context for attacking religion and what Dawkins (2010) has described as 'the uniquely ridiculous nature of religious belief'. In this context evolution has become something of a 'cause célèbre' (Moreland and Reynolds 1999), probably because the claims of some who support young earth creationism seem to deny so much empirical evidence that it is easy to rhetorically associate their literal interpretation of scripture (and so support for positions that seem untenable to many outside their communities) with irrational thinking. Given this, a belief in a creator (and the possibility of eternal salvation, etc.) is targeted *by association* as irrational as well. This argument is akin to claiming that Newton's mechanics should be dismissed as the product of a primitive and superstitious mind because of his apparent commitment to numerology (Pesic 2006), or that Kepler's laws should be rejected because he used horoscopes, or that Linus Pauling's fixation on the imagined health benefits of megadoses of vitamin C should lead us to discount his findings in chemistry.

Richard Dawkins, probably the most high profile of the philosophical materialists to campaign on this issue, sets out the debate as being a clash between modern, rational, evidence-based science and illogical, outdated superstitions – although of course when scientists are interviewed about their beliefs, it is found that 'believers, atheists and agnostics' all use rational argument to justify their positions (Falcão 2008, p. 1261). These two distinct threads to the evolution debate are set out in Table 3.3.

The educational position presented in Table 3.3 would be adopted or at least accepted by the vast majority of scientists and science educators. From that perspective it is important to explore how to facilitate effective teaching of evolution, even among communities where religious or other cultural beliefs are inconsistent with the science. The metaphysical position presented in Table 3.3, however, whilst being strongly held and advocated by some scientists, is not a consensus view of the scientific community. However, it becomes important to science educators because (a) science education in many countries puts increasing emphasis on teaching about the nature of science and (b) the claims of the materialistic naturalists are often high profile in media discussions of areas of science such as evolution. Indeed, it has been reported that, in part, the

Table 3.3 Evolution is not only a major issue in science education, but has become something of a cause célèbre for a group of scientists who consider science should supplant religion

Level of debate	Focus of debate	Issue	Role of evolution	Status
Educational	Effective teaching of scientific ideas	Science education is concerned with teaching people about the scientific models and theories of the material world. Sometimes learners fail to appreciate, or even reject, scientific ideas because of commitments drawing upon religious and other cultural beliefs	Evolution by natural selection is an especially important scientific theory, which is fundamental to an understanding of modern biology. Rejection of evolution is a major impediment to developing a scientific understanding of the living world, including many aspects of medicine and health, and of ecology	Close to consensus position in science and science education
Metaphysical	Nature of science	Science is a modern, rational way of understanding the world based on logical analysis of evidence that is superior to more primitive modes of thought based upon superstition and supernatural entities. Science education should persuade people to abandon their primitive beliefs and adopt a more rational approach to the world	Evolution is an important 'battleground' because it is one of the most visible areas where common superstitions (e.g. religious beliefs) lead to people denying evidence and rejecting the rational approach of science. Evolution provides a valuable case study that illustrates just how problematic religious ideas are in obstructing education that should be promoting rationality	The views of a pressure group within the scientific community, which are not shared by scientists generally

development of the intelligent design movement (see Chap. 4) was a response to the atheistic gloss on evolution in the writings of Dawkins (Alexander 2008).

At the start of the chapter, the rather unsavoury matter of Prof. Reiss' departure from his role as education director for the Royal Society (RS) was raised. Reiss is a highly respected science educator with a background in evolutionary biology. He accepts natural selection as the best scientific account of the origin of species and is concerned that learners in schools and colleges should learn about evolutionary theory. Given that Reiss is unambiguously pro-evolution, and pro-teaching evolution, his dismissal seems a miscalculation and overreaction by the RS. Reiss's comments were reported in the press:

> In his speech, Reiss said that while creationism had no scientific basis, science teachers risked alienating pupils who believed in the idea by dismissing it out of hand. "They should take the time to explain how science works and why creationism has no scientific basis", he said.
>
> (guardian.co.uk 2008)

Reiss's comments were, however, also misreported (as supporting teaching of creationism in science) by some media outlets immediately after the speech, and there were then public reactions to those reports by people who had not heard the talk. Reiss issued a clarifying statement, but the RS asked for his resignation – apparently on the grounds that he had uttered comments capable of being misinterpreted in ways that reflected badly on the RS. It is in the nature of controversial issues that statements are readily misconstrued (deliberately or otherwise), so one might expect the RS to have taken a more principled approach and to have supported its officer.

However, it seems that Reiss was a victim of the tension within science about the nature of science itself and whether it should encompass an atheistic, materialist worldview. A number of Fellows of the Royal Society had already criticised Reiss's appointment before he made any public pronouncements because he was a committed Christian and an ordained minister of the Anglican Church. Even though Reiss does not dress as a priest when working as a science educator, and does not use his academic positions to push theological views, some Fellows objected that such an openly devout Christian was unsuitable for a high-profile office in a scientific society (Vallely 2008). So, there was a campaign to remove Reiss from his position at the RS because *some other scientists considered his personal beliefs incompatible with science*.

The behaviour of the RS in capitulating to its materialist Fellows illustrates that the science and religion issue is a complex one. It also demonstrates that where there are conflicts, they can originate from either direction: (some) people of religion rejecting science or (some) people of science being intolerant of religion.

3.7.2 Implications for Education

Reiss has argued that:

> It is perfectly possible for a science teacher to be respectful of the worldviews that students occupy, even if these are scientifically limited, while clearly and non-apologetically helping them to understand the scientific worldview on a particular issue.
>
> (Reiss 2009, p. 783)

Perhaps what some of Reiss's critics found unacceptable, but seems sensible and appropriate from an educational perspective, is that the teacher's aim should be limited to helping students *understand* the scientific ideas and the arguments and evidence for them. It is not the science teacher's job to ask learners to *commit* to accepting those ideas (Taber 2013b) – even when the teacher finds the case for them as overwhelmingly convincing. Science teaching should not be about persuading learners to adopt beliefs, but about providing a solid understanding of core scientific ideas that students can then make up their own minds about. Science should certainly not be learnt as dogma – and indeed when scientific principles come to be taken for granted as if articles of faith, they can act as obstacles (Bachelard 1940/1968) to scientific progress (e.g. the existence of the ether, the central dogma of molecular biology and the noble nature of the inert gases).

In recent years there has been research in a number of countries exploring students' notions of the relationship between science and religion (Francis et al. 1990; Fulljames et al. 1991; Hansson and Redfors 2007; Taber et al. 2011a, b). Given the complexity of the intellectual 'landscape' around this issue, it is not surprising that students fail to appreciate many of the subtleties of the different positions adopted. Moreover, and of some concern, there is some sense that learners' perceptions often tend to demonstrate particular awareness of the less conciliatory positions. Learners tend to recognise those more fundamentalist religious positions that reject scientific accounts and are likely to assume that atheistic and scientistic perspectives are the norm in science. One US study reported that high school students could be sufficiently troubled by the perceived incompatibility of science and religion 'that a significant percentage of our most motivated and capable students feel they may be deterred from a science career' (Esbenshade 1993, p. 336). This is surely an important topic that should be a priority for further research in different national contexts (Reiss 2008).

Science educators need to adopt sensitivities to students and colleagues, bearing in mind that there can be deep convictions to both religious and materialistic worldviews. Those working in science education also need to be aware that the subtleties and complexities of the 'science and religion' issue are unlikely to be accessible to many of the learners coming to our classes, who are increasingly likely to unhelpfully consider 'science *and* religion' in terms of 'science *versus* religion'.

References

Alexander, D. R. (2008). *Creation or evolution: Do we have to choose?* Oxford: Monarch Books.

Alexander, D. R. (2009). After Darwin: Is intelligent design intelligent? In M. S. Northcott & R. J. Berry (Eds.), *Theology after Darwin* (pp. 22–40). Milton Keynes: Paternoster.

Allen, N. J., & Crawley, F. E. (1998). Voices from the bridge: Worldview conflicts of Kickapoo students of science. *Journal of Research in Science Teaching, 35*(2), 111–132. doi:10.1002/(SICI)1098-2736(199802)35:2<111::AID-TEA3>3.0.CO;2-V.

Bachelard, G. (1940/1968). *The philosophy of no: A philosophy of the scientific mind.* New York: Orion Press.

Barbour, I. G. (2002). *Nature, human nature, and God.* London: Society for Promoting Christian Knowledge.

Berkes, F. (1993). Traditional ecological knowledge in perspective. In J. T. Inglis (Ed.), *Traditional ecological knowledge concepts and cases* (pp. 1–9). Ottawa: International Program on Traditional Ecological Knowledge International Development Research Centre.
Billingsley, B., Riga, F., Taber, K. S., & Newdick, H. (2014). Secondary school teachers' perspectives on teaching about topics that bridge science and religion. *The Curriculum Journal, 25*(3), 372–395. doi:10.1080/09585176.2014.920264.
Capra, F. (1975). *The Tao of physics: An exploration of the parallels between modern physics and eastern mystcism*. Berkeley: Shambhala Publications.
Cobern, W. W. (1994). *Worldview theory and conceptual change in science education*. Paper presented at the National Association for Research in Science Teaching, Anaheim.
Coll, R. K., Lay, M. C., & Taylor, N. (2008). Scientists and scientific thinking: Understanding scientific thinking through an investigation of scientists views about superstitions and religious beliefs. *Eurasia Journal of Mathematics, Science & Technology Education, 4*(3), 197–214.
Cooper, L. N. (1984). Source and limits of human intellect. *Leonardo, 17*(1), 40–45.
Cray, D., Dawkins, R., & Collins, F. (2006, November 5). God vs. science. *Time*. Retrieved from http://www.time.com/time/printout/0,8816,1555132,00.html.
Dagher, Z. R. (2009). Epistemology of science in curriculum standards of four Arab countries. In S. BouJaoude & Z. R. Dagher (Eds.), *Arab States* (Vol. 3, pp. 41–60). Rotterdam: Sense Publishers.
Dawkins, R. (1995/2001). *River out of Eden: A Darwinian view of life* (Science Masters edn.). St. Helens: The Book People Lfd.
Dawkins, R. (2010, March 20). The faith trap. *The Washington Post*. Retrieved from http://newsweek.washingtonpost.com/onfaith/panelists/richard_dawkins/2010/03/the_faith_trap.html.
Esbenshade, D. H. (1993). Student perceptions about science & religion. *The American Biology Teacher, 55*(6), 334–338.
Falcão, E. B. M. (2008). Religious beliefs: Their dynamics in two groups of life scientists. *International Journal of Science Education, 30*(9), 1249–1264. doi:10.1080/09500690701765863.
Fergusson, D. (2009). Darwin and providence. In M. S. Northcott & R. J. Berry (Eds.), *Theology after Darwin* (pp. 73–88). Milton Keynes: Paternoster.
Fisher, L. (2004). *Weighing the soul: The evolution of scientific beliefs*. London: Weidenfeld and Nicolson.
Francis, L. J., Gibson, H. M., & Fulljames, P. (1990). Attitude towards Christianity, creationism, scientism and interest in science among 11–15 year olds. *British Journal of Religious Education, 13*(1), 4–17.
Freeman, M. M. R. (1992). The nature and utility of traditional ecological knowledge. *Northern Perspectives, 20*(1), 9–12.
Fulljames, P., & Stolberg, T. (2000). Consonance, assimilation or correlation?: Science and religion courses in higher education. *Science & Christian Belief, 12*(1), 35–46.
Fulljames, P., Gibson, H. M., & Francis, L. J. (1991). Creationism, scientism, Christianity and science: A study in adolescent attitudes. *British Educational Research Journal, 17*(2), 171–190.
Gilley, S., & Loades, A. (1981). Thomas Henry Huxley: The war between science and religion. *The Journal of Religion, 61*(3), 285–308.
Gould, S. J. (2001). *Rocks of ages: Science and religion in the fullness of life*. London: Jonathan Cape.
Grumett, D. (2009). Natural theology after Darwin: Contemplating the vortex. In M. S. Northcott & R. J. Berry (Eds.), *Theology after Darwin* (pp. 155–170). Milton Keynes: Paternoster.
guardian.co.uk. (2008). *Reiss resigns over call to discuss creationism in science lessons*. Retrieved April 14, 2013, from http://www.guardian.co.uk/science/2008/sep/16/michael.reiss.resignation.
Habgood, J. (2002). *The concept of nature*. London: Darton, Longman and Todd Ltd.
Halloway, A. (2008, September 23). Reiss resigns as Royal Society stifles debate on evolution. Retrieved March 27, 2013, from http://creation.com/reiss-resigns-as-royal-society-stifles-debate-on-evolution.

Hansson, L., & Redfors, A. (2007). Physics and the possibility of a religious view of the universe: Swedish upper secondary students' views. *Science & Education, 16*(3–5), 461–478. doi:10.1007/s11191-006-9036-8.

Hewitt, D. (2000). A clash of worldviews: Experiences from teaching aboriginal students. *Theory Into Practice, 39*(2), 111–117. doi:10.1207/s15430421tip3902_8.

Humphreys, C. (2004). *Can scientists beleive in miracles?* (Lecture). Cambridge: Faraday Institute. http://www.st-edmunds.cam.ac.uk/faraday/Multimedia.php

Inglis, J. T. (1993). *Traditional ecological knowledge concepts and cases*. Ottawa: International Program on Traditional Ecological Knowledge International Development Research Centre.

Long, D. E. (2011). *Evolution and religion in American education: An ethnography*. Dordrecht: Springer.

McCalla, A. (2006). *The creationist debate: The encounter between the bible and the historical mind*. London: Continuum.

Moreland, J. P., & Reynolds, J. M. (1999). Introduction. In J. P. Moreland & J. M. Reynolds (Eds.), *Three views on creation and evolution*. Grand Rapids: Zonderzan.

Morris, B. (2006). *Religion and anthropology: A critical introduction*. Cambridge: Cambridge University Press.

Pesic, P. (2006). Isaac Newton and the mystery of the major sixth: a transcription of his manuscript 'Of Musick' with commentary. *Interdisciplinary Science Reviews, 31*, 291–306. doi:10.1179/030801806x143268.

Popper, K. R. (1934/1959). *The logic of scientific discovery*. London: Hutchinson.

Reiss, M. J. (2008). Should science educators deal with the science/religion issue? *Studies in Science Education, 44*(2), 157–186. doi:10.1080/03057260802264214.

Reiss, M. J. (2009). Imagining the world: The significance of religious worldviews for science education. *Science & Education, 18*(6), 783–796. doi:10.1007/s11191-007-9091-9.

Sadler, T. D. (Ed.). (2011). *Socio-scientific issues in the classroom: Teaching, learning and research* (Vol. 39). Dordrecht: Springer.

Sagan, C. (1985/2006). *The varieties of scientific experience: A personal view of the search for God*. New York: Penguin Books.

Taber, K. S. (2009). *Progressing science education: Constructing the scientific research programme into the contingent nature of learning science*. Dordrecht: Springer.

Taber, K. S. (2013a). Conceptual frameworks, metaphysical commitments and worldviews: The challenge of reflecting the relationships between science and religion in science education. In N. Mansour & R. Wegerif (Eds.), *Science education for diversity: Theory and practice* (pp. 151–177). Dordrecht: Springer.

Taber, K. S. (2013b). *Modelling learners and learning in science education: Developing representations of concepts, conceptual structure and conceptual change to inform teaching and research*. Dordrecht: Springer.

Taber, K. S., Billingsley, B., Riga, F., & Newdick, H. (2011a). Secondary students' responses to perceptions of the relationship between science and religion: Stances identified from an interview study. *Science Education, 95*(6), 1000–1025. doi:10.1002/sce.20459.

Taber, K. S., Billingsley, B., Riga, F., & Newdick, H. (2011b). To what extent do pupils perceive science to be inconsistent with religious faith? An exploratory survey of 13–14 year-old English pupils. *Science Education International, 22*(2), 99–118.

Thagard, P. (2008). Conceptual change in the history of science: Life, mind, and disease. In S. Vosniadou (Ed.), *International handbook of research on conceptual change* (pp. 374–387). New York: Routledge.

Vallely, P. (2008, October 11). Religion vs science: Can the divide between God and rationality be reconciled? *The Independent*. Retrieved from http://www.independent.co.uk/news/science/religion-vs-science-can-the-divide-between-god-and-rationality-be-reconciled-955321.html.

Westfall, R. S. (1971). *The construction of modern science: Mechanisms and mechanics*. Cambridge: Cambridge University Press.

Chapter 4
Representing Evolution in Science Education: The Challenge of Teaching About Natural Selection

Keith S. Taber

4.1 Introduction

This chapter will consider the difficulties of effectively teaching about evolution. Firstly, the importance of teaching evolution as part of school science or college courses in biology is established. It will be argued that evolutionary theory cannot be sensibly omitted from any authentic science or biology curriculum. Then difficulties in teaching about the topic will be considered. The argument made here is that there are a number of features of natural selection that make teaching this topic challenging for many teachers but that it is useful to separate these into two major categories. The chapter will discriminate (a) those features which are similar to problems in teaching other 'difficult' science topics (such as force and motion or ionic bonding or photosynthesis) and which relate to intellectual challenges students face in learning about abstract and counterintuitive ideas from (b) those issues particular to topics such as evolution where people in many societal groups actively deny the science and oppose its teaching in schools and colleges. Natural selection is not only a theory that many students find difficult to grasp but also one that many learners have been told is false and perhaps even wicked. All teachers of evolutionary theory will face the first set of challenges, but for those working in particular countries or areas, there will be the additional problem of being asked to teach something that some students find morally objectionable.

K.S. Taber (✉)
Faculty of Education, University of Cambridge, Cambridge, UK
e-mail: kst24@cam.ac.uk

B. Akpan (ed.), *Science Education: A Global Perspective*,
DOI 10.1007/978-3-319-32351-0_4

4.2 The Importance of Evolution in Biology

It is generally accepted within the scientific community that evolution is a very important topic in science and that natural selection is a key theory in biology. That position will be strongly adopted here, although it is acknowledged that many people, including a very small proportion of scientists (Hameed 2010), do not accept natural selection describes a genuine process. It is important then to consider the nature of science, to explain how there can be such 'qualified consensus'.

4.2.1 How Do We Know What Scientists Think?

Before making any statements about the canonical claims of science, it is sensible to offer a caveat. The very nature of science is both dynamic and largely decentralised. The science that is currently considered of merit is – in principle – what the scientific community accepts. Yet there are no regular formal polls where scientists are asked to vote for or against particular ideas being accepted as canonical science. Moreover, although science is democratic in the sense that scientific results and claims are meant to be evaluated in their own terms rather than accepted on the authority of particularly influential scientists, that does not mean that every person who has qualified as a scientist is considered equally qualified to judge any scientific question.

Inevitably modern scientists are specialists, and if one was interested in knowing about the currently valued theories of superconductivity, or the best synthetic route for producing a particular new drug, or the best available understanding of hemispheric lateralisation in the brain, then we would find that (i) most scientists would not be well placed to advise us as the matter was outside their particular areas of specialist expertise, and (ii) even within the relevant subfield, some particular researchers would be considered to have the experience and insight to offer more authoritative views than others. So, to find out whom we should ask, the person outside the field would rely on advice from those within the field suggesting who they thought were the most important experts. There is clearly a sense of bootstrapping here: the people in a field decide who they consider to be in their field (Kuhn 1996) and then who has greater authority within it.

Scientists working within any particular field are usually only able to set up their own labs or be awarded research funds once they are recognised as well prepared to contribute to that field, meaning they have been inducted into the accepted theories, approaches and methodologies in the field (Kuhn 1996). Scientific fields tend to develop through the establishment of specific research programmes that are based upon particular (so-called hard core) commitments established when the programme is initiated (Lakatos 1970), and so scientists working in a field come to take those commitments for granted as long the programme is considered to be making progress.

Science is now so specialised that new researchers are not expected to 'get up to speed' by personally working their way through all the arguments and prior studies that had led the field to its current position – that simply would not be feasible. This is similar to how we cannot expect school students to rediscover all the great ideas of science by open-ended and unguided discovery learning (Driver 1983), as it would take them centuries! This feature of the nature of science does however mean that there tends to be built-in bias in a scientific field that protects key commitments in that field to ensure scientific results fit with, rather than question, those core assumptions. Indeed, one prominent philosopher of science has described the auxiliary theory built up around a field's most precious ideas as acting as a 'protective belt' for those key ideas, offering peripheral 'refutable' (and modifiable) aspects of theory cocooning the hard core ideas (Lakatos 1970).

This means that those working in a scientific field, and so considered experts, tend to necessarily be rather more committed to the key ideas in their field than a naive notion of the critical, open-minded scientist might suggest. The corollary to this is that often – although not always (Kuhn 1996) – the strongest critics of particular scientific ideas tend to be those who are outside the particular field and so have more distance on the scientific questions. External critics may be readily dismissed by those in a field as – by definition – they are not the experts who best know the core topics and the latest research.

We should not be surprised that experts working in evolutionary theory seldom raise questions about the fundamental notion that living things evolve into different species over time. Those scientists who reject evolution (e.g. Morris 1985) tend to have backgrounds in other fields and so lack the expertise and depth of knowledge of evolutionary biologists. The non-scientist has a choice between accepting the minority claims of those who lack expertise in the topic or the opinions of the many experts working in the field who may be so well socialised into the ways of thinking in that field that they find it difficult to recognise potential merits of alternative views (Thagard 1992).

4.2.2 The Status of Scientific Literature

This discussion of the scientific community may seem a diversion, as readers might consider that scientific knowledge is found in publications. Of course individual scientists will be subject to foibles and human limitations, but (it could be argued) scientific knowledge is actually to be found in 'black and white', in the pages of research journals. This is certainly a widely shared notion. However, this is a problematic idea in two senses.

For one thing, there is an argument that publications only contain representations of the knowledge of their authors and that really only people (not journals or books or computers) know things. So, in reading a scientific report, we are not acquiring knowledge that is unambiguously located in the report and can be unproblematically transferred to our brains but rather interpreting *a representation* of someone else's knowledge through our own prior understanding of the topic (Taber 2013a). We are used to thinking about *student* learning in these terms (Taber 2011a), but of course it is a general feature of human learning that applies just as much to teachers and indeed to scientists.

Perhaps an even more serious problem is that the primary scientific literature – generally seen as where scientific knowledge is reported – is vast and often contradictory. Published reports have survived a process of peer review, which means that other scientists in the field considered them to make a genuine contribution to knowledge (which may sometimes just mean they raise an interesting new hypothesis) and so to move forward the understanding of current issues in the field. However, that does not mean that all the claims in published papers are correct: rather simply that they seem to be based on reasonable interpretations of data that had been collected and analysed in sensible ways informed by a theoretical perspective that is considered viable in the field (Taber 2014a). Sometimes there are alternative perspectives related to a phenomenon under active consideration within a field (Lakatos 1970), and so papers from different perspectives might offer contrary accounts of nature, despite each being considered to be making valuable contributions to debate in the field.

Sometimes papers published decades ago are considered classics and still worth reading (Garwin and Lincoln 2003), whilst most others from the same period have long since been seen as surpassed and are now considered largely irrelevant. Usually only those working in the field will know which are which, although citation records (the extent to which papers are cited in other more recent papers) can offer a useful indication. Many newly published papers will make only modest contributions and will be cited very little in future: whilst a few will – in time – come to be seen as seminal. However, even those working in a field cannot always accurately predict which papers will stand the test of time.

The observer outside the field may fare better looking at secondary sources. Reviews of research published in research journals are likely to offer synthetic accounts, albeit sometimes from one of a number of alternative perspectives or research programmes operating in a field. Textbooks at least attempt to offer an overview of current thinking for a non-specialist, but school-level textbooks are inevitably written by non-experts. School textbook authors are often primarily teachers rather than researchers (which potentially helps them build effective pedagogy into their writing as they are more likely to appreciate the appropriate level of treatment), but even if a school textbook was written by an active scientific researcher, it is likely to have a broad scope (e.g. the whole of biology) such that most of the text would concern fields outside the author's own particular expertise. Commonly school texts are only very indirect representations of current scientific thinking.

There are mechanisms by which the scientific community and the general population can be kept informed of major developments in different fields, such as the well-established and prestigious general science journals *Nature* and *Science* and through science journalism, for example, magazines such as *New Scientist* and *Scientific American*. However, it is important for educators to realise that deciding what counts as current scientific consensus or orthodox thinking is not always a clear-cut matter. This is something that should also be borne in mind when curriculum developers and teachers are encouraged to include more on the nature of science in the curriculum. The processes by which science progresses and by which some kind of consensus position emerges are complex and nuanced and may not be readily appreciated by learners unless carefully developed teaching models of the nature of science are adopted (Taber 2008).

It seems then that statements about what science currently tells us, or what scientists currently 'think', need to be measured and qualified as these are not usually straightforward matters. To some extent the classroom teacher will often rely upon some curriculum authority or examining board to set out a representation of scientific ideas as target knowledge for learners of a particular age. Indeed, to some extent, the responsibility for deciding what counts as scientific knowledge rests with curriculum authorities (such as government education ministries or organisations charged with making such decisions in a particular national context). However, when a topic *is* included in the curriculum, the teacher still has a responsibility to offer learners some sense of its status within science, and this may not be a straightforward matter.

4.2.3 Evolution Is Fundamental to Modern Biology

Despite this caveat about the complications that arise when evaluating the current status of scientific knowledge, there are ideas that are generally recognised as being fundamental to the sciences. In chemistry, to offer an example where there is little or no controversy, a core idea is that the structure and behaviour of substances can be explained by theories that assume that matter has a particulate nature at a tiny, submicroscopic scale (Taber 2013b). This idea – let us call it atomic theory – has become so well established, and been found to be so useful, that virtually all of modern chemistry relies upon theoretical explanations built upon this basic notion. In most areas of chemical research, it would be very hard to make any progress without accepting and applying this principle. There may perhaps be some scientists that reject the 'atomic hypothesis', but if so they are surely few in number and have no influence in mainstream chemistry. Within chemistry, the particulate model of matter, the atomic hypothesis, takes on the status of a paradigm of 'normal science' (Kuhn 1996) – a set of ideas so widely accepted that they completely dominate the field compared with alternative views.

In a similar way, a great deal of modern biological thinking assumes evolution: the idea that the types of living things found today, the different species, are modified from ancestors that were quite different and that different species found today are related through descent. That is, individuals from different species share common ancestors, albeit ancestors that lived a very long time ago and are separated from their modern descendants by a very great many intermediate generations. This notion of evolution is generally understood to be possible through a mechanism known as natural selection, first proposed in outline by Charles Darwin and Alfred Russel Wallace (1858), and since developed through a 'modern synthesis' with findings from genetics, into a theory of descent with modification that is very widely accepted by biologists (Mayr 1991). Indeed, like the atomic hypothesis in chemistry, evolution by natural selection has become a core part of the paradigm of modern biology.

Professional biologists would overwhelmingly agree that evolution by natural selection is a central and generally accepted theory in biology. However, there is a small minority of biologists who would disagree and suggest that instead it was simply one model or perspective and that – in their view – it was not sufficiently supported by the available evidence to be accepted. Whilst that is very much a minority view, the analysis above suggests that decisions about which ideas in science are sufficiently well accepted to be considered canonical are complicated by the dynamic nature of science and the complex structure of the scientific community. Unfortunately for the science teacher, there is no website representing 'official science' that is kept updated with a list of the currently approved models, theories, laws, etc., that make up scientific knowledge. As suggested above, science just does not work in that way.

The view taken here, and the position recommended to all science educators, is that evolution by natural selection is the canonical scientific explanation for the origin of different species of living things and a core theoretical principle of modern biology. However, the absence of a formally approved canon of scientific knowledge offers scope for those who have issues with a particular scientific idea (such as evolution) to identify scientists and scientific publications that appear to cast doubt on both the merits and the status of that idea. This is certainly so in the case of evolution. As there are well resourced and highly committed organisations actively advocating against evolutionary ideas, such examples are regularly put into the public domain where it is hoped they will influence people (including students and teachers) to question or reject evolution and its status as widely accepted scientific knowledge.

Although evolution by natural selection may not seem to the layperson to be obviously relevant to many issues in biology, it has become so central to key explanatory schemes of how living things come to be the way they are that – like the atomic hypothesis in chemistry – it has become an integral part of the nature of biology as understood today. Perhaps to the school or college student, evolution is seen as just one topic among many and has little to do with understanding other topics such as, say, digestion. Yet to the modern biologist who has been trained to adopt an evolutionary mind-set, evolution has *everything* to do with digestion: the structure of the alimentary canal, the presence of specific digestive enzymes, the nature of the blood supply, the incidence of appendicitis, and so forth are all understood in terms of the evolutionary journey through which an organism's anatomy, physiology, and biochemistry came to have the form they have today.

Indeed, one key evolutionary thinker, Theodosius Dobzhansky (1973), went so far as to publish an article aimed at biology teachers entitled 'nothing in biology makes sense except in the light of evolution'. One could seek to quibble with the absolute inclusiveness of the claim (really, nothing at all?), but most biologists would feel that, if anything, the argument has become stronger in the years since Dobzhansky's article was first published.

Like the atomic theory in chemistry, evolution has acted as a major integrative theory in biology, which has allowed results from across a whole science to be understood within a common theoretical framework. Indeed there is even an argument that the theory of natural selection helped facilitate the transition of biology from being nature study ('natural history') to a mature science. Darwin set out on the Beagle as an amateur naturalist, but through his life's work, he did more than anyone to establish biology as a suitable discipline for a professional scientist.

4.3 The Importance of Evolution in Science Education

Given the importance of evolution to biology, there is a very strong case for considering it to form the basis of an essential topic in the science curriculum. This argument can be made at two levels. For one thing, evolution can be seen as an important topic in its own right, simply in terms of reflecting the pattern of scientific activity in the discipline. Evolutionary studies are a substantive part of biological research and provide a major area of activity for those who might decide to enter into professional work in the life sciences. Evolutionary theory offers accounts of the diversity of the biota, which might be considered an important question for biologists to be concerned with. So, as an important topic, evolution should be included in the school/college biology/science curriculum alongside other important topics. In particular, a school science curriculum that omitted evolution is ignoring one of the most important topics in the subject.

However, if we accept that very little in modern 'biology makes sense except in the light of evolution', then we can go beyond this argument to suggest that evolution has a stronger claim on its place in the science curriculum than many other biological topics. Ecological relationships cannot be understood, at least in the way they are understood in modern biology, except in terms of evolution; the geographical distribution of different species cannot be understood, at least in the way it is understood in modern biology, except in terms of evolution; and so forth. From this perspective we might argue that (i) if there is pressure on the curriculum, and only a limited number of biological topics can be included, then evolution should have the highest priority, as it is a more important topic than other biological topics and (ii) if evolution is seen as a key underpinning of a modern understanding of biology, and a core theoretical perspective for understanding other topics, then evolution should have a central role in the biology curriculum, and should be used as an organising theme, introduced as an early topic that is then drawn upon (and reinforced and developed) in learning other topics.

This position is consistent with the proclamations of many organisations concerned with science education and the public understanding of science. For example, the US National Science Teachers Association (NSTA) 'strongly supports the position that evolution is a major unifying concept in science and should be included in the [school] science education frameworks and curricula. Furthermore, if evolution is not taught, students will not achieve the level of scientific literacy they need' (National Science Teachers Association 2013).

4.4 Impediments to Learning About Evolution

There is a large body of research into the nature of learning in science and into the ideas about scientific topics that have been elicited from learners before and after teaching (Duit 2009; Taber 2009). Much of this work is based on a constructivist perspective that sees learning as an active process of meaning-making, which is channelled by the existing state of a learner's knowledge and understanding (Driver et al. 1994; Fensham et al. 1994; Mintzes et al. 1998). As with many other topics, research suggests that learners tend to develop their own notions around evolution which often do not fit the scientific accounts being taught in the curriculum (Wood-Robinson 1994).

4.4.1 The Challenging Nature of Natural Selection as Target Learning

There are a number of features of evolution as a topic that make it challenging for most students. For one thing, natural selection is a complex theory. Now the target knowledge presented in a school curriculum is not usually at the level of the current frontiers of scientific knowledge but rather based on simplifications suitable for the age and ability of the learners – with some of the less essential detail and confusing complications omitted. Such simplified versions of scientific ideas do however have to be what Jerome Bruner (1960) referred to as 'intellectually honest' simplifications that retain the essence of the more sophisticated scientific models. An optimal level of simplification offers a version of the scientific account that students can access and make sense of, yet which is still good enough to provide a sound basis for later progression through more advanced learning (Taber 2000). The challenge of developing such simplifications is greater in some topics than others.

Evolution by natural selection is problematic because although it is intellectually satisfying for the learner who has mastered it, and indeed often appears a 'simple' idea to experienced biologists, it only fully makes sense once a range of different ideas are coordinated together into a complex scheme. These ideas relate to how genes inform the development of individual characteristics, how genes are passed through hereditary, how occasional 'copying errors' occur leading to mutations, variations of characteristics within species, the failure of some (most in many species) offspring to themselves reproduce, the relationship between surviving to maturity and reproductive success on the fit between individual characteristics and environment, the possibility of geographical separation of distinct breeding populations within the same species, and so on (Taber 2009).

A second problem with natural selection is that it is counterintuitive for many learners because we lack direct experience of key features (the timescales over which natural selection occurs are so far from human experience), and our experience of the world generally reflects discrete and quite distinct species (Ruse 1987/1993). The former point is important because it is difficult to appreciate the sheer number of generations separating, say, the last common ancestors of humans and chimpanzees and their modern descendants.

Yet this timescale is important given that natural selection works with myriad chance events. Each fertilisation event reflects a successful union between particular packets of genes that could so easily have been different (given the number of sperm likely to be potentially able to join with a particular egg, if we consider mammalian fertilisation as an example). Within any particular environment, individuals born with the best *potential* characteristics for reproductive success are only *slightly* more likely to *actually have* reproductive success than their peers – as they still risk predation, starvation, drowning, poisoning, etc., even if at a slightly reduced level of risk compared to those peers. As the evolutionary theorist Stephen Jay Gould (1991) pointed out, so much in evolution is contingent: there is a limit to the extent your genes will prevent you from being in the wrong place when a hungry predator, a landslide, a volcanic eruption, or a tsunami, arrives.

The second point is perhaps more of an impediment. The idea that there are certain discrete 'natural kinds' of living things seems to be part of folk biology in diverse cultures (Medin and Scott 1999) and an idea spontaneously developed from a very young age (Keil 1992) – suggesting that there may even be a genetic predisposition to forming a cognitive bias towards recognising natural kinds in the world. So, ironically, evolution may have predisposed us to see the world in terms of a discontinuity of living things that is an impediment to appreciating evolution.

Darwin and Wallace (1858) recognised how the diversity of life on earth could be related through descent from common ancestral forms, and there is now a vast evidence base to support this view from comparative anatomy, palaeontology, molecular biology, etc. However, their insight was based on years of close engagement with samples, from myriad species in diverse habitats, and consideration of fossil forms. Darwin famously represented the process as a great bush of life, yet most people only experience a small part of what is in effect a single transverse section of that bush (see Fig. 4.1).

From the scientific perspective – the development of the 'bush' of life as shown in Fig. 4.1 – the species we see today are temporary islands of stability within an inherently dynamic picture of life that (when considered at the geological scale) is always in flux. Yet the environment we experience presents us only with the apparent discontinuity of the biota at one moment in earth's history, not the continuity highlighted by the scientific perspective. This is perhaps accentuated if, as has been suggested, evolutionary change does not tend to be uniform, but instead proceeds through a series of punctuated equilibria (Gould and Eldredge 1977, 1993/2000)

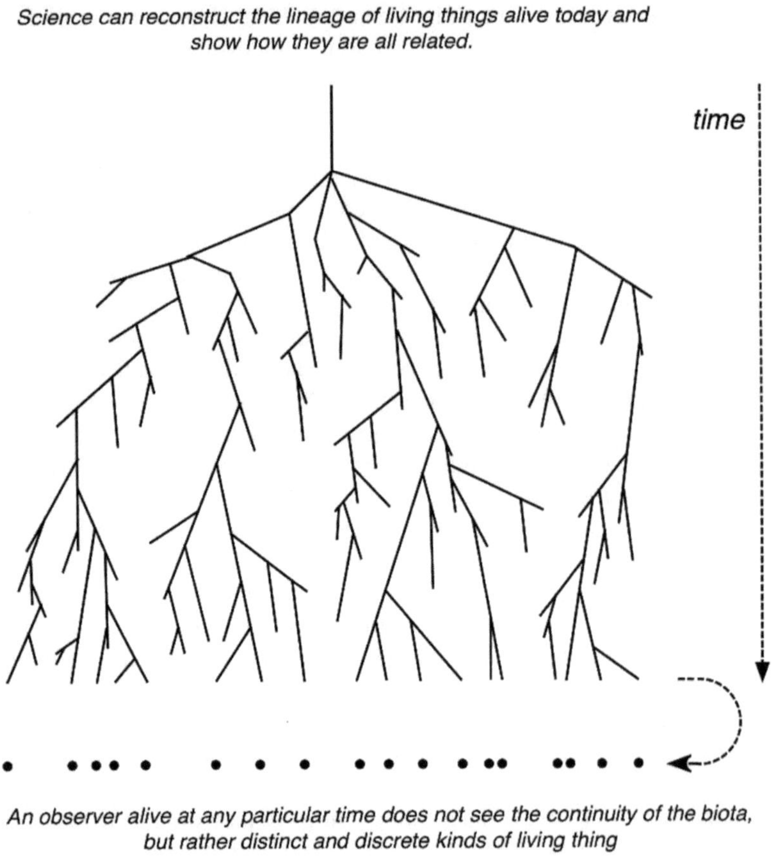

Fig. 4.1 The scientific perspective on the evolution of living things considers 'deep time', whereas the everyday experience of learners is limited to a 'snapshot' of the species alive at one geological moment

such that significant changes tend to occur only during a small proportion of the evolutionary history of any species, as the fossil record will predominatnly reflect the periods of equilibrium.

An additional common problem for teaching concerns the status of natural selection as a scientific idea. Evolution by natural selection is *a theory*. This means that it is not something scientists are setting out as a proven fact, as an absolutely certain account, but rather as an explanatory narrative that is conjectural and open to revision. However, that should not be understood to mean that natural selection is a low-status idea within science. To scientists, theories are well-developed systems of concepts that are strongly supported by evidence and which have been found to have considerable explanatory and predictive power. Although scientific knowledge is *in principle* provisional, some theories (such as natural selection) are so well supported that they come close to being considered as if factual for most purposes. Yet, technically, they have not been – and nor can they ever be – 'proved'.

A modern understanding of science does not admit the possibility of absolute proof of general ideas (Popper 1934/1959). All scientific generalities are – strictly – logically underdetermined by available evidence and inevitably reliant upon fallible interpretation (Kuhn 1996; Lakatos 1970). That is not considered a failing of science but rather a reflection on the limits of human knowledge. All scientific knowledge is subject to these constraints: science produces models and theories and then selects and develops those found to be most useful in the light of further testing.

This principled limit to the status of scienific knowledge, which is now well accepted among philosophers of science, creates a major challenge for science teachers when they are asked to help learners understand the nature of science (Hodson 2009). This is especially so for learners who are still developing the intellectual ability to effectively cope with complex and uncertain information – and this is likely to include most school pupils and many college level students (Perry 1970). The teacher has to help learners appreciate that scientific knowledge is always, technically at least, provisional and open to being revised, whilst claiming that science has reliable apparatus for developing robust and trustworthy accounts of the world.

What *is* known is that commonly learners have developed much less sophisticated understandings (Driver et al. 1996), such that they do not appreciate the nature, and in particular the status, of scientific theories. So, secondary-age learners, for example, may simply consider a theory to be an idea a scientist had – little more than a hunch or a guess (Taber 2006) – rather than the outcome of extensive development of a formal perspective that is consistent with evidence and coherent with other key ideas in science. Lacking a sophisticated understanding of the nature of science often means that students do not appreciate how (i) acknowledging that theories are not 'proven' is *not* the same as (ii) considering them as just hunches awaiting testing (Taber et al. 2015).

To return to a comparison used earlier, the particulate nature of matter that is so central to modern chemistry is, like natural selection, technically only theoretical knowledge. Yet, like the evolutionary modern synthesis, it provides a basic framework of understanding relied upon by effectively all scientists working within the discipline. Its theoretical status does not undermine its central importance in science. By its nature, science develops abstract generalised knowledge that is inherently theoretical. Without theory, science is reduced to natural history collecting and loses its explanatory power. Not all scientific theories are well developed and strongly supported by evidence: but just like particle theory, evolution by natural selection is considered 'reliable' knowledge (Ziman 1978/1991).

4.4.2 Teaching Evolution and Conceptual Change

This analysis suggests that teaching about natural selection is always likely to be challenging for teachers, as it is both a complex theory and an area where students' existing learning is often likely to be suboptimal for constructing good understanding of the scientific ideas. Each of these issues is very common in science education.

When teaching complex ideas, teachers have to scaffold learning in ways that help learners build up an overview of a topic with support until they are in a position to see how the components of the theory fit together in a coherent and logical way. Teaching from a constructivist perspective is a challenge for teachers – who have to both work out how to teach from the students' starting points and offer optimally guided instruction (Taber 2011a) that structures content into manageable 'learning quanta' (Taber 2005) and supports personal construction of knowledge with suitable 'scaffolding' (Wood 1988).

Teaching science therefore involves the teacher in building metaphorical conceptual bridges or ladders between students' current knowledge and the scientific concepts prescribed in the curriculum (Leach and Scott 2002). Teachers have to understand the learner's starting point and work out how to proceed stepwise using various pedagogical tools to help the learner make sense of scientific ideas. Those tools include models, demonstrations, analogies, metaphors, examples, thought experiments, etc. There is much that can go wrong in this process, as evidenced by the extensive literature describing students' failures to learn – or at least failing to learn what was intended rather than developing their own alternative conceptions of the subject matter. A range of types of learning impediments may occur (Taber 2005): for example, students may lack expected prior knowledge, or may fail to see how it relates to new instruction; students may misinterpret teaching through existing concepts that are inconsistent with scientific knowledge; or they may make creative, but unhelpful, connections between instruction and unrelated prior knowledge.

A particular problem with the usual pedagogic tactics teachers commonly adopt is that when they are used to make evolution accessible, they may encourage misconceptions. A lot of the language used tends to personify nature and/or imply teleology – that evolution is seeking to do something specific as if there is foresight at work or specific target states built into the process (Ruse 1986/1993; Zohar and Ginossar 1998). Darwin himself used personification as a device in explaining natural selection (Beer 1986).

There is also a tendency for students who have no principled objection to evolution to find Lamarckian models of evolution (where individual organisms change during their lifetime in response to environmental conditions and then pass on these acquired characteristics to their offspring) more feasible. This is despite Lamarck's ideas being considered to have been 'easily and repeatedly refused by all writers on the subject of varieties and species' and 'finally settled' by Wallace's time (Wallace 1858/2003). The author of this chapter has interviewed students who have explained that inheritance of acquired characteristics is what they have been taught in school science, although the author suspects this is rather how they personally made sense of what they were taught (Taber 2014b).

There are many other topics where learners' everyday experience fails to provide sufficient background as the foundation for building up scientific knowledge (so teachers need to provide demonstrations, analogies, models, etc.) or where scientific ideas seem to be counterintuitive (the relationship between force and motion, for example, (Gilbert and Zylbersztajn 1985)) or to be inconsistent with folk-theories (that exercise can produce energy, for example, (Solomon 1992)), and approaches have to be developed to help learners to appreciate the scientific models. That this is seldom a simple matter, and that the best strategy may vary across science topics, is reflected in the extensive literature discussing how to best teach for conceptual change (Vosniadou 2008). Natural selection is therefore going to be inherently a difficult topic to teach effectively, without any consideration of potential clashes due to learners' religious commitments.

4.5 Worldview Commitments and Learning About Evolution

Yet, in addition to the inherent problems with natural selection as a complex and counterintuitive idea, those charged with teaching evolution often have the additional complication that learners in some classes will reject evolution on principle (Hokayem and BouJaoude 2008) because they consider the notion is contrary to core commitments that make up part of their worldview (see the discussion of metaphysical commitments and worldview in Chap. 3).

Some communities reject evolution because of commitments to alternative general ideas about (a) the origins of the biota and/or (b) the potential consequences of evolutionary ideas for the nature of human beings. Although it is difficult to generalise, because different communities that object to evolutionary ideas hold different beliefs, problems commonly arise due to commitments to accounts in religious Scriptures as being technically accurate (rather than offering theological or metaphorical truths – again see Chap. 3). Particular problems among some Christian communities may relate to beliefs that:

(i) God created the world through a number of discrete acts of creation, bringing into being different classes of animals and plants through their own special creation event.
(ii) The creation occurred sometime in the last 10,000 years.
(iii) Human beings are more than just a particular type of animal, having a different relationship to God than other living things.
(iv) Death was not initially inevitable for humans and (sic, other) animals – there was no death until sin entered the world.
(v) God renewed his covenant with humanity following a worldwide flood, from which were saved specimens of the types of creature he had created previously.
(vi) The cosmos beyond the earth was created as a perfect realm which is unchanging.

4.5.1 *The Creation of Living Things*

Scriptural accounts of the creation of the world in Genesis (in the Jewish Torah and Christian Old Testament) refer to God creating different types of living things as part of a staged 6-day programme of creation. There are very different traditions in Christianity about how such accounts are to be read, and in many Christian traditions they are understood as poetic or metaphorical accounts conveying deep theological truth. However, during the twentieth century, a number of Christian denominations in the US popularised the notion that these accounts should be taken as true technical accounts of the creation of the world (McCalla 2006).

From this perspective the main types of living things alive today are descended from the different discrete types of living thing originally created by God. The evolutionary notion that all living things on earth may have developed from the same, much simpler, ancestor organism is completely inconsistent with the beliefs of those holding a religious commitment to Genesis as a literal account.

Usually there is no objection to the idea that within the main groups of living things created by God, different descendent populations might diverge to give different variations (sometimes referred to as microevolution) but always within the bounds of the general type of organism God originally created. The notion that animals may have descended from fundamentally different types (known as macroevolution) is completely excluded by this perspective.

4.5.2 *The Dateline*

Jesus, the founder of Christianity, is a historical figure, so there is little dispute over when he lived on earth and carried out his ministry (about 2000 years ago). The Christian Scripture offers his genealogy in the form of an unbroken male line back to Adam, the first man. Based on this (and various clues in the Old Testament), it is possible to produce a timescale for the creation of the world and Adam. One scholarly calculation that was influential in the eighteenth century suggested that the creation occurred about 4000 years before Jesus's birth (starting on October 23, 4004 BC), and whilst this rather precise date has since fallen into dispute, many of those who today reject evolution because of its inconsistency with Christian Scriptures consider that the earth is no more than about 10,000 years old.

Clearly a figure of 10,000 years is completely contrary to the scientific view that the earth is something like 4,500,000,000 years old (and the Universe much older). The much shorter timescale is inconsistent with the time needed for evolution to occur through natural selection.

4.5.3 *The Special Relationship*

A particular problem that some groups find with evolution is the idea that humans evolved over millions of years from earlier hominids that in turn had themselves evolved from earlier non-hominid species. Evolutionary theory suggests, that in biological terms at least, there is nothing special about humans marking us off as separate from the rest of the biota. Scientifically, we are primates, and mammals, and indeed animals, and only as special as any other particular species.

This is completely contrary to the notion that human beings were marked out as a special creation by God. Scriptural accounts are read to imply that humans were always intended to have a special relationship with God and seen as distinct from the rest of creation. In some religious traditions, humans have immortal souls that survive death whereas other animals do not. In some traditions, the rest of the biota is seen as provided by God for mankind's use – a perspective that may not always encourage strong ecological thinking (although in some religious traditions, people are also considered to be in stewardship of nature and responsible for it to God).

Perhaps a major area of contention is the issue of the boundary between humans and non-humans. In some religious traditions, this is unproblematic, because it is taught that God created man separately from other kinds. However, the scientific account offered through evolution by natural selection suggests that if we were able to move back generation by generation, there would be no sudden discontinuity between humans and their prehuman ancestors, but rather there has been a slow process through which creatures developed that we now consider human. On the evolutionary model, there would be many generations of ancestors where there would be no clear basis for considering them definitely human or definitely prehuman. The scientific account presents challenges for those who see humans as in a special relationship with God (Rachels 1990): for example, does a severely mentally damaged newborn child have an immortal soul, whilst a modern chimpanzee does not?

Darwin was certainly aware of the potential difficulties of his evolutionary ideas conflicting with religious views on the special status and nature of humans. Darwin only made a brief allusion to the question of human origins in his 'Origin of Species' (1859/1968), leaving the topic for a later book (Darwin 1871/2006) only published once his evolutionary ideas had been widely discussed and come to be generally accepted in many quarters.

4.5.4 *The Fall*

Another tenet of some 'fundamentalist' traditions in Christianity is 'the Fall'. According to Scripture, Adam and Eve disobeyed God's direct command in eating from the tree of knowledge. This event is considered as man inviting sin into a world that God had made perfect (Williams 2001). In this tradition, there was no death prior to the Fall (Messer 2009; Moreland and Reynolds 1999), and all of the creatures of the earth had coexisted peacefully (and so were all herbivores); but afterwards prey-predator relationships developed. The scientific interpretation of the fossil record is inconsistent with the idea that at one time all animals were plant eaters. The Fall is often considered very important in Christian thinking as it is related to the idea of salvation in Jesus Christ and the possibility of eternal life with God after death.

4.5.5 The Flood

The Genesis account does not only record the creation but also a later cataclysmic event where God sent a worldwide flood to punish humanity for its evil ways, only saving one family (Noah and his three sons, and their wives, who were forewarned to build a great ark) who would be the ancestors of all humans in the world after the deluge. God also had Noah and his family save stock of all the animals he had created to repopulate the animal world as well. Many of those who adopt the scriptural account as an accurate technical account consider this event as very significant in human history and see the act of saving representatives of the different animals God had created in keeping with the idea of them having been created as distinct types unchanged from the creation, through the deluge, to the present day (McCalla 2006).

The scientific account of earth's geology suggests there have been enormous changes in the face of the earth since its formation due to seismic activity and plate tectonics, changes in the atmospheric composition, etc., whereas many of those who reject evolution feel that the earth is basically unchanged since its creation, apart from the powerful effects of the great flood.

4.5.6 The Heavens

Many who take scriptural texts as offering a literal account of the creation and history of the world find Scripture to disconfirm scientific ideas about cosmology. An obvious point of contention is scientific accounts of the earth as orbiting a second-generation sun, because the earth is composed mainly of elements that were created in the nuclear furnaces of earlier suns and which – according to scientists – were not present in the Universe until they were formed in stars. The scriptural account of creation is often read as allowing 6 days for the whole creation process and does not allow for billions of years between the formation of first-generation stars and the formation of the earth.

Moreover, scriptural verses can be read to imply that heavenly bodies are unchanging, so the notion of stars themselves passing through a kind of life-cycle and then being destroyed in supernovae is seen by some as completely contrary to Scripture (Morris 2000).

4.6 Young-Earth Creationism

These scriptural perspectives are often associated together in the notion of young-earth creationism (YEC), a worldview that excludes the possibility of cosmic evolution, geological timescales, or evolution of species through natural selection. From the scientific perspective, such a position is untenable given the vast evidence base from diverse fields such as astronomy, geology, palaeontology, comparative anatomy and molecular biology.

However, of course, scholarship into the nature of science reminds us that evidence never unproblematically leads to particular definite conclusions. Rather, any empirical evidence is always interpreted within some theoretical framework or other. This is one reason why few scientific experiments or observations can be seen as completely refuting a hypothesis (Lakatos 1970): conclusions always depend upon evidence, *plus* its interpretation. If we reject some aspect of the theoretical framework, we can reinterpret the evidence. So, when Galileo (Galilei 1610/1989, p. 35) suggested that Jupiter had its own satellites ('four planets hitherto never seen'), based on 'observations recently made, with the benefit of a new spyglass', some of those who were not open to such a possibility (because it seemed to be contrary to scriptural teaching) rejected his evidence on the basis that they did not accept the validity of his instrumentation – the telescope.

Similarly today, various arguments are used to fit scientific evidence into different interpretative frameworks by those who reject the scientific interpretations. There are museums in the USA where dinosaurs are displayed as being contemporaneous with humans rather than part of an earlier evolutionary epoch. To many working in science education that seems ridiculous, but surveys suggest that *most people* in the USA accept this idea (McCalla 2006). Indeed some even suggest that such creatures are only not around today because – for whatever reason – Noah did not take them in his ark.

Similarity of anatomy and biochemistry across species can be understood by those adopting YEC as evidence for God having used optimal designs that therefore were very similar across different parts of His creation. Methods that date rocks and suggest some rocks are millions or billions of years old are based on flawed assumptions, YEC adherents would claim: we cannot know that radioactive decay rates have been constant during earth's history, as we did not have the technology to test this during most of the earth's 10,000 years.

Ultimately, regardless of how superior the scientific account may seem to science teachers in terms of fitting a diverse and extensive evidence base, such arguments are of limited value in persuading those who find evolution to rely upon or suggest ideas that are directly contrary to matters they take as central to their religious convictions. If you 'know' that the earth is young, and evolution does not occur, because this is seen as central to your religious faith, then no amount of argument from scientific evidence is relevant (Long 2011). In some parts of the world, children are being told by their parents and Church elders that evolution is false and often that it is an evil idea that leads otherwise decent people on the road to eternal damnation (see below).

Leaders of YEC movements are aware that science has a vast evidence base it uses to persuade people of the worth of evolutionary theory, and so they invest time and scholarship into addressing scientific arguments, looking for flaws, identifying minority dissenting voices from within the scientific community, and offering alternative interpretations of scientific evidence that can appear convincing from within the YEC worldview. The YEC movement does not need to provide young people with convincing creationist accounts of all possible scientific evidence but just enough examples of how the evidence can make sense from a YEC perspective so that when they meet evolutionary evidence in school or college, they are convinced that there must be a perfectly good explanation for that data that fits with their own convictions.

4.7 Moral Objections to Evolution

As well as arguments for rejecting evolution based on inconsistencies between religious commitments and the details of the scientific account, there is also what might be considered a 'secondary' line of argument that evolution and championing of evolution are not simply incorrect according to religious teaching but actually represent something that is (from this perspective) inherently bad or evil.

From a scientific perspective, a theory can be more or less supported by evidence but cannot be morally good or bad. However, people may draw implications for behaviour based on scientific theories, and so theories may be seen to be associated with ideological positions that others judge as morally desirable or undesirable. Certainly some commentators see evolutionary ideas as dangerous or morally wrong, and this seems to be based upon at least three distinct issues:

(i) Evolutionary ideas lead to people questioning the authority of Scripture and so doubting articles of their faith.
(ii) Evolution is part of an inherently atheistic and materialist worldview that denies the existence of God Himself.
(iii) Evolution supports values and ideologies at odds with the moral teaching of religion.

Issue (i) is clear from the discussion of possible interpretations of religious Scripture above. Issue (ii) is less clear-cut. Evolution itself is not inherently atheistic and indeed, even Darwin – who found much evidence to bring into doubt the Christian account of a personal, loving God – did not see natural selection as an absolute reason for excluding the existence of a creator God (Mandelbaum 1958).

This issue is complicated by the stance taken by a minority of scientists who are atheist materialists, who consider that their perspective should be the proper basis for science itself and who seem happy to encourage debate on the basis of setting acceptance of scientific accounts of origins against what they see as irrational and primitive supernatural alternatives. Although this group is not representative of the scientific community (see Chap. 3), they do have a high public profile in some countries and so may often *be thought to* represent the scientific view.

4.7.1 Ideological Positions Associated with Evolutionary Ideas

Issue (iii) concerns the *implications* that some might consider follow from accepting natural selection. It has been suggested that opposition to evolution within Muslim communities is generally of this kind as the Qur'an is not usually considered to specifically exclude evolution (Hameed 2010).

From a scientific perspective, the theory of evolution (in common with any other scientific theory) does not tell people how to behave, but such theories can inform ideologies and lead to questioning of cultural traditions. So, for example, if all living creatures are related by descent, and species are not absolute, the tradition of not eating other human beings (something taboo in most human cultures), but eating other mammals, could be questioned.

Evolution has certainly been *associated* in the past with eugenics (Bowler 1983/1989), and so with suggesting it might be acceptable to not allow those with inheritable diseases or of severely low intelligence levels to reproduce. As always, science cannot offer a view of what is right or wrong but only help inform us of what is technically possible and what the likely consequences of different actions might be.

Yet some opponents of evolution have stretched the argument to make claims that belief in evolution is responsible for various things seen (from their perspective) as undesirable, including Nazism, communism, fascism, romanticism, homosexuality, promiscuity, imperialism, teenage pregnancy, divorce, public unrest and so forth (Berry 2009; Hameed 2010; Yahya 2008). One popular writer and speaker against evolution went so far as to describe evolution as 'the philosophy [sic] underlying all the evils of the world' (Morris 2000, p. 18).

Whilst this is nonsensical from the scientific perspective, there *are* some people who will use arguments about evolution to justify behaviours or opinions others find undesirable. Many learners in science classes, especially at school level, will not be well placed to make distinctions between the science and the ideology, and if they come from communities where they are warned that evolution is an immoral and dangerous 'philosophy', then they may understandably tend to be very wary of classroom teaching about evolution.

4.8 Creation Science

Some of those who reject evolution and oppose the teaching of evolution attempt to locate their arguments within a scientific perspective or at least to claim that their argument is based on scientific evidence (McCalla 2006). In many parts of the USA, there has been a campaign to teach about creationist views in schools as a counter to the teaching of evolution. The US constitution does not allow the teaching of religion in state schools, but the argument has been made that (a) evolution is only a theory and not known definitively to be true and (b) that there are alternative interpretations of the scientific evidence that should also be taught. Point (a) is correct but would apply to any other scientific theory: plate tectonics, the role of enzymes in digestion, flux cutting as a mechanism in electromagentism, the particulate nature of matter, etc. There have not been major campaigns to have alternatives to these other ideas taught on the basis that they are only theories and not definitive knowledge.

Creationists will tend to marshal evidence to support their alternative views, although this sometimes involves scant regard for well-accepted scientific principles: adherents of so-called creation science have been said to 'play fast and loose with the facts of geology and biology' (Mandelbaum 1958, p. 381). So, the theory of punctuated equilibria, that suggests evolutionary change tends to be uneven, is presented by creationists as scientists acknowledging that the fossil record does not provide evidence of modification *and therefore* provides no support for natural selection (Morris 1985).

As an example, one book written by an author who taught science at a British university (Pimenta 1984, p. 29) argues that because all matter was created from hydrogen (*all* atoms of which, the reader is told, contain neutrons), it is reasonable to suppose that all bodies in the Universe contain hydrogen deep within them (which does not follow), which is liable to be sufficiently heated by radioactivity to give rise to sudden events 'equivalent to millions of subterranean hydrogen bombs' (which certainly would not follow). This (non-feasible) violent mechanism is mooted to explain the current appearance of the earth with its apparent geological history, despite a recent creation. According to Pimenta, radioactive methods that date rocks to great age 'cannot be valid' *because* time only began about 6000 years ago. He suggests that the rate of decay of elements has been shown to have changed significantly in recent centuries and that when it is used to persuade people of evolution, it amounts to 'a satanic ploy' (p. 238).

These efforts to present alternative interpretations of scientific evidence, and sometimes scenarios completely disregarding scientific evidence, may be well resourced – explaining the museums in the USA that present geological and paleontological material arranged in accordance with YCE interpretations. In general, however, these approaches have made few official inroads into state education – although that does not mean that classroom presentations always cover evolution according to the curriculum (Long 2011). Elsewhere, copies of beautifully illustrated hardback books (volumes of an 'Atlas of Creation') reflecting an antievolutionist stance have been distributed to thousands of schools in some countries, from an Islamic organisation in Turkey (Hameed 2008). This material claims that Darwin's theory was derived from his imagination – which in a sense is inevitably true of course (Taber 2011b) – and had no basis in 'scientific evidence or findings'; rather because science was 'fairly primitive' in Darwin's time, people did not recognise 'the full extent of the ridiculous and unrealistic nature of his assertions' (Yahya 2008).

4.8.1 Intelligent Design

In recent years a new strand of thinking has developed, known as intelligent design or ID. The ID movement is not formally linked to any religious organisation and accepts the geological evidence for the age of the earth and much of the evidence for *some* aspects of evolution. However, ID adherents argue that there are aspects of the structure and organisation of living things that demonstrate a kind of irreducible complexity that is inconsistent with being formed through random events in natural selection (Behe 1996, 2007). The scientific account of natural selection argues that complex structures such as the mammalian eye or a wing capable of supporting flight developed in small steps – a proto-eye simply offering a gross indication of light intensity and direction; the precursor to the wing just supported gliding between branches, and not actual flight, etc. (Dawkins 1988).

However, ID supporters argue that there are some complex structures at cellular level that only offer any advantage to the organism once they are fully formed (Behe 1996). The bacterial flagellum was a choice example: a structure composed of specific subsystems which all had to be present and properly integrated to function but which individually offered no obvious value to the organism. The ID argument runs that given such complexity could not have been provided by natural selection (which would not favour the commitment of resources to building structures that only have a viable use many generations later), then such structures demonstrate that organisms have at some level been designed by an intelligence. According to ID, evolution occurs but cannot be the whole story: rather evolution is helped and steered by some guiding intelligence. The official ID stance does not identify such intelligence with a God (rather than perhaps a very advanced alien genetic engineer), but that is the association that is available to those who wish to adopt it.

Because ID accepts most of the scientific account, and looks to adopt scientific evidence, its proponents have claimed it should be seen as an alternative scientific account and so considered in courses teaching evolutionary theory. However there are a number of objections to ID. One is that it adopts non-scientific (non-testable) hypotheses (Alexander 2009). That is, science should look for naturalistic explanations and not invoke God or other teleological arguments to cover gaps in scientific knowledge. This is not an argument that necessarily excludes God but assumes that scientific explanations must be based on evidenced mechanisms rather than conceding that some natural phenomena may not have natural explanations. This is a position that most religious scientists would adopt (Alexander 2009) as part of 'methodological naturalism', the idea that within the work of science, only natural mechanisms and explanations are adopted (see Chap. 3).

ID has also been seen by many working in science education as an attempt to offer a version of creationism that might be admitted into the science curriculum. However, ID has been widely rejected by the scientific community, and many organisations concerned with science and science education have taken public positions opposing the teaching of ID in science classrooms. The Association for Science Education in the UK issued a statement on ID to the effect that:

> it is clear to us that Intelligent Design has no grounds for sharing a platform as a scientific 'theory'. It has no underpinning scientific principles or explanations to support it. Furthermore it is not accepted as a competing scientific theory by the international science community nor is it part of the science curriculum … Intelligent Design has no place in the science education of young people in school.
>
> (Association for Science Education 2007)

4.8.2 Responding to Creationism in the Classroom

It has been recognised that science teachers working with students who reject evolution on non-scientific grounds face a particular challenge, as no amount of argument or appeal to evidence is likely to be effective when the whole idea of evolution seems contrary to deeply held beliefs. Advice to engage with students' viewpoints (Reiss 2008) has alarmed some scientists (see Chap. 3) and may make some science teachers uneasy due to their limited preparation for dealing with religious questions.

A useful perspective may be to keep in mind that science is not meant to be about belief and the teacher's job is not to persuade students to believe in evolution by natural selection or indeed any other theory (Taber, in press). The teacher's job is to help students understand (i) the scientific model and (ii) the evidential basis for that model. Teachers are likely to make more headway in helping learners from creationist communities understand natural selection if it is presented as a theory to be understood and critiqued and not as a true account they are being asked to believe.

Such a strategy will clearly be more viable in classrooms where science is generally taught from a perspective informed by a modern view of the nature of science, so scientific knowledge on all topics is presented as reliable – but also conjectural, theoretical and inherently open to review. Students in classes that regularly learn about historical scientific models which were once widely accepted, but become replaced as new evidence became available, will be used to critiquing scientific ideas and will appreciate that this is important for scientific progress. In such a context teachers can invite questions and objections to natural selection (as they would with other topics) but ensure that these are all dealt with in terms of the scientific evidence. The aim must not be to demonstrate that evolution by natural selection is 'true' but rather to show why it is currently considered the best available scientific account.

4.9 Conclusion

This chapter has considered the challenges involved in teaching about evolution. Natural selection is a difficult and counterintuitive idea for many learners, and teaching about the theory is commonly misunderstood by students. There is much research into learners' ideas in science and how to address them that can offer teachers some guidance on how to develop presentations of evolution that will allow learners to construct their own understandings in keeping with scientific ideas.

However, this is complicated in many national contexts where learners may understand evolution to be contrary to religious teaching. Sometimes learners may have even been told that evolutionary theory itself is in some sense an evil idea that corrupts society and undermines faith. The science education community in many parts of the world has held firm to the idea that evolution should be taught and that it should not share the science classroom with presentations of creation science or alternatives such as ID. This seems a sensible policy: science teachers should teach the currently accepted scientific theories, emphasising both that they are theory but also that they are strongly supported by empirical evidence. However, it is much less clear how to effectively respond to the reactions of those students who themselves bring creationist beliefs into the classroom, and to do so in ways that both do justice to science and show appropriate respect for the values and views of the learners. Arguably, science teaching that is informed by a perspective from the history and philosophy of science, and where learners appreciate the nature and status of scientific models and theories, is more likely to support students in learning about natural selection, without them feeling they are being asked to accept ideas contrary to their own convictions.

Acknowledgement The author would like to thank Dr. Mark Winterbottom for advice given during the preparation of this chapter.

References

Alexander, D. R. (2009). After Darwin: Is intelligent design intelligent? In M. S. Northcott & R. J. Berry (Eds.), *Theology after Darwin* (pp. 22–40). Milton Keynes: Paternoster.

Association for Science Education. (2007). *Science education and intelligent design: A statement from the Association for Science Education*. Association for Science Education, Hatfield.

Beer, G. (1986). 'The face of nature': Anthropomorphic elements in the language of the origin of species. In L. J. Jordanova (Ed.), *Languages of nature: Critical essays on science and literature* (pp. 212–243). London: Free Association Books.

Behe, M. J. (1996). *Darwin's black box: The biochemical challenge to evolution*. New York: Simon & Schulster.

Behe, M. J. (2007). *The edge of evolution: The search for the limits of Darwinism*. New York: Free Press.

Berry, R. J. (2009). Biology after Darwin. In M. S. Northcott & R. J. Berry (Eds.), *Theology after Darwin* (pp. 4–21). Milton Keynes: Paternoster.

Bowler, P.J. (1983/1989). *Evolution: The history of an idea* (Rev. ed.). Berkeley: University of California Press.
Bruner, J. S. (1960). *The process of education*. New York: Vintage Books.
Darwin, C. (1859/1968). *The origin of species by means of natural selection, or the preservation of favoured races in the struggle for life*. Harmondsworth: Penguin.
Darwin, C. (1871/2006). The descent of man, and selection in relation to sex. In E. O. Wilson (Ed.), *From so simple a beginning: The four great books of Charles Darwin* (pp. 767–1248). New York: W W Norton & Company.
Darwin, C., & Wallace, A. R. (1858). On the tendency of species to form varieties; and on the perpetuation of varieties and species by natural means of selection. *Proceedings of the Linnean Society, 3*, 45–62.
Dawkins, R. (1988). *The blind watchmaker*. Harmondsworth: Penguin.
Dobzhansky, T. (1973). Nothing in biology makes sense except in the light of evolution. *The American Biology Teacher, 35*(3), 125–129.
Driver, R. (1983). *The pupil as scientist?* Milton Keynes: Open University Press.
Driver, R., Asoko, H., Leach, J., Mortimer, E., & Scott, P. (1994). Constructing scientific knowledge in the classroom. *Educational Researcher, 23*(7), 5–12.
Driver, R., Leach, J., Millar, R., & Scott, P. (1996). *Young people's images of science*. Buckingham: Open University Press.
Duit, R. (2009). *Bibliography – students' and teachers' conceptions and science education*. Kiel: IPN – Leibniz Institute for Science and Mathematics Education.
Fensham, P. J., Gunstone, R. F., & White, R. T. (1994). *The content of science: A constructivist approach to its teaching and learning*. London: Falmer Press.
Galilei, G. (1610/1989). Sidereus Nuncius (A. van Helden, Trans.). In A. van Helden (Ed.), *Sidereus Nuncius or the Sidereal Messenger* (pp. 25–86). Chicago: University of Chicago Press.
Garwin, L., & Lincoln, T. (Eds.). (2003). *A century of nature: Twenty-one discoveries that changed science and the world*. Chicago: University of Chicago Press.
Gilbert, J. K., & Zylbersztajn, A. (1985). A conceptual framework for science education: The case study of force and movement. *European Journal of Science Education, 7*(2), 107–120.
Gould, S. J. (1991). *Wonderful life: The burgess shale and the nature of history*. London: Penguin.
Gould, S. J., & Eldredge, N. (1977). Punctuated equilibria: The tempo and mode of evolution reconsidered. *Paleobiology, 3*(2), 115–151. doi:10.2307/2400177.
Gould, S. J., & Eldredge, N. (1993/2000). Punctuated equilibrium comes of age. In H. Gee (Ed.), *Shaking the tree: Readings form nature in the history of life* (pp. 17–31). Chicago: University of Chicago Press.
Hameed, S. (2008). Bracing for Islamic creationism. *Science, 322*(5908), 637–1638. doi:10.1126/science.1163672.
Hameed, S. (2010). Evolution and creationism in the Islamic world. In T. Dixon, G. Cantor, & S. Pumfrey (Eds.), *Science and religion: New historial perspectives* (pp. 133–152). Cambridge: Cambridge University Press.
Hodson, D. (2009). *Teaching and learning about science: Language, theories, methods, history, traditions and values*. Rotterdam: Sense Publishers.
Hokayem, H., & BouJaoude, S. (2008). College students' perceptions of the theory of evolution. *Journal of Research in Science Teaching, 45*(4), 395–419.
Keil, F. C. (1992). *Concepts, kinds and cognitive development*. Cambridge, MA: MIT Press.
Kuhn, T. S. (1996). *The structure of scientific revolutions* (3rd ed.). Chicago: University of Chicago.
Lakatos, I. (1970). Falsification and the methodology of scientific research programmes. In I. Lakatos & A. Musgrove (Eds.), *Criticism and the growth of knowledge* (pp. 91–196). Cambridge: Cambridge University Press.
Leach, J., & Scott, P. (2002). Designing and evaluating science teaching sequences: An approach drawing upon the concept of learning demand and a social constructivist perspective on learning. *Studies in Science Education, 38*, 115–142.

Long, D. E. (2011). *Evolution and religion in American education: An ethnography*. Dordrecht: Springer.
Mandelbaum, M. (1958). Darwin's religious views. *Journal of the History of Ideas, 19*(3), 363–378. doi:10.2307/2708041.
Mayr, E. (1991). *One long argument: Charles Darwin and the genesis of modern evolutionary thought*. London: Allen Lane.
McCalla, A. (2006). *The creationist debate: The encounter between the Bible and the historical mind*. London: Continuum.
Medin, D. L., & Scott, A. (Eds.). (1999). *Folkbiology*. Cambridge, MA: The MIT Press.
Messer, N. (2009). Natural evil after Darwin. In M. S. Northcott & R. J. Berry (Eds.), *Theology after Darwin* (pp. 139–154). Milton Keynes: Paternoster.
Mintzes, J. J., Wandersee, J. H., & Novak, J. D. (Eds.). (1998). *Teaching science for understanding: A human constructivist view*. San Diego: Academic.
Moreland, J. P., & Reynolds, J. M. (1999). Introduction. In J. P. Moreland & J. M. Reynolds (Eds.), *Three views on creation and evolution*. Grand Rapids: Zonderzan.
Morris, H. M. (1985). *Scientific creationsim* (2nd ed.). Green Forest: Master Books.
Morris, H. M. (2000). *The long war against God: The history and impact of the creation/evolution conflict*. Green Forest: Master Books.
National Science Teachers Association. (2013). *NSTA position statement: The teaching of evolution*. Retrieved 8 Apr 2013, from http://www.nsta.org/about/positions/evolution.aspxacce
Perry, W. G. (1970). *Forms of intellectual and ethical development in the college years: A scheme*. New York: Holt, Rinehart & Winston.
Pimenta, L. R. (1984). *Fountains of the deep*. Chichester: New Wine Press.
Popper, K. R. (1934/1959). *The logic of scientific discovery*. London: Hutchinson.
Rachels, J. (1990). *Created from animals: The moral implications of Darwinism*. Oxford: Oxford University Press.
Reiss, M.J. (2008, September 11). *Science lessons should tackle creationism and intelligent design*. guardian.co.uk
Ruse, M. (1986/1993). Teleology and the biological sciences. In M. Ruse (Ed.), *The Darwinian paradigm: Essays on its history, philosophy and religious implications* (pp. 146–154). London: Routledge.
Ruse, M. (1987/1993). Biological species: Natural kinds, individuals, or what? In M. Ruse (Ed.), *The Darwinian paradigm: Essays on its history, philosophy and religious implications* (pp. 97–117). London: Routledge.
Solomon, J. (1992). *Getting to know about energy – in school and society*. London: Falmer Press.
Taber, K. S. (2000). Finding the optimum level of simplification: The case of teaching about heat and temperature. *Physics Education, 35*(5), 320–325.
Taber, K. S. (2005). Learning quanta: Barriers to stimulating transitions in student understanding of orbital ideas. *Science Education, 89*(1), 94–116.
Taber, K. S. (2006). Exploring pupils' understanding of key 'nature of science' terms through research as part of initial teacher education. *School Science Review, 87*(321), 51–61.
Taber, K. S. (2008). Towards a curricular model of the nature of science. *Science & Education, 17*(2–3), 179–218. doi:10.1007/s11191-006-9056-4.
Taber, K. S. (2009). *Progressing science education: Constructing the scientific research programme into the contingent nature of learning science*. Dordrecht: Springer.
Taber, K. S. (2011a). Constructivism as educational theory: Contingency in learning, and optimally guided instruction. In J. Hassaskhah (Ed.), *Educational theory* (pp. 39–61). New York: Nova. Retrieved from https://camtools.cam.ac.uk/wiki/eclipse/Constructivism.html
Taber, K. S. (2011a). The natures of scientific thinking: Creativity as the handmaiden to logic in the development of public and personal knowledge. In M. S. Khine (Ed.), *Advances in the nature of science research – concepts and methodologies* (pp. 51–74). Dordrecht: Springer.
Taber, K. S. (2013a). *Modelling learners and learning in science education: Developing representations of concepts, conceptual structure and conceptual change to inform teaching and research*. Dordrecht: Springer.

Taber, K. S. (2013b). Revisiting the chemistry triplet: Drawing upon the nature of chemical knowledge and the psychology of learning to inform chemistry education. *Chemistry Education Research and Practice, 14*(2), 156–168. doi:10.1039/C3RP00012E.

Taber, K. S. (2014a). Methodological issues in science education research: A perspective from the philosophy of science. In M. R. Matthews (Ed.), *International handbook of research in history, philosophy and science teaching* (pp. 1839–1893). Dordrecht: Springer.

Taber, K. S. (2014b). *Student thinking and learning in science: Perspectives on the nature and development of learners' ideas*. New York: Routledge.

Taber, K. S. (in press). Knowledge, beliefs and pedagogy: How the nature of science should inform the aims of science education (and not just when teaching evolution). *Cultural Studies in Science Education*. doi:DOI: 10.1007/s11422-016-9750-8

Taber, K. S., Billingsley, B., Riga, F., & Newdick, H. (2015). English secondary students' thinking about the status of scientific theories: Consistent, comprehensive, coherent and extensively evidenced explanations of aspects of the natural world – or just 'an idea someone has'. *The Curriculum Journal, 26*(3), 370–403.

Thagard, P. (1992). *Conceptual revolutions*. Oxford: Princeton University Press.

Vosniadou, S. (Ed.). (2008). *International handbook of research on conceptual change*. London: Routledge.

Wallace, A. R. (1858/2003). On the tendency of varieties to depart indefinitely from the original type. In A. Berry (Ed.), *Infinitive tropics: An Alfred Russel Wallace anthology* (pp. 52–62). London: Verso.

Williams, P. A. (2001). *Doing without Adam and Eve: Sociobiology and original sin*. Minneapolis: Fortress Press.

Wood, D. (1988). *How children think and learn: The social contexts of cognitive development*. Oxford: Blackwell.

Wood-Robinson, C. (1994). Young people's ideas about inheritance and evolution. *Studies in Science Education, 24*(1), 29–47. doi:10.1080/03057269408560038.

Yahya, H. (2008). *Atlas of creation* (13th ed., Vol. 1). Instanbul: Global Publishing.

Ziman, J. (1978/1991). *Reliable knowledge: An exploration of the grounds for belief in science*. Cambridge: Cambridge University Press.

Zohar, A., & Ginossar, S. (1998). Lifting the taboo regarding teleology and anthropomorphism in biology education—heretical suggestions. *Science Education, 82*(6), 679–697. doi:10.1002/(sici)1098-237x(199811)82:6<679::aid-sce3>3.0.co;2-e.

Chapter 5
History and Philosophy of Acidity: Engaging with Learners by a Different Route

John Oversby

5.1 Introduction

Too often, young novice chemists engage with chemical reactions and then learn the facts off by heart, deadening any future interest. The material here is relevant to lower secondary school science (11–14 years old) and for all abilities. The theme of acidity is commonly treated in lower secondary schools, with occasional reference to 'everyday' science such as treating wasp and bee stings. Here, I aim to set the topic firmly in a historical and philosophical context. It owes its origin to work carried out for the History and Philosophy in Science Teaching (HIPST) European Commission project.

The progression of ideas about acidity described in many books about the history of chemistry begins with the uniquely dangerous, and possibly fatal, classification based on taste. It moves on through its effect, at first, on homemade medicines that were useful as indicators of acidity and finally searching for the magic ingredient (oxygen or hydrogen?). It is an amazing story that covered many centuries, but finished nearly 200 years ago with Sir Humphrey Davy. Yet these ideas are still commonly in use today in the upper reaches of secondary schools. Even some research chemists continue to base some of their thinking about acidity in organic materials on hydrogen atoms that can be replaced by a metal. While the story can be ended at the Davy idea of replaceable hydrogen atoms, it leaves open development to other ideas.

Much of the language we use about acidity, such as neutralisation, comes from medieval power politics where great powers neutralised each other through pitting equally strong armies against each other. They were thought to be equivalent to each other, and the notion of equivalence is fundamental to salt preparation. Although, at

J. Oversby (✉)
Independent Researcher, The University of Reading, Reading, UK
e-mail: oversby61@gmail.com

B. Akpan (ed.), *Science Education: A Global Perspective*,
DOI 10.1007/978-3-319-32351-0_5

first sight, the topic appears to be just a description of phenomena, the philosophical problem of characterising a chemical class boundary through the reactions of imperfect acids is solved through induction to realise the intangible concept of the ideal or perfect acid.

Philosophical considerations of 'explanation' based on cause and effect drive the search for the magic ingredient that is at the heart of all acids, even giving an element its name of oxygen from the Greek 'sour maker'. This search drew evidence from practical knowledge such as combustion of sulphur and phosphorus to make acrid (from Latin: sharp) smoke, suggesting that acids contained something that made them acidic. Of course, the role of water in taste was not considered (e.g. sulphur oxide was named as sulphuric acid). Nevertheless, the oxygen containing theory was very comforting, until the chemists discovered an acid that did not contain oxygen, hydrochloric acid, when chlorine was shown to be an element. There followed a period of 'accommodation' where two kinds of acids were proposed, ones containing oxygen and ones containing hydrogen. The story stops here in this study, as it did for many historical chemists who never knew about the contribution of Arrhenius.

In this section, I will give a personal outline of ideas from philosophy of science.

Theory	An overarching term for a general explanatory and causal idea that links concepts, models and laws	The particle theory of matter explains macroscopic properties and behaviour through the properties and behaviour of individual and collections of submicroscopic particles. Entities such as atoms, molecules and ions are central for chemists. Each of these is modelled mathematically, as visuals including drawings and animations, and by behaviours of the models such as vibrations and flexing
Concept	An abstraction that represents a general class and contains what are considered as essential features of that class	This idea is characterised by the lack of 'a' or 'the' in front of the noun. So, an acid refers to an example, but acid refers to an idealised idea about acids. Concepts also enable us to distinguish between instances and non-instances, based on these essential features. Some concepts are clear, while others are described as fuzzy, that is, it is very difficult to be certain about which features are actually essential. Concepts usually link with each other, such as acid, base, salt and neutralisation to form schema
Model	A model is a representation of an event, object, system, process or idea	Mental models are representations and are only known, partially, by the individual. To communicate mental models, they are expressed as drawings, diagrams, animations, physical objects, mathematically, gestures and orally and by electronic systems (usually as computer software). Each expressed model is only a partial expression of a mental model. Mental and expressed models interact with each other, so that, for example, construction of a displayed formula for a compound may produce novel developments in mental models. Models can also be categorised as static or dynamic

(continued)

Law	A scientific law describes a regular pattern but without implying any causal effect	Acids effervesce with metals such as magnesium and zinc is an example of a chemical law. Laws, typically, apply in specific and limited circumstances, such as the previous example applying in solution. While laws may imply a causal effect, and may give rise to explanatory elaboration, the law itself is without a mechanistic cause
Phenomenon	A phenomenon is a part of the natural world that excites interest, observation with or without instruments, description and investigation and, where possible, explanation	The behaviour of acids in general is a phenomenon that characterises what we call acids and can be further subdivided into sub-phenomena, such as effervescence with certain metals, forming salts with many metal oxides
Data	Data is specific information, measurements or variables collected	The most common use of data is fine-grain evidence collected by measurements, usually in physical science investigations. A broader interpretation includes coarse-grained evidence such as drawings, chemical formulae and affective expressions such as interest and motivation, themselves also coarse grained. Sometimes it is possible to extract fine-grained data from coarse-grained data
Idealisation	Idealisation is a deductive process of extracting essential features and characteristics to produce an abstract idea	Creation of the idea of ideal acid is usually made by investigating the behaviour of many specific examples of acids and then refining the central features of acid. This is described below in the case study
Simplification	Simplification means removing complexity or elaboration	Simplification is a common process in chemistry during the creation of the ideal. It can involve removing unnecessary concepts or entities. In acidity, one example is given by 'an acid contains hydrogen'. This also illustrates a danger of oversimplification, since methane also contains hydrogen but is not usually considered an acid. This can lead to modification of the simplification, such as 'an acid contains hydrogen that is replaceable by a metal'. The value of simplification is tested by its use in causal explanations and its acceptance by the community of practitioners
Deduction	Deduction is the process of creating a conclusion from a premise by reasoning	If an ideal acid is a material that forms salts when reacted with many metal oxides, then when hydrochloric acid reacts with magnesium oxide to form magnesium chloride, it is behaving typically as an acid
Inductive reasoning	Creation of a general proposition or idealisation by taking specific examples or instances and extracting essential characteristics	Many metal oxides react with specific acids to form salts. By inductive reasoning, we create the notion of a base that includes the examples of these metal oxides

5.2 The Case Study

Introducing History and Philosophy of Science (HPS) in the school chemistry curriculum is a relatively recent activity for curriculum designers. There are significant reasons for this. There are few recognised paradigm (fundamental ideas) changes in chemistry compared, for example, with physics. The usual ones quoted include the phlogiston idea, although this only lasted for 50 years in history, and quantum mechanics, which is, perhaps, more properly learned in physics.

There are few historical objects that have lasted in chemistry, compared with physics which has abundant examples. Partly this is because chemistry widely used glass apparatus which is fragile and easily destroyed. Partly it is because some of the equipment, such as the balance, was considered to be so commonplace that they were simply not worth saving or recording. Partly it is because the furnaces that were used were destroyed. Finally, partly it is because many of the early chemical discoveries were based on observing phenomena and not on using measuring instruments.

The structure of the chemistry curriculum in schools is such that phenomenological investigations abound at junior levels and quantitative considerations, such as calculations based on chemical formulae and chemical equations, are treated and assessed in a rule-based (algorithmic) way. The topic is initially about acidity, a well-defined topic in historical terms. I have chosen to use much of the traditional classroom material, since it is largely parallel to the historical line. Where it differs from normal teaching is in making explicit historical and philosophical thinking.

The origins of ideas about acidity are largely lost in the mists of history. An Arabic alchemist, abir ibn Hayyan Geber, had made a wide variety of acids around AD 750. His acids included sulphuric, nitric, citric and malonic. Many alchemists also knew about the effect of these and other acids on limestone and marble. Robert Boyle, in 1670, knew that an acid gives hydrogen with a metal such as zinc, or iron, although he would not be sure about the chemical nature of hydrogen. Boyle also invented the use of everyday indicators such as syrup of violets, a household medicine, for testing acidity. The term alkali came from the Arabic for the ashes left in the roasting pan. Boyle knew that acids and alkalis, or bases, could neutralise each other, and chemists/alchemists previously had studied the salts formed. Based on alchemy, many chemists were searching for magic ingredients. Lavoisier had known that many acidic substances, such as sulphur oxide and phosphorus oxide, had been formed by burning the non-metal in his newly named oxygen (Greek: acid maker) and that acids therefore contained oxygen. Lavoisier understood sulphur oxide to be sulphuric acid (French: acide sulphurique) and that its solution in water, which we now call sulphuric acid, was sulphuric acid hydrate. Not knowing the function of the water as a chemical here, it is not surprising that Lavoisier did not appreciate what was going on. In 1818, Davy showed that hydrochloric acid was made up of only hydrogen and chlorine with no oxygen. This led to the idea that there were two kinds of acid, the oxoacids and the hydracids. This was resolved later in the century when it was proposed by Arrhenius that all acids in water gave off hydrogen ions, more or less.

The action of creating a class of substances known as acids is a process of induction. Although many acids have similar properties (turning vegetable dyes red or orange or yellow, reacting with zinc and iron, reacting with carbonates, making salts with bases), there are enough exceptions to make the classification unclear. In this case study, I discuss the notion of an ideal or perfect acid, which has all the acidic features and which we create based on the imperfect acids by their reactions. This is a challenging idea. I finish the cognitive aspect by exploring whether an acid always has oxygen or whether it might have hydrogen. This is a modest paradigm shift in chemistry, not always recognised by philosophers. Finally, I use various reflective devices to explore both their understanding of the chronology of discovery and the philosophical processes that have taken place.

I have used links to drama in producing play scripts of historical events and to English in the form of newspapers to present historical information. There are strong links to history in the cultural background of the scientific exploration. A homework box for students to record their thinking about idealisation on the outside provided an insight into their progress, and a card sort enabled them to discuss their view about the chronology of ideas about acidity.

5.3 History of Acidity

Alchemists were aware of many acids from AD 750 onwards, although it is not clear from their writings how much they were explicit about their nature as a class of chemicals.

Acids were first classified by taste (Latin: acetum sour), and acids are probably the only class of chemicals to be identified this way.

'The conceptual division of certain substances into acids and bases was already evident in the Middle Ages, the terms "acid", "base", and "salt" occurring in the writing of medieval alchemists. Acids were probably the first to be recognized, apparently because of their sour taste: The English word "acid", the French word "acide", the German "Sauer", and the Russian "kislota" are all derived from words meaning "sour".'

Boyle popularised the use of vegetable dyes as acid indicators, around 1670. Most of these dyes were already known as home medical remedies, such as syrup of violets. This link between medicine and chemistry was more common than we commonly imagine. Many early researchers in chemistry worked in pharmacies or were Professors of Medicine. Boyle also published his knowledge that acids gave a flammable gas (now known as hydrogen) with some metals such as zinc and iron.

Nicholas Lemery tried to explain (1680) how acids tasted sour by imagining that they had very small particles with points on them. This is a cause and effect explanation.

Lavoisier used his oxygen theory of acids to create a new naming (nomenclature) system with some of his fellow chemists (1787) to aid learning in a systematic way.

Davy, in 1810, showed that chlorine was an element and that hydrochloric acid contained no oxygen. He created the idea that all acids contained hydrogen, but it took many years before textbooks adopted this. Wilson, writing in 1856, described two kinds of acids, oxoacids and hydracids.

Arrhenius, in 1884, put forward the thesis that acids form hydrogen ions when in water, using electrical conductivity equipment invented for the new telegraph industry in 1840 by Kohlrausch.

The progression of ideas about acidity moved from a descriptive basis for a class of chemicals (the idea of an ideal or perfect acid) to a causal effect based on acids containing oxygen and then hydrogen. The idea that it was a component of the acid that made it acidic changed when Arrhenius proposed that a new species, the hydrogen ion, was the cause of acidity.

5.4 Pedagogy

Discussions. Inclusive discussions are common in English classrooms. However, good whole-class discussions may have interference from poor behaviour. Alternatively, well-structured group discussions may be more inclusive. Clear roles for group members need to be established through agreed ground rules.

Role play. Good role play has clear direction concerning the roles and sufficient time for the participants to think about their roles. A danger is that participants identify too much with their role. The teacher needs to give them time and space to step out of role, perhaps by an individual reflection activity about the other roles.

Teacher-generated plays. These were included in the student newspapers to provide evidence for the kind of discussions that were taking place at the time. Writing them at an appropriate literacy level was a challenge. One teacher-generated play on Boyle's use of indicators was used with significant success. A second play to discuss alternative views about acids, using a Galileo discourse style involving two attendees at one of Boyle's lectures, and a listening pie-seller, generated some valuable reflective discussion.

At the end of the lesson on ideal acids, they were shown a box containing a bottle of ideal acid. Opening the box, they saw the bottle was empty, because, of course, the ideal acid only exists in the imagination; it is not real. For homework, they created a box of their own with their learning about acids and ideal acid on the outside.

I was interested in whether the students could construct a chronology of discoveries about acids. This was done by a card sort where the random cards had to be placed in chronological order through discussion in small groups. This was not based on memory but on rational discussion. Most groups were able to do this.

5.5 Research Evidence

Some evidence was collected by means of a word association test, student notebooks and field notes.

Fig. 5.1 Word cloud

The word cloud in Fig. 5.1 results from asking one class to record all the words they associated with 'acid'. The responses ranged from 5 to 37 words. The text from all students was pasted into Wordle which outputs a word cloud, with the font size representing the frequency of use of the word. The word cloud demonstrates the largely negative view of 'acid' by this group of students. The collection of words was made at the start of the module and took 5 min.

The student boxes showed that the great majority understood the features of the ideal acid. My analysis suggests that a minority understood the nature of idealisation.

5.6 Reflections

There were a number of factors that influenced the study. Some of these were inevitable and out of control of the researcher, while others could be changed.

Since the teachers had expressed their view that their personal knowledge of HPS was limited, this provided a base for their work, such that the teachers saw this as a strong learning activity for them. In the first place, they claimed that they would simply enact the strategy in their classes, but it was clear that local circumstances and their personality led them to adapt the programme.

One teacher was a graduate chemist, and the other was a graduate physicist. Both had been teachers for many years.

The lessons were quite short, 50 min with the usual delays in students arriving late from another lesson. This influenced time for discussions and scene-setting and may have resulted in these sections being too short, in retrospect. Shorter lessons are becoming more common as head teachers try to squeeze more into the curriculum as a result of government interventions.

The lesson formats were influenced by pedagogical preferences of the two teachers, one taking the approach of question-answer sessions with the whole class and the other giving more structured instructions for activity. In both cases, the students were following the lead of the teachers, limiting their own inquiries. The shortage of time in each lesson, and pressures to complete the work in a short unit, led us to provide extra material in the form of student newspapers. While we took every effort to make the reading as simple as possible, it remained a challenge for some of the students.

Dealing with younger students, who may lack maturity and skills to carry out activities required, put lots of pressure on simplifying the ideas.

5.7 Conclusion

I have described embedding HPS within a traditional framework for a sequence of lessons on acidity. I have provided evidence for this to be an effective alternative to engaging 11–14-year-old students with this, otherwise dry, topic.

5.8 Website

History and Philosophy in Science Teaching (HIPST) project, funded by the European Commission http://hipstwiki.wetpaint.com/

Part II
Science and National Development

Chapter 6
Science Education for National Development: Indian Perspective

Sudhakar C. Agarkar

6.1 Introduction

At the time independence in 1947, India inherited the education system set up by the British colonial rulers. School education focussed on three Rs (reading, writing and arithmetic), and higher education was available only to a small number of students living mainly in urban areas. The industrial development was in its infancy, and Indian industries produced only a few products of daily use indigenously. Hence, republican India embarked on self-reliance as well as sustainable and equitable growth by improving the status of science education in the country. India, it is widely known, has a long tradition of science and technology. By revitalising its tradition, it has committed itself to the task of national development (MIB 2003). One can identify six different components in this context: (1) scientific policy resolution, (2) improvement in school science education, (3) identification and nurture of talent, (4) establishment of research and development (R&D) institutions, (5) encouragement in industrial growth and (6) spread of scientific literacy. We will now discuss each of these components in detail.

6.2 Scientific Policies

The first scientific policy resolution was drafted in 1958. It has been revised thrice (1983, 2003 and 2013) taking into account the feedback and the needs of the countrymen. It would be useful to go through some of the pertinent aspects of these

S.C. Agarkar (✉)
VPM's Academy of International Education and Research, Thane, Maharashtra State, India
e-mail: sudhakar.agarkar@gmail.com

B. Akpan (ed.), *Science Education: A Global Perspective*,
DOI 10.1007/978-3-319-32351-0_6

policy documents that shaped the developments of science and technology in the country.

6.2.1 Scientific Policy Resolution, 1958

The policy resolution of 1958 clearly stated that "the key to national prosperity lies in combination of three factors: technology, raw material and capital, of which the first is perhaps the most important of all. But technology can only grow out of the study of science and its applications" (DST 2004a).

6.2.2 Technology Policy Statement, 1983

The basic objectives of the technology policy brought in 1983 were to attain technological competence and self-reliance and to reduce vulnerability, particularity in strategic areas, making the maximum use of indigenous resources (DST 2004b). It focussed on the use of traditional skills and capabilities, making them commercially competitive and to ensure the correct mix between mass production technologies and production by the masses (DST 2004b).

6.2.3 Scientific Policy Resolution, 2003

The policy framed in the light of new millennium emphasised that the fruits of science reach every citizen of India. It proposed to create food, agricultural, nutritional, environment, water, health and energy security of the people on a sustainable basis. In addition, it emphasised the need to vigorously foster scientific research in universities and other academic, scientific and engineering institutions and attract the brightest young persons to careers in science and technology (DST 2004c).

6.2.4 Science Technology Innovation Policy, 2013

The scientific policy resolution unveiled in January 2013 is called science technology innovation policy as it focuses on innovation (DST 2013). The key features of this policy are (1) promoting the spread of scientific temper among all sections of society; (2) enhancing skills for applications of science among the young from all social sectors; (3) making careers in science, research and innovation attractive enough for talented and bright minds; (4) establishing world class infrastructure for research and development (R&D) for gaining global leadership in some select frontier areas of science; (5) positioning India among the top five global scientific powers by 2020; (6) linking contributions of science research and innovation system with the inclusive economic growth agenda and combining priorities of excellence and relevance; (7) creating an environment for enhanced private sector participation in R&D; (8) enabling conversion of R&D output with societal and commercial applications by replicating hitherto successful models, as well as establishing of new PPP (public–private partnership) structures; (9) seeking S&T-based high-risk innovation through new mechanisms; (10) fostering resource-optimised cost-effective innovation across technology domains; (11) triggering in the mindset and value systems to recognise, respect and reward performances which create wealth from S&T-derived knowledge; and (12) creating a robust national innovation system.

6.3 Science Education in Schools

Policies mentioned above had to be translated into practice. It was achieved by setting up appropriate infrastructure within the country. Naturally, the first attention was given to teaching of science in schools. This task was tackled through a variety means undertaken simultaneously.

6.3.1 Curriculum Development

Science as a discipline of formal study received low importance in the schools established in the precolonial period. During the British Raj, too, hardly any effort was made to teach science at school level (Sharma 1975). In the post-independence period, however, special efforts were made to teach science to all. The first attempt to include science as a subject of study in school education was made by Secondary Education Commission (1952–1953). The commission recommended teaching of simple principles of science to primary school children followed by a diversified course for high schools. Accordingly, general science was taught to the students up to grade 8, and three disciplines (physics, chemistry and biology) were taught to students who opted for science stream in grades 9 and 10.

The report of the Education Commission (1964–1966) made a lasting impact on school science education in the entire country (Govt. of India 1966). The report entitled "Education and National Development" laid great emphasis on making science and mathematics important elements in school curriculum and recommended teaching of these subjects on a compulsory basis to all the pupils during the first 10 years of schooling. The National Council of Educational Research and Training (NCERT), an apex body in education, undertook the task of framing national curricula for school subjects. Following the recommendations of the National Education Commission in 1966, NCERT came out with a National Curriculum Framework to teach science to all in 1968.

The Government of India decided to take stock of the school education once again and brought out the National Policy on Education (NPE) in 1986 (Govt. of India 1986). Instead of a discipline-focused curriculum, NPE suggested integrated approach of teaching of science where all its branches are taught in a coherent manner bringing out their interrelationships and social relevance. Towards the end of the century, the Ministry of Human Resource Development (MHRD), Government of India, decided to take a fresh look at school education. NCERT, once again, brought out a National Curriculum Framework for School Education in 2000 (NCERT 2000). The framework highlighted the importance of integrating indigenous knowledge with school science curriculum and urged the educators to teach science in such a way that the future citizens of India are competent enough to meet the challenges of globalisation. The recent National Curriculum Framework prepared in 2005 (NCERT 2005) goes beyond the classrooms and brings out the importance of out-of-school experiences. It also emphasises linking science and mathematics with life skills and highlights the need to inculcate values among the students.

6.3.2 Instructional Material

Based on the guidelines given by education policy documents and the framework prepared by NCERT, the task of preparing instructional material was undertaken.

Textbooks Department of Science and Mathematics (DESM) in NCERT prepared textbooks and made them available for the entire country. Schools following the syllabus of the Central Board of Secondary Education (CBSE) accepted these books and planned their science teaching accordingly. Some state governments translated them into regional languages and made them available to their school system. The majority of the states established their own units for textbook production. These units prepared books based on the guidelines of NCERT and made them available at a reasonable cost. These books were distributed free of cost for the economically weaker sections of the society.

Laboratory kit Science being an experimental subject needs to be taught through activities. However, the school laboratories are often inadequately equipped. In order to overcome this problem, a laboratory kit for schools was designed by the NCERT and was made available to the system. Training courses were conducted to acquaint the teachers with the kit for its effective utilisation.

Teachers' handbook Government as well as voluntary agencies undertook the task of preparing handbooks for teachers. These handbooks provide inputs in (a) classroom management, (b) effective teaching of school science, (c) assessment of students' learning, (d) assignments to be given to students for understanding of concepts and (e) out-of-school activities to support school science education.

Popular science books With a view to explain science concepts in a simple language to students, popular science books are brought out in large numbers. These books might take the format of science quiz, fiction or non-fiction stories, biographies of scientists and answers to children's questions (Lagu 1978). Many of them are written in local languages. It must be noted that there is a great demand for such books. State governments encourage the writing of such popular books by giving away Best Literature Awards in Science and Technology.

Digital resources With the advent of technology, it has become possible to design digital resources for teachers, students and parents to support school science education. In this context, a programme undertaken by Homi Bhabha Centre for Science Education is worth mentioning. Supported by the Rajiv Gandhi Science and Technology Commission, Government of Maharashtra, the project entitled Open Educational Resources for Schools (OER4S) was undertaken to make available resources free of costs to all the stakeholders of school education, namely, students, teachers and parents in regional language (www.mkcl.org/mahadnyan). At the national level, the Government of India has come out with a portal named "Sakshat" that provides such resources in ample numbers. A low-cost tablet (named Akash) has been developed that can access the digital materials uploaded to the website. In addition to resources on the website, space-based resources are also being used to support science education (Bhatia and Joshi 2006). A dedicated satellite named EDUSAT was launched by ISRO in October 2007. Being in the geostationary orbit, this satellite is available all the time for Indian teachers and students. Through this satellite, two-way video as well as audio links can be established to support school science education.

6.3.3 Teacher Preparation

It is well known that the teacher plays a crucial role in curriculum transactions. Hence, due attention was given to teacher preparation. It involves pre-service as well as in-service training courses for teachers. Professional training along with basic qualification is mandatory for the job of a school teacher in India. Primary teachers are expected to get their Diploma in Education (D.Ed.) after Higher Secondary School Certificate (H.S.S.C.) examination. For teaching science at the secondary level, teachers are expected to obtain a degree of Bachelor of Education (B.Ed.) after a degree in science. A large number of teacher training institutions are established in the country to provide in-service training to prospective teachers. The syllabus in these institutions has been revamped to prepare teachers capable of teaching school science effectively.

Through pre-service courses, a teacher gets general guidelines for classroom interactions. These inputs are seldom sufficient to deal with changing curriculum and societal expectations. Practising teachers, therefore, need to undergo refresher courses during their service period. Hence, State Councils of Educational Research and Training (SCERTs) in different states of the country arrange in-service training courses for science teachers. In recent years District Institutes of Education and Training (DIET) have been established to take care of educational needs of schools within the district. In addition, teacher associations and voluntary agencies also come forward to conduct in-service training courses for practising science teachers.

The National Council of Teacher Education (NCTE) as an advisory body in teacher education came into existence in 1973. It was then made a statutory body by the Act of Parliament in 1993. The main aim of NCTE is to achieve planned and coordinated effort of teacher education throughout the country, regulate and maintain norms and standards in the teacher education systems. The mandate given to NCTE is very broad and covers the whole gamut of teacher education programmes. From its establishment, it is trying to maintain certain standards and bring in uniformity in the curriculum for pre-service training of teachers. In addition, it is also preparing guidelines for conducting in-service training courses. It has initiated training programmes for practising teacher educators so that a large number of teachers could be trained on a regular basis.

6.4 Identification and Nurture of Science Talent

Systematic efforts are being made to search and nurture science talent among young students throughout the country. A brief account of these activities follows.

6.4.1 National Talent Search Scheme

The National Council of Educational Research and Training (NCERT) launched a scheme called National Science Talent Search Scheme (NSTSS) in 1963. In the initial period, these scholarships were awarded to students for pursuing education only in basic sciences up to doctoral level. NSTS scheme underwent a change in 1976 and was extended to include social sciences along with engineering and medicine. It was renamed as National Talent Search Scheme (NTSS). Apart from financial assistance, summer camps and seminars are also arranged for the students who are selected through this scheme. Over the last 40 years (1963–2013), a large number of students, especially from poor economic background, have benefited from this activity. It has enabled the country to increase the manpower in science and technology to a considerable extent.

6.4.2 *KVPY*

A scheme called Kishore Vaigyanik Protsahan Yojana (KVPY) was launched in 1999 to encourage scientific creativity among post-school students. Funded by the Department of Science and Technology (DST), Government of India, the scheme is meant to attract highly motivated students for pursuing basic science courses and career in research. Managed by the Indian Institute of Science, the candidates are selected through written test, project work and personal interview. There are three streams denoted by SA, SX and SB: stream SA is for students enrolled in grade XI, stream SX is for students enrolled in grade XII, and stream SB is for those who have registered for the undergraduate programme (B.Sc.). The selected candidates are given generous scholarship and contingency grant to pursue their higher studies in science (www.kvpy.org.in).

6.4.3 *Inspire Award*

A new scheme was launched in 2010 by the Department of Science and Technology to tap the science talent existing in young students. The scheme popularly known as INSPIRE (Innovation in Science Pursuit for Inspired Research) encourages school and college students to come up with innovative ideas in science. The scheme has three components: SEATS (Scheme for Early Attraction of Talent), SHE (Scholarships for Higher Education) and AORC (Assured Opportunity for Research Career). As the name suggests, the first part of the scheme (SEATS) attempts to attract talents at the early age (10–15 years). The students with potential are given a financial assistance of Rs. 5,000 each to convert their ideas into workable projects. This financial support ensures that even poor children have a scope to show their talents. Once chosen for the award, these students are given an opportunity to interact with practising scientists in the country. The second part of the scheme (SHE) is for the students in a higher age group (17–22 years) who are pursuing their education in science or technology. The third part of the scheme (AORC) is for those in the age group of 27–37 years – students pursuing doctoral studies or for those who have entered into profession recently. Thus, one can see that these schemes attempt to attract students at different ages to science-related activities and professions (www.inspire-dst.gov.in/award.html).

6.4.4 *Ignite Scheme*

A National Innovation Foundation (NIF) was established as an autonomous unit of the Department of Science and Technology, Government of India, in March 2000 to provide institutional support to grass-root innovators and outstanding knowledge holders in informal sectors. NIF came out with a scheme entitled "Ignite" for the creative youths of India. It attempts to identify hidden talent from among the community through submission of write-ups. Any child irrespective of educational qualification can submit the entry form for this competition. The best entries are recognised and supported. In a short time, the scheme has reaped good fruits. It could collect about 20,000 new ideas and file about 650 patents, of which 40 have been granted (www.nif.org.in/ignite).

6.4.5 International Olympiads

Olympiad competitions are held internationally to spot the talents among students in different disciplines. The Government of India decided to send the teams to international science Olympiads and entrusted the responsibility of selecting teams to the Homi Bhabha Centre for Science Education (HBCSE). Acting as the nodal agency, HBCSE has been identifying and preparing students for international Olympiad competitions in physics, chemistry, biology, astronomy and junior science for the last decade. A large number of students from secondary and higher secondary schools appear for national level competitive examinations. Training cum selection camp is held at the premises of HBCSE in Mumbai for the toppers of the national competitions. The performance of Indian teams in international competitions has been encouraging. India also hosted a few international Olympiad events like International Biology Olympiad, International Astronomy Olympiad and International Junior Science Olympiad (www.olympiads.hbcse.tifr.res.in).

6.4.6 Science Exhibitions

Organisation of school level science exhibitions was initiated in the 1970s to provide opportunities to innovative teachers and students to display their talents. After a slow start, the activity has now spread to the entire country. NCERT suggests the theme of the exhibition every academic year, and students from different schools prepare exhibits/working models pertaining to the theme. Starting from taluka level (small administrative block of a district), the exhibitions are arranged at district, state, national as well as at international levels. Through this activity a large number of innovative ideas from students and teachers have surfaced.

The National Council of Science and Technology Communication (NCSTC), Department of Science and Technology, Government of India, initiated a children's science movement in 1993. Under this scheme, NCSTC arranges National Children's Science Congress (NCSC) annually for the children between 10 and 17 years from all over the country. The main objectives as advocated by NCSTC (www.ncstc.dst.gov.in) are to provide a forum to the young scientists to pursue their natural curiosity and to quench their thirst for creativity by experimenting on open-ended problems and to make them feel that science is all around and one can gain knowledge as well as solve many problems by relating the learning process to the physical and social environment of the neighbourhood. Over the last 20 years, this activity has taken a good shape involving a large number of teachers and students from both urban and rural schools. This activity has stimulated scientific activities among the students and enabled them to learn the scientific methodology involving questioning, hypothesising, experimenting, analysis of data, conclusion and communication of results. This mode of exhibition is offering an opportunity to the students to actually play the role of a scientist at school level. The majority of the students find it both challenging and enjoyable.

6.5 R&D Institutions

The Ministry of Human Resource Development (MHRD) of the Government of India gave adequate attention to the development of institutions of higher learning and centre for research and development both in science and in technology. The University Grants Commission (UGC), an apex body looking after higher education in the country, was encouraged to start new colleges with facilities for science education and research. In order to ensure that students receive quality education in science, efforts were made to equip the college laboratories with adequate equipment. Apart from the establishment of science colleges in the country, the MHRD also initiated the establishment of special institutions of higher learning and research in science and technology.

6.5.1 IITs/IISERs

To provide quality education in science and technology and to carry out fundamental research in these areas, a few premier institutions were set up by the MHRD. Establishment of Indian Institute of Technology (IIT) can be mentioned as an example in this context. Initially there were five IITs located in different parts of India. Now the number has increased fourfold covering many cities in the country. On the lines of IITs, MHRD came forward to start institutions for science education called Indian Institute of Science Education and Research (IISER). These institutions, having facilities for teaching as well as for research in the frontier areas of science, attempt to prepare students to face the challenges of the future.

6.5.2 Council of Scientific and Industrial Research (CSIR)

The Council of Scientific and Industrial Research (CSIR) was established in pre-independence period in 1942 to provide scientific and industrial R&D that maximises the economic, environmental and societal benefits for the people of India. It presently has more than 50 institutions all over the country working in the area of basic and applied sciences. Some of the notable institutions of CSIR are National Chemical Laboratory, Pune; National Institute of Oceanography, Goa; and the Centre for Cellular and Molecular Biology, Hyderabad (www.csir.res.in). These institutions are providing opportunities to young graduates to pursue their career in scientific research.

6.5.3 Department of Atomic Energy (DAE)

Department of Atomic Energy (DAE) was set up by the Government of India as early as 1954. DAE, as the name suggests, takes care of research and development activities related to atomic research. It presently has a large number of institutions under its purview which are categorised as research institutions, supported institutions and industrial units. Since the beginning, the Government of India embarked on the peaceful uses of atomic energy. DAE has steered the programme successfully and has demonstrated capacity to manage nuclear reactors successfully. In spite of opposition from certain sections of the society, DAE has emerged as an umbrella body for basic and applied research within the country over the past five decades. It has embarked on technology transfer to the society as well as to industries (www.dae.nic.in).

6.5.4 Indian Space Research Organisation (ISRO)

Indian Space Research Organisation (ISRO) set up in 1969 takes care of activities related to space science. It works in the areas of satellite launching, weather forecasting, space explorations, etc. As mentioned earlier, it has made available a dedicated satellite for education named EDUSAT. In addition, ISRO provides support to science education through its Cluster Resource Centre and Educational Portal (www.isro.org). The objective of ISRO is to develop space technology and its application to various tasks of national and international interests. Accordingly, it has successfully put into operation two major satellite systems, namely, the Indian National Satellites (INSAT) and the Indian Remote Sensing (IRS) satellites. Over the years, it has also developed various launch vehicles, like the Polar Satellite Launch Vehicle (PSLV), the Geosynchronous Satellite Launch Vehicle (GSLV), etc. for launching satellites. Recently, ISRO has launched a capsule that can be used for sending human beings into the space.

6.5.5 Indian Council of Agricultural Research (ICAR)

The Indian Council of Agricultural Research (ICAR) is an autonomous organisation under the Department of Agricultural Research and Education (DARE), Ministry of Agriculture, Government of India. The council is the apex body for coordinating, guiding and managing research and education in agriculture including horticulture, fisheries and animal sciences in the entire country. With 99 institutions and 53 agricultural universities, spread across the country, this is one of the largest national agricultural systems in the world. The ICAR has played a pioneering role in ushering Green Revolution and subsequent developments in agriculture in India through its research and technology development that has enabled the country to increase the production of food grains by 4 times, horticultural crops 6 times, fish 9 times (marine 5 times and inland 17 times), milk 6 times and eggs 27 times since 1950–1951, thus making a visible impact on the national food and nutritional security.

6.5.6 Indian Medical Council (IMC)

The Medical Council of India was established in 1934 under the Indian Medical Council Act, 1933. It has a main function of establishing uniform standards of higher qualifications in medicine and recognition of medical qualifications in India and abroad. The number of medical colleges had increased steadily during the years after independence. It was felt that the provisions of Indian Medical Council Act were not adequate to meet with the challenges posed by the very fast developments and the progress of medical education in the country. As a result, in 1956, the old Act was repealed and a new one was enacted. This was further modified in 1964, 1993 and 2001. The objectives of the council are as follows: maintenance of uniform standards of medical education, recommendation for recognition/derecognition of qualifications of medical institutions of India or foreign countries, permanent registration/provisional registration of doctors with recognised medical qualifications and reciprocity with foreign countries in the matter of mutual recognition of medical qualifications. A variety of research is being carried out in preventive as well as curative methods in the country by this organisation (www.mciindia.org).

6.6 Industrial Growth

Over the past six decades, there has been a notable growth in Indian industries. With the opening up of the economy in 1991, a large number of multinationals have made their headway into the country. At the same time, Indian industries are making their presence felt in other countries. Some of the notable industrial areas are automobiles, pharmaceuticals and IT-based services. The manpower trained in educational and R&D centres could be absorbed in these industries. It not only offers incentives to youths to study science and technology but also provides opportunities for further innovation and development.

6.6.1 Automobiles

The automobile industry in India has grown tremendously over the last few decades. To begin with, there were collaborative ventures. Slowly, however, home grown automobile products are available in the market. The presence of some of the well-established auto industries in India clearly shows the growth in this area. Not only are they fulfilling the domestic demands but are also exporting the automobiles to developed as well as developing countries.

6.6.2 Health Care and Pharmaceuticals

Health-care industries have shown a tremendous growth in India. Medical tourism has now become good business for people in India. Many of the hospitals in the country receive patients from other countries for medical treatment as it is both cheap and reliable. Likewise, the pharmaceutical industry has also seen an unprecedented growth over the past few years. Apart from satisfying the needs of the country, it also supplies drugs to many other countries both developed and developing. It began with the multinational companies setting up their units here. Slowly, Indian pharmaceutical companies have also grown. Recently, there was news item allowing Indian pharmaceutical companies to manufacture life-saving drug that is patented by a developed nation. Indian companies can now sell it to the customer at one-fourth of its original price (TOI 2012).

6.6.3 IT Services

India has seen tremendous growth in information technology (IT) sector. It has two components: IT services and BPO (business process outsourcing). Growth in both fields has been significant. IT, thus, has become an employment and revenue-generating sector for the country. The growth of IT sector is still on and would contribute substantially to economic progress of the country. Apart from many multinational IT companies opening their offices in India, the Indian companies provide services to a large number of countries all over the world. IT export now forms a major source of foreign exchange for India (TOI 2012).

6.7 Scientific Literacy

In addition to teaching of science formally in schools, deliberate efforts are made to spread scientific literacy among the masses through informal modes. Both governmental and non-governmental organisations are involved in this important task. Here is a brief account of this mission.

6.7.1 *NCSTC*

The National Council of Science and Technology Communication (NCSTC) has been set up by the Government of India under the Department of Science and Technology (DST) to oversee the spread and popularisation of science in the society (http://dst.gov.in/scientific-programme/s-t_ncstc.htm). Towards this end NCSTC undertakes a variety of activities like (a) research in thrust areas of science and technology communication; (b) development of scripts, films, video and radio programmes; (c) training of teachers and voluntary workers in science communication; (d) development of science journalism; (e) giving awards and recognition to outstanding science communicators; (f) coordinating with state councils and S&T networks; (g) developing capacity through science communication; (h) arranging field programmes; (i) organisation of children's science congress; and (j) promoting international cooperation in science communication.

6.7.2 *NCSM*

The National Council of Science Museums (NCSM) is established by the Government of India with a mandate of taking science to the doorstep of the society. Towards this end it has set up science centres and science museums in different parts of the country. These centres/museums provide hands-on opportunities to people to perform simple and interesting experiments. Apart from making available working models, these centres/museums arrange science-based lectures and competitions regularly. Spread all over India, they have become places of informal learning of science both for children and for adults. Starting from metropolitan cities, it has extended its wings to small towns in the country. In order to take care of rural population, NCSM has set up mobile science exhibitions. These exhibits are taken from one place to another to create scientific literacy among rural folks (www.ncsm.gov.in).

6.7.3 Voluntary Efforts

In India one finds a large number of voluntary agencies that undertake activities for popularisation of science. They often arrange popular lectures, lecture demonstrations and question-answer sessions. They also occasionally arrange competitions like science quiz and talent search examinations. More useful service is offered by these organisations through printed materials. For example, Kerala Sastra Sahitya Parishad, apart from bringing out a monthly magazine, publishes books and pamphlets on relevant issues in Malayalam, an official language of the state of Kerala (Janardhanan 2002). Similarly, Marathi Vidnyan Parishad in Mumbai brings out a monthly magazine and books with scientific articles written in lucid and simple Marathi (official language of the state of Maharashtra).

6.8 Challenges

The foregoing description brought out the fact that through deliberate efforts, India could achieve considerable progress in improving the base of science and technology in the country. It has successfully developed manpower which can face the challenges of the modern world. At the same time, it has successfully spread scientific and technological literacy among the masses so that they can make use of modern gadgets profitably. In spite of these successes, there are many challenges that the country faces. It would be appropriate at this juncture to discuss some of the challenges and suggest some solutions.

6.8.1 Equity

In India school education is on the concurrent list of the state as well as central governments. Although major responsibility of education lies with the state government, central government also runs some schools like Navodaya Vidyalaya (model school for rural gifted children) and Kendriya Vidyalaya (school for the wards of the employees of the central government). Within the state one finds four types of schools: (i) schools run by local self-governments like the Zilla Parishad or Municipal Corporation, (ii) schools run by the State Government Education Departments or Tribal Development Department (Ashram Schools), (iii) aided schools run by private organisations and (iv) unaided schools managed totally by private bodies. One notices tremendous variation in these schools in terms of the facilities and preparation of students attending these schools. Students belonging to poor socio-economic status invariably attend schools run by local self-governments. Those who can afford to pay heavy fees go to private unaided schools. Therefore, there is a great disparity in the educational opportunities available to the students. Achieving equity in educational opportunities, therefore, poses a great challenge.

Studies were conducted at HBCSE to understand the difficulties faced by first-generation learners and to design appropriate remedial measures. These studies revealed that lack of learning prerequisites, non-availability of educational opportunities and inadequate communication competence were responsible for the poor performance of socially deprived students. Attempts made to compensate for these lacunae were found to enhance student's scholastic attainment in science and mathematics significantly (Kulkarni and Agarkar 1985). Findings of this study need to be implemented on a large scale.

Hodson (2003) warns us that this is a time for action. Action will have to be taken to make school science curriculum relevant to the needs, interests and aspirations of young citizens. He also cautions us that if current social and environmental problems are to be solved, we need a generation of scientifically and mathematically literate citizens. To achieve this goal, these subjects need to be taught through activities and problem-solving assignments. Lack of proper resources in the school, however, precludes teacher from doing so. Some government as well as voluntary agencies took this issue on a priority basis. They developed low-cost kits along with manuals to perform experiments in schools. An educationally relevant laboratory programme has been designed and implemented successfully on a pilot level in selected schools (Agarkar 1998). There is a need to undertake large-scale programmes to implement activity-based teaching methods in the classroom.

6.8.2 Teacher Training

India has a good and well-established institutional set-up for providing pre-service training to the teachers. Even then many of the schools do not have trained teachers to teach science. It is imperative that training opportunities are provided for untrained teachers without delay. The Open Universities and Open Schools in India are trying hard in this regard. Subject teaching is undergoing paradigm shift world over (Cheng et al. 2002). It is important that prospective teachers are prepared to adjust with this shift effectively. Appropriate modification in the curriculum for pre-service education of teachers is called for. The role of a teacher is changing in the modern society. A teacher is now expected to be a knowledge manager rather than knowledge disseminator (Tan 2001). Pre-service training is expected to prepare the teacher to play this role effectively.

Although some in-service training programmes are arranged both by government and non-governmental organisations, there is hardly any policy of training worked out for the entire country. The number of teachers in India is more than four million (NCERT 1998). It is almost impossible to train these teachers if one limits to direct mode of interaction. Instead, efforts will have to be made to design training modules for teacher educators who in turn would train the teachers (COMSEC 2000). Encouraging teachers to undertake action research projects would probably help empower them to some extent. Provision for life-long education of teachers using modern technology has to be explored to prepare teachers to play leadership role in the future (Agrawal 2002). The IT facilities within the country need to be exploited for school education. It is high time that the space-based resources are also used to improve the teaching of science and mathematics.

6.8.3 R&D in Science Education

A few decades ago, science education was not recognised as a discipline for serious research in India. However, as the science and technology education took shape in the country, some institutions started taking interest in conducting research in this field. NCERT has a special wing named as the Department of Education in Science and Mathematics (DESM) which conducts research in school curriculum, teacher education and mode of evaluation. The University Grants Commission has identified some of the education departments as progressive institutions and has given them a status of Institute for Advanced Study in Education (IASE). They have a faculty and research students to conduct research in science education. Institutions engaged in scientific and technological research are also coming forward to undertake research and developmental activities in science education. One such institution is the Homi Bhabha Centre for Science Education (HBCSE), established in 1974 as a constituent unit of the Tata Institute of Fundamental Research (TIFR). Since its inception HBCSE has been conducting research in curriculum development, teacher education, cognitive science, learning hurdles, etc. Based on the research findings, the centre has designed methods and materials to be used by the system (www.hbcse.tifr.res.in).

There is a big challenge to translate laboratory findings into classroom practices. At the same time, there is a need to design methods and materials to improve the teaching of science taking into account the constraints of resources in the schools. Heterogeneity of Indian classrooms poses challenges to the teacher. Research will have to be conducted to understand diversities in sociocultural background of students and to benefit them through inclusive education. Another area of attention is assessment since the present mode of assessment entirely focuses on written mode, which tests only the information. Proper evaluation methods are to be worked out to test skills and attitudes of students.

Through various intervention projects, innovative methods of teaching are being developed. These methods when tried on a small scale are found to be successful. However, these innovations remain at the local level or die off as soon as funding ceases. There is, therefore, an urgent need to ensure that these innovations are critically assessed and tried on a large scale. A central assessment body might be a good solution in this context. This body, apart from assessing the utility of innovations, should be encouraged to undertake research and development work in science and mathematics education. The tradition of conducting research in these areas is now well established in the country. What is needed is the continuous flow of finance and regular meetings of active researchers to achieve good quality research output.

6.8.4 Collaborations

Coordination among educational institutions, research organisations and industries is the need of the day. There has been an attempt to develop links between educational institutions and industries in recent years. These efforts are to be strengthened so that people from industries spend time in schools and students in turn get an opportunity to work in industries. Similarly, organic linkages have to be established between educational and research organisations. This collaboration should be on an institutional basis so that a direct contact between scientists and students is established. This linkage should aim at exposing students to scientific research in their formative years (Limaye 2006).

6.9 Conclusion

From the foregoing discussion, it is clear that the notable industrial and economic progress of India can be attributed to systematic use of science at all levels. The education system in the country attempted to prepare adequate manpower for RD activities and for industries. Setting up of adequate number of R&D institutions within the country gave them opportunities to innovate. At the same time, industrial growth within the country enabled people to practise those innovations leading to economic development. The Indian model of social and economic progress is, thus, based on the use of science and technology. Any developing country can follow this model to achieve social as well as economic progress in a short time.

International cooperation in the field of science education has proved useful in the past (Ramaraju 1999). There is still a wide scope for international cooperation in this area, at least, at the regional levels. Collaborative cross-cultural research projects could be planned both with developed and with developing countries. Moreover, exchange of teachers could prove to be an effective way to acquaint them with global problems and solutions. The League for the Exchange of Commonwealth Teachers (LECT) under its Teachers' International Professional Development Programme (TIPD) has a scheme to arrange such exchanges. In this scheme a team of practising teachers from one country spends extended period of time in another country to interact with teachers, teacher educators and students. Such an interaction is found to be beneficial in enhancing teachers' professional knowledge apart from providing them first-hand experiences on how to tackle culture-specific problems in the classroom (Agarkar and Bedekar 2004). Arranging such visits frequently would certainly improve collaboration among educators from different countries.

References

Agarkar, S. C. (1998). Relevant laboratory programme for secondary schools: Problems and solutions. *School Science, XXXV*(3), 46–52.

Agarkar, S. C., & Bedekar, V. V. (2004). Visit of British teachers to India: A report. *CASTME Journal, 24*(1), 17–26.

Agrawal, M. (2002). Developing educational and social leadership in teachers in the context of school curriculum in India. *Staff and Educational Development International, 6*(3), 246–259.

Bhatia, B. S., & Joshi, S. R. (2006). Science communication: needs and experiences, In S. Agarkar, S. Limaye, & R. Pertzborn (Eds.), *Proceedings of the Indo-US workshop on utilization of space-based resources to enhance science education in India*, Mumbai: HBCSE, TIFR.

Cheng, Y. C., Tsui, K. T., Chow, K. W., & Mok, M. (2002). Searching for paradigm shift in subject teaching and teacher education. In Y. C. Cheng, K. T. Tsui, K. W. Chow, & M. Mok (Eds.), *Subject teaching and teacher education in the new century* (pp. 3–30). Dordrecht: Kluwer Academic Publishers.

COMSEC. (2000). *Training of trainers in science and technology education* (Asian ed.). London: Commonwealth Secretariat.

DST. (2004a). *Scientific policy resolution 1958*. Available at http://dst.qov.in/StPolicy/st policy 1958.htm

DST. (2004b). *Technology policy statement*. Available at http://dst.qov.in/StPolicy/st policy 1983.htm

DST. (2004c). *Science and technology policy*. Available at http://dst.gov.in/StPolicy/st policy 2003.htm

DST. (2013). *Science and technology policy*. Available at http://dst.gov.in/StPolicy/st policy 2013.htm

Govt. of India. (1966). *Education and national development* (pp. 389–421). New Delhi: Ministry of Education.

Govt. of India. (1986). *New education policy*. New Delhi: Ministry of Education.

Hodson, D. (2003). Time for action: Science education for an alternative future. *International Journal of Science Education, 25*(2), 645–666.

Janardhanan, K. R. (2002). Popularization of science. In S. C. Agarkar & V. D. Lale (Eds.), *Science, technology and mathematics education for human development* (Vol. 2, pp. 63–65). Mumbai: HBCSE.

Kulkarni, V. G., & Agarkar, S. C. (1985). *Talent search and nurture among the underprivileged*. Mumbai: HBCSE.

Lagu, R. G. (1978). *How and why in science*. Mumbai: Oxford University Press.

Limaye, S. (2006). Scientists: A resource for science education. In S. Agarkar, S. Limaye, & R. Pertzborn (Eds.), *Proceedings of the Indo-US workshop on utilization of space-based resources to enhance science education in India.* Mumbai: HBCSE, TIFR.

MIB. (2003). *India 2003: A reference annual*. New Delhi: Publication Division, Ministry of Information and Broadcasting, Govt. of India.

NCERT. (1998). *Sixth all Indian educational survey: National tables* (Vol. III, p. 3). New Delhi: Publication Department, NCERT.

NCERT. (2000). *National curriculum framework for school education* (pp. 58–62). New Delhi: Publication Department, NCERT.

NCERT. (2005). *National curriculum framework for school education* (pp. 58–62). New Delhi: Publication Department, NCERT.

Ramaraju, R. (1999). *Impact of international cooperation on selected fields of Indian education* (pp. 115–185). Mumbai: Himalaya Publishing House.

Sharma, R. C. (1975). *Modern science teaching* (pp. 2–5). Delhi: Dhanpatrai and Sons.

Tan, K. S. (2001). Addressing the lifelong learning needs of teachers: A continuing teacher education framework. *Asia Pacific Journal of Teacher Education and Development, 4*(2), 173–188.

TOI. (2012, March 14). *The Times of India.*

Chapter 7
Improving Basic Science and Mathematics Education in Southeast Asia: The Role of SEAMEO RECSAM

Robert Peter Devadason

7.1 Introduction

Most of the countries in the Southeast Asian region are at various stages of economic development and have included bold visions to reach developed status in the near future. One of the important contributors towards this end is the notion that investments in science, mathematics and technology at the basic education level will prepare its citizens to be able to capitalise on the many high-income jobs of the future which are scientific and technology literate dependent. Some of the evidences can be gleaned from the number of countries participating in international assessments in science and/or mathematics (like Trends in International Mathematics and Science Study, TIMSS, and the Programme for International Student Assessment, PISA) and the preparation of national education blueprints that emphasise science and mathematics education (for a Malaysian example, see MOE 2013). Some countries, in addition, also have institutions specially dedicated to the promotion of science and technology like the Institute for the Promotion of Science and Technology (IPST) in Thailand and the National Institute for Science and Mathematics Education (NISMED) in the Philippines. Malaysia is so serious about its commitment on science and mathematics education that it commissioned a study on the policy to increase the number of science stream students in schools (MOE 2012) and has decided to implement a number of the recommendations emanating from the study.

Non-governmental agencies within each country whose scope and focus towards basic science and mathematics education vary have emerged and are playing an increasing role. For instance, the Association of Science and Mathematics Education Penang (ASMEP) undertakes various association-related activities (Akpan 2009) to spur interest in science and mathematics to a large extent within the state of Penang

R.P. Devadason (✉)
Ipoh Campus, Institute of Teacher Education, Ministry of Education Malaysia,
Hulu Kinta 31150, Perak, Malaysia
e-mail: drobertp40@gmail.com

B. Akpan (ed.), *Science Education: A Global Perspective*,
DOI 10.1007/978-3-319-32351-0_7

and to a lesser extent Malaysia and beyond. The Academy of Sciences Malaysia (ASM) is mainly an advocacy body in promoting science and technology in the country. Science industries and museums of a country on the other hand support the emphasis of science and technology of the government by serving as channels for non-formal and informal science and technology education.

Supports towards promotion of basic science and mathematics education also come from international agencies at various degrees of involvement. For instance, the United Nations Educational, Scientific and Cultural Organisation (UNESCO) and the International Science, Technology and Innovation Centre (ISTIC) for South-South Cooperation are international organisations that mainly provide advocacy support to individual countries. Funding agencies like the World Bank and Asian Development Bank have continued to provide support towards capacity building in science and technology mainly in the form of financial assistance. Japan, for instance, has, in addition, given technical assistance. The International Council of Associations for Science Education (ICASE) also serves in the capacity of advocacy.

With all these attempts by various players in each country at various levels of implementation, what has a regional centre like the Southeast Asian Ministers of Education Organisation Regional Centre for Education in Science and Mathematics (SEAMEO RECSAM) to offer? This chapter begins with a description of RECSAM within the context of SEAMEO and Malaysia where it resides. The major activities of RECSAM and its activities are explained within the framework of the various forms of cooperation. It concludes with a section on challenges and future directions.

7.2 SEAMEO RECSAM as a Regional Centre

Cooperation efforts between nations have been in existence for quite a number of years. These efforts, whether as financial aid or technical cooperation, have normally been from developed nations to developing nations. de Sa e Silva (2009) reports that developing countries have themselves cooperated with each other and that it has been in existence, at least, since the early 1950s. Collaborations could be between two countries or between groups of countries sharing commonalities like religion or proximity to one another. It is thus little wonder that the leaders of countries in Southeast Asia saw it strategic to form the Southeast Asian Ministers of Education Organisation (SEAMEO) as one formal mechanism to help one another.

SEAMEO was established on 30 November 1965 with the aim of promoting cooperation in education, science and culture in the Southeast Asian region. Its membership consists of 11 countries, namely, Brunei Darussalam, Cambodia, Indonesia, Lao People's Democratic Republic (Lao PDR), Malaysia, Myanmar, Philippines, Singapore, Thailand, Timor Leste and Vietnam. Eight associate member countries and three affiliate members as well as a partner country support

SEAMEO. These members support SEAMEO in various ways including providing funding and technical support.

SEAMEO carries out its cooperation activities through the 20 regional centres established in various member countries. The centre's operational budget, capital and annual recurring cost are undertaken by the respective host countries. The Regional Centre for Education in Science and Mathematics (RECSAM), formed in 1967, is one of the first centres formed after the formation of SEAMEO. It was formed with the premise that developing countries need an increasing supply of human resource that rely on science and technology and a strong basics on science and mathematics will contribute towards that end. According to SEAMEO RECSAM:

> *The purposes of the Centre shall be to assist the Member States to improve education in science and mathematics in their respective countries and to that end, to undertake research and development, to provide training, to serve as an information centre and clearinghouse as well as to engage in other related activities within and outside Malaysia.* (SEAMEO RECSAM 2012a, p. 4)
>
> *Training of teachers in science and mathematics had high priority throughout the region and that member countries should support a regional centre devoted to this function, in view of the fact that there is a shortage of facilities in the region for such a purpose. .... Help member countries to improve the teaching of science and mathematics in order to lay the foundation for meeting the technically and scientifically trained manpower requirements of the region for national development.* (SEAMEO RECSAM 1976, p. 5)

RECSAM is located in Penang, Malaysia, which is increasingly seen as an emerging donor country (Devadason 2010). The centre enjoys significant support from the government of Malaysia although it was given autonomous status in 1995. For instance, the operating budget, the provision of the site and buildings and its maintenance are heavily financed by the Malaysian government. Azian et al. (2010) provide details of this mechanism. The Enabling Instrument of RECSAM (SEAMEO RECSAM 2012a) is the legal document to guide its operation.

Since its inception in 1967, SEAMEO RECSAM has assisted in the major development of educational manpower for the advancement of science and mathematics education at both primary and secondary school levels for its member countries. RECSAM has been continuously offering full scholarships to teachers and educators in the region to take part in training to improve their skills and make them more adaptable to the changing educational environment. At the moment, well over 19,000 teachers and key educators in science and mathematics have benefitted from RECSAM's programmes and activities. RECSAM has made significant steps in its purposes and roles, as it is currently nearing the end of its Ninth Five-Year Plan (SEAMEO RECSAM 2010) which is for the period July 2010 to June 2015.

The Centre Director is the head of RECSAM and reports to the Governing Board which meets every year to endorse activities proposed by the centre. RECSAM has three divisions: the Research and Development Division, the Training Programme Division and the Administration Division. The academic activities of the centre are carried out by a number of seconded, permanent and contract academic staff. Seconded staff are government educators of member countries deployed to

RECSAM. They work for about 3–5 years at RECSAM before returning to their country. The total number of staff approved in the organisational structure of RECSAM is 98 including 22 academic staff. The Administration Division has seven executives taking charge of finance, publication/multimedia, marketing, administration, International House, library, information/documentation and ICT support units. The support staff assist in the technical, administrative and general duties.

7.3 SEAMEO RECSAM's Scope of Activities

RECSAM actively pursues six major regional functions that contribute to the improvement of science and mathematics education. They include a focus on research and development; capacity building; regional conferences, seminars and workshops; networking among experts and institutions; consultancy services; and clearing house for information. These foci have been in place since the early days of the centre's existence (Cruz 1980).

7.3.1 Research and Development

RECSAM conducts a number of research activities, and they can be conveniently divided into research that informs policy and research that informs pedagogy. Since the early days of its existence, RECSAM has contributed much towards regional level research (for a historical context, see Clements and Ellerton 1996). Singh and Ellerton (2013) while reporting on mathematics education research have argued that RECSAM had provided an effective collaborative research environment towards improving mathematics instruction in the region. Research activities as well as innovative capacity building programmes may result in the development of resources. They serve as catalyst to bridge the theory-practice gap as well as serve to translate innovations to fit indigenous needs.

Research informing pedagogy may be conducted in the region and the host country or even confined to neighbouring schools within the vicinity of RECSAM. A book that does secondary analysis of the TIMSS 2007 results (Ong and Gonzalez 2012) and another book with the same theme for TIMSS 2011 results (Ong et al. 2014) are being published, not to repeat reports published by IEA or the individual countries but to add value to their work in the form of secondary analysis reports. The book by Devadason et al. (2010) documents the state of the use of ICT in the teaching of science and mathematics among the member countries. A recent national survey research on teachers' use of the Geometer's Sketchpad in Malaysian secondary schools was conducted by Warabhorn et al. (2012). The purpose of this survey was to investigate how far mathematics teachers in secondary schools used GSP for instructional purposes. The findings of this study was to help the academics of RECSAM and other training providers offer more focused training on GSP or other dynamic geometry software as well as to inform policy (for the Ministry of Education (MOE) Malaysia to consider scope and depth of support for the use of GSP for instruction). The research on lesson study (Cheah 2012) which was conducted on a number of schools within the vicinity of RECSAM demonstrates the effort by the researcher to adopt and adapt this successful Japanese professional development practice to Malaysian classroom situation.

An example of an on-going research essentially informing policy takes the form of the development of Southeast Asian Regional Standards for Mathematics Teachers Project (SEARS-MT). This project involves the development of a set of standards for the twenty-first century mathematics teacher for member countries. The standards, when complete, will serve as a guideline for teachers' self-development as well as help departments in charge of teachers' development to adapt and adopt them to provide systematic development paths for the mathematics teacher. The SEARS for Science Teachers (SEARS-ST) project began in early 2014 and was expected to be completed in early 2015. The Southeast Asian Regional Assessment (SEARA) was a project to consider a regional assessment system for the member countries. Various modes were considered: develop a home-grown system, outsource the development of such a system or use existing systems like the IEA's TIMSS and OECD's PISA.

RECSAM publishes and coordinates the publication of teaching and learning resources for the benefit of teachers and teacher educators. Products developed during the regular courses, which are normally in the form of lesson plans, are by default compiled and kept in the library as resource materials although there have been attempts at publishing them (e.g. see Dominador et al. 2014, which is a booklet on a past regular course topic with lesson plan exemplars). At the regional level, a teacher's guidebook integrating climate change issues in Southeast Asian schools (Azian 2010) was developed in collaboration with six other SEAMEO centres. Translation (from French to English) of the book *Teaching Science in School: La main a la pate Resource Materials for the Primary Classroom* (Foo 2007) has also been carried out for the benefit of English-speaking readers. "The Training of Trainers Manual for Promoting Scientific and Technological Literacy (STL) for all" (Lee 2008) is another resource developed for teachers and teacher educators.

7.3.2 Capacity Building Programmes

Four main modes of training are offered by RECSAM, namely, regular courses, in-country courses, customised courses and training workshops. The focus of these courses is to serve as catalyst to improving the quality of science and mathematics teachers in the region. Both pedagogical content knowledge and knowledge of content are emphasised depending on needs. Emphasis is also given to school-based professional development. ICT is used extensively during the course to serve as examples of its use for instruction as well as provide ample opportunity to the participants to explore ICT-based tools for teaching and learning. For fiscal year

2013/2014 (SEAMEO RECSAM in press), 1,568 participants were trained impacting not only the 11 member countries but also beyond.

Regular courses are funded by the Malaysian government through a yearly Special Educational Development Fund (SEDF) which is to be used to conduct activities for the member countries through full scholarships. The past 4 years saw four 1-month regular courses being offered to member countries every fiscal year (fiscal year begins on 1 July and ends 30 June the following year), thus potentially benefitting a maximum of 77 participants (two participants per country for three of the courses and one for the fourth course) per fiscal year. Additional participants may participate in these courses as fee-paying participants. The four regular courses conducted for fiscal year (FY) 2013/2014 were RC-SS-138-1: Linking Secondary Science Learning to Everyday Life and Societal Needs; RC-PM-138-2: Making Real Life Connections and Developing Mathematical Ideas in Primary Classrooms; RC-PM-138-3: Developing Mathematical Thinking through Game-based Learning in Primary Classrooms; and RC-SS-138-4: Lesson Study: Improving Instructional Practices in Secondary Science Classrooms (SEAMEO RECSAM in press).

The SEDF is also used to fund a maximum of six in-country courses per fiscal year although some in-country courses are funded by the course hosting country. In-country courses are typically of 5-day duration (30 contact hours) conducted in the requesting member country by two academic staff assigned by RECSAM. They were introduced as a solution to concurrently address funding issues, increasing the centre's reach to more participants as well as customising the course specifically to the needs of the recipient country roughly based on a number of themes suggested by the centre. RECSAM uses the SEDF to finance the two staff to be sent to the recipient country, while the latter will take responsibility for the local costs such as venue, participants' accommodation and meals and other costs. In-country courses were held in Indonesia and Lao PDR (SEAMEO RECSAM in press) for the 2013/2014 fiscal year.

Customised courses are initiated and paid for by the recipient country or a sponsoring agency. The content of the courses is "customised" according to the needs of the requesting agency. It also serves as an important income-generating activity for RECSAM. Two customised courses for Indonesian educators, Twenty-First Century Skills in Secondary Science: Enhancing Higher Order Thinking Skills and Integrated Values Education and Twenty-First Century Skills in Secondary Mathematics: Enhancing Higher Order Thinking Skills and Integrated Values Education, were conducted from 21 October to 15 November 2013. The Directorate General Management of Basic Education, Ministry of Education and Culture, Indonesia, sponsored the 30 participants.

The Third Country Training Programme (TCTP) courses are a class of customised courses which generally involves one or more funding organisations or country to sponsor participants in developing countries to participate in courses conducted by an organisation which itself is located in a developing country. Malaysia, through its Malaysian Technical Cooperation Programme (MTCP), cooperates with a number of external funding partners to offer such courses. In lieu of RECSAM's location in Malaysia, Malaysia uses RECSAM as its venue to offer courses on science and mathematics education. Currently, two other funders cooperate with Malaysia: Japan International Cooperation Agency (JICA) and Colombo Plan Secretariat. JICA funds professional development of African teachers, while Colombo Plan Secretariat which is located in Sri Lanka funds professional development of teachers from Colombo Plan countries.

Workshops are another type of income-generating activity conducted by RECSAM, and they are normally of duration between 1 and 3 days and sometimes up to 5 days. Since these workshops are of very short duration, it may not be feasible to open them to participants from beyond Malaysia. However, workshops are normally strategically held to coincide with other courses of longer duration to provide access to course participants from beyond Malaysia as well as to ensure financial viability. Workshops are also a way to attract high-quality facilitators from around the world (including from Malaysia and the region) who may not be able to participate in courses of longer duration due to cost and time constraints. The "Inspiring Science: Comprehensive support for contemporary developments in science education" facilitated by Dr Mark Windale from Sheffield Hallam University and held from 29 October to 1 November 2012 and the "Web 2.0: A new wave of innovation for teaching and learning science" by Dr C. Annamalai, an academic from the Centre and held from 24 to 25 October 2012, serve as examples of workshops. A regional workshop for science educators from developing countries "Innovative Teaching and Learning of Science through Inquiry-Based Science Education (IBSE)" was organised by SEAMEO RECSAM in cooperation with International Science, Technology and Innovation Centre (ISTIC) for South-South Cooperation under the auspices of UNESCO from 1 to 5 October 2012. It was attended by 41 teachers from 15 countries.

At an even more local level are two-hour colloquia organised for teachers, teacher educators and researchers from around the states of Penang and Kedah (they are located near RECSAM premises) whenever academic staff from RECSAM and experts from within the region or beyond that visit the centre are available to share their research or training expertise. Colloquia are held for free.

A number of countries in the SEAMEO region strategically send its teachers and teacher educators to do their postgraduate degree in a developed country. RECSAM realised that two of the major hindrances to teachers and teacher educators in the region having access to higher degree programme in a developed country were high cost and low proficiency of English. In collaboration with Deakin University, a way out to this predicament was to conduct part of the programme in RECSAM which was very much cheaper as well as provide opportunities for participants to reach the minimum English proficiency level set by the university. A postgraduate diploma was presented by RECSAM to those who successfully completed the programme and for those in addition who obtained the minimum English proficiency level continued into the master's degree programme in Deakin University. This programme enabled a group of Indonesian participants to obtain master's in education degrees (in Science and Mathematics) from Deakin University in 2010.

7.3.3 Convener of Regional Conferences, Seminars and Workshops

The International Conference on Science and Mathematics Education (CoSMEd) and Search for SEAMEO Young Scientists (SSYS) are two biennial events convened by RECSAM. CoSMEd was first convened in 2005 with the intent of providing an avenue for educators, researchers and experts in science and mathematics education to converge on the centre and exchange knowledge on current issues. CoSMEd was held for the fifth time as CoSMEd 2013 with the theme "Empowering the Future Generation through Science and Mathematics Education". Based on feedback from past conferences, CoSMEd has evolved to satisfy both researchers and classroom practitioners.

The SSYS is a regional congress conducted in the form of a science exhibition and science congress for the youth to embark on scientific and technological research projects. It is one of the projects to benefit students directly. SSYS in its current form accommodates both science and mathematics categories (see Ng 2012 for detailed description and selected projects of SSYS). Sponsorship in terms of registration fee waiver are given to all official delegates from SEAMEO member countries; teacher and student delegates from Cambodia, Lao PDR, Myanmar, Vietnam and Timor Leste are also provided with return airfare. The first SSYS regional congress was held in 1997; the 9th SSYS Regional Congress was held in March 2014 with the theme "Disaster Risk Reduction for Sustainable Development". Robottom (2013) argues that SSYS is an example of an event that attempts to find a balance between the "approach of grounding science education in science-related problems within local communities, and, on the other, the development of scientific rigour in contemporary science education".

Occasionally, regional seminars or workshops are held to serve as a platform to discuss major issues that may have implications for the region. For instance, RECSAM collaborates with ICASE to jointly host regional seminars. They are normally held after the World Conference on Science and Technology Education (WorldSTE) which is organised by ICASE. In these seminars, the committee members of ICASE present papers and discuss important issues brought up during the WorldSTE. An example is the Joint RECSAM-ICASE Regional Seminar on "The Way Forward for Science and Technology Education: Implication for Policy Makers" which was held from 16 to 19 February 2009. The outcome of this seminar was a road map formulated to improve the quality of science and mathematics education in Southeast Asia which significantly contributed to the Southeast Asian Regional Assessment (SEARA) and Southeast Asian Regional Standards for Mathematics and Science Teachers (SEARS-MT and SEARS-ST) projects which are currently in progress. A second Joint RECSAM-ICASE Regional Seminar on "Innovations in Science and Technology Education (STE) for the twenty-first century" was held from 16 to 17 February 2011 at RECSAM, and it discussed among others some of the issues brought up during the World Conference STE 2010 held in Tartu, Estonia.

7.3.4 Networking Among Experts and Institutions

Networking among experts and institutions in the region and beyond is continuously undertaken through various channels. The usual way is through the acquaintances made during the conduct of many activities of the centre as mentioned above. More serious networking may result in memorandum of understanding (MOU) signed with interested institutions. For instance, an MOU was signed with Thepsatri Rajabhat University (TRU) and Kamphaeng Phet Rajabhat University (KPRU) for academic collaboration on 14 May 2012. This MOU allowed for each party to collaborate and develop initiatives and educational programmes, research and staff professional development projects and also to set up a mechanism for academic collaboration between both parties upon consultation and meeting. Another common way is for academic staff of the centre to participate in international science and/or mathematics education conferences like the fourth World Conference on Science and Technology Education (WorldSTE 2013) which was held in Sarawak, Malaysia, from 29 September to 3 October 2013. These and other activities allow for RECSAM to build a network of experts and institutions on science and mathematics education from within the region and beyond.

7.3.5 Engagement in Consultancy Services

RECSAM staff are also frequently involved in consultancy services. The top leadership of the centre are often invited to MOE Malaysia's national level meetings pertaining to science and mathematics education. RECSAM's academic staff are also called by schools, whether within Malaysia, within the region and outside the region to provide input and advice on innovative practices of the centre. For instance, RECSAM staff have been invited by schools around the vicinity to help them implement lesson study as a professional development activity (Cheah 2012). At a regional level, RECSAM's academic staff have provided consultancy services to countries like the Philippines, Indonesia, Thailand, Lao PDR and Malaysia in implementing climate change issues in their curricula (Aligaen 2013).

7.3.6 Serve as Clearing House for Information

One important activity of RECSAM is to serve as a clearing house for information. Clearing house activities are performed through three main ways: a well-stocked library, the publication and dissemination of regional level research through two journals, and the publication and dissemination of capacity building exercises as well as research and development efforts in the form of books and monographs.

RECSAM has a specialised library stocked with over 30,000 books and other materials (including digital materials and journals) on science and mathematics content, pedagogy and research. Among the resources are the many publications of the centre, whether by its experts or participants, since the early years of its founding. While most of its users use the resource when they visit the centre or attend events in the centre, the online public access catalogue (OPAC) allows for public online access to its list of resource materials.

RECSAM publishes two journals: the *Learning Science and Mathematics*, which is an online journal that primarily caters to the needs of teachers and teacher educators, while the other, *International Journal of Science and Mathematics Education in Southeast Asia*, caters primarily for science and mathematics education researchers that is published in June and December each year. Both are peer reviewed with the *International Journal of Science and Mathematics Education in Southeast Asia* enjoying the advice of very notable academics in the field and reviewed by an international panel of reviewers.

Results of significant activities of the centre are published in the form of reports, monographs and books. Selected reviewed papers of CoSMEd 2013 were compiled into two books (Devadason et al. in press; White et al. 2014), while a research project on children's connectedness to nature was published as a project report (Hazura et al. 2013). These publications are either given away free, made available for download, or sold.

7.4 Cooperation Efforts

The previous section provided a glimpse of the breadth and depth of activities conducted by the centre. However, the activities of the centre have to be relevant to the needs of the member countries. RECSAM's activities continue to follow the guiding principle of the founding members of RECSAM:

> *The Regional Centre should give thought to strengthening and supporting existing national institutions where possible, and that it should complement and supplement existing national programmes in order not to duplicate or compete with them. Each SEAMEO country should expect to get these benefits from the Centre's programmes that are not attainable at comparable costs.* (SEAMEO RECSAM 1976, p. 7)

Although there are many commonalities among the SEAMEO member countries as "South" states (perhaps Singapore as an exception), there are indeed significant differences between each country as well as significant differences within each country to make it difficult for aid to be not only relevant but also effective and efficient (Fredriksen 2011).

It is thus imperative that a high level of collaboration with clients from member countries is necessary so as not to duplicate or compete with their efforts but to complement them. As a centre formed to serve the needs of the region, a number of mechanisms and practices have been put in place to enhance effectiveness of regional cooperation and maximum consultation through the involvement of various stakeholders. This high level of collaboration can be seen at various levels of communications with member countries. Formal channels put in place include the SEAMEO Council, SEAMEO Secretariat, High Officials, Governing Board, Centre Directors as well as some internal mechanisms which will be explained in this section.

The SEAMEO Council is the policy-making body of SEAMEO and consists of Ministers of Education of each member country as well as representatives from associate and affiliate members, and they normally meet every 2 years during the SEAMEO Council Conference to take policy decisions. The SEAMEO Secretariat is the executive arm of the Council and it is located in Bangkok, Thailand. The SEAMEO Secretariat executes decisions made by the Council. SEAMEO carries out its activities through 20 specialist regional centres/units that undertake training and research programmes in various fields of education, science and culture, and SEAMEO Secretariat is the coordinating body for these centres. SEAMEO activities follow guidelines provided for in the SEAMEO Charter (SEAMEO 1968).

The High Officials are the highest level civil servants in the education ministry of each country. They meet formally once a year through the High Officials Meeting (HOM). They deliberate on projects and issues which may require their support and commitment in terms of human resource and finance as well as policy that had been brought up by the SEAMEO centres. They give feedback and, if required, endorse selected projects and issues to be brought to the attention of the SEAMEO Council Conference for approval by the education ministers.

The Centre Directors meet formally once a year during the Centre Directors Meeting (CDM) where they discuss issues at the centre level. The meeting also acts as a focal point for aid providers, especially from the "North" countries to propose North-South Cooperation activities for SEAMEO.

RECSAM is an autonomous organisation, and its governance as stipulated in the Enabling Instrument of RECSAM (SEAMEO RECSAM 2012a) is provided by the Governing Board. Its membership consists of a nominee from each member country. They meet formally once a year to provide input, deliberate and approve the activities of the centre.

All major activities of the centre including training programmes and research are formalised into Five-Year Development Plans. These activities are developed through a regional workshop organised for this purpose where the following are invited: Governing Board Members and their representatives, Ministry of Education officials from member countries as well as consultants from associate member countries through a regional workshop. RECSAM is currently in the final phase of its Ninth Five-Year Plan which began in 1 July 2010 and expected to end on 30 June 2015. This regional workshop serves as an important channel to thrash out issues on programme relevance and potential impact to member countries. They are formally endorsed by the SEAMEO Council after going through the CDM and HOM. The development of RECSAM's Tenth Five-Year Plan for 1 July 2015 to 30 June 2020 is currently in progress.

From the operational point of view, the detailed design of the regular courses are finalised about a month before the commencement of the courses. These course outlines undergo refinement through input from external consultants who are from the university or teacher training institutes located nearby.

Maximum consultations with member countries are made when conducting research activities that have implication for the region. For instance, the SEARS-MT project working paper was brought to the GBM for input and approval of funds. This working paper further went through the CDM, the HOM and the SEAMEO Council before implementation. From its operational point of view, the SEARS-MT involves phases where experts from the Ministry of Education of the member countries are called upon to give feedback on the final form this standard should take. In the case of SEARS-MT, experts from two associate member countries (Japan, Australia) too participated in drafting the MT framework.

7.5 Challenges and Future Directions

While the centre participates as a regional centre through a variety of activities and enjoys support from the different countries, it continuously faces a number of challenges, the most important of which is financial sustainability.

7.5.1 Financial Sustainability

The centre is hosted by the Malaysian government and as such receives a yearly grant through the Ministry of Education which is revised from time to time. It is used for staff remunerations and operation of the centre. Besides the yearly grant, the government also gives a yearly Special Educational Development Fund (SEDF) which is used to provide scholarships for science and mathematics educators from the member countries (regular courses and in-country courses). With the granting of autonomous status by the government in 1995, the centre was able to generate its own income to supplement the yearly grant given by the government to cover shortfalls and expand the extent of professional services it can provide. However, a number of events these past few years are making this mode of funding support extremely difficult to sustain. First is the drastic increase in salary of Malaysian staff as the country progresses into a high-income country. Second is the increase in transport, food and accommodation cost throughout the region. Third is that the centre's buildings are ageing and require more expensive and frequent maintenance and so are the support staff that are drawing higher salaries due to seniority in service. Though the centre is an autonomous organisation, the rate at which costs have escalated can hardly be addressed by pursuing higher income from self-revenue-generating activities alone. It has to depend on the government although it is the author's opinion that in the long run the centre needs to wean itself from relying too much on the host government, at least in terms of financial support. Gone are the days when the centre enjoyed funding and technical support from developed countries (SEAMEO RECSAM 1976) apart from the host country as RECSAM is now seen as residing in a country at the upper end of the developing country spectrum.

The rise in cost of living is one of the main factors that have been instrumental in shaping the duration of the regular courses funded by the government of Malaysia. These courses in the past used to be of 3-month duration which was reduced to 6 weeks and then eventually to the current combination of 4-week course and in-country courses. Azian et al. (2010) provide more details on this issue.

It is interesting to note that not all centres operate similarly. For instance, SEAMEO Regional Training Centre (RETRAC), which is mandated to improve educational management for the region, is able to generate enough supplementary income through its short courses on English language which has high commercial value. Meanwhile, SEAMEO Regional Language Centre (RELC) which is located in Singapore operates a profitable hotel which is located in the tourist area as it enjoys a very high occupancy rate. SEAMEO Tropical Medicine (TROPMED), Malaysia, is supported by the Institute of Medical Research (IMR), Malaysia, and in lieu of its location within a larger facility as well as access to existing experts, operational cost would be expected to be much lower than for a physically separate centre like RECSAM.

Income-generating activities can be divided into academic and non-academic categories. Academic activities like customised courses, workshops, seminars and conference not only help generate income for the centre but also extend the scope of its mandate for the region. In recent years there have been attempts to generate income through professional activities by making use of ICT, that is, through the distance learning mode. This was actually touted by a Governing Board Member as one direction to go (SEAMEO RECSAM 2012b, p. 21). It remains to be seen how effective this mode will turn out to be. Non-academic income-generating activities include the use of the International House as well as rental of facilities. The location of the International House is not strategic enough for it to generate income as generated by the hotel in SEAMEO RELC. In its current mode of operation, a tension exists between how much emphasis is to be placed on income generation and the focus on its core mandate. Too much emphasis on income generation may reduce the energy spent on improving science and mathematics in the region.

As one of the oldest and most established regional centres, RECSAM has attracted limited foreign funding agencies like Japan International Cooperation Agency (JICA) and Colombo Plan Secretariat who have partnered with the Ministry of Foreign Affairs (MOFA), Malaysia, through its Malaysian Technical Cooperation Programme (MTCP) to sponsor in-service courses for participants from beyond the region like Africa and the Asia-Pacific region under the TCTP. The positive experience of making use of RECSAM has contributed to these funding agencies continuing their service with the centre.

The centre perhaps, too, has not mastered the ability to consistently draw upon external funding agencies to sponsor its mandated activities. This is an issue that is being acknowledged to be a significant problem among most of the centres so much so that one of the more successful centres experienced in sourcing for funds has been tasked to provide mentorship to the other centres in the near future.

7.5.2 Hiring and Retaining of Able Academics

The organisational structure of RECSAM allows for 22 academic staff including 2 deputy directors and 4 senior specialists. It is desired that academic staff have at least a master's degree in the relevant field as well as classroom teaching experience on top of being fluent in English. This condition greatly reduces potential resource persons from a number of member countries. Specialists are normally seconded from the respective education ministries of the member countries and depending on the policies of the government, the secondment term varies from about 3 to 5 years. They are expected to use the experience gathered in RECSAM to bring back to their own country. This does not occur in practice. The centre often has to hire contract staff to fill the many vacancies.

The salary of academic staff is pegged at a scale slightly higher than that offered to education staff in Malaysia, and it is adjusted accordingly whenever Malaysian staff salary is adjusted. RECSAM in recent years has never been able to attract specialists from Singapore and Brunei as their staff receive a salary much higher than that offered in Malaysia. It is a pity because there is much that specialists from these English-speaking countries, especially from Singapore, can offer to the region. Realistically, it is difficult to hire specialists from most of the countries as English is not even considered a second language except for the Philippines, Indonesia and sometimes Vietnam and Thailand. Gone are the days when governments of developed countries were liberal in funding long-term experts to participate in the centre's programmes. The exception is Japan, which in recent years has renewed sending its experts for a year. Over the years, the centre has evolved a number of strategies to tap into their expertise. Among them include the sabbatical leave programme and the foreign fellow programme which allows for experts from around the world to contribute to RECSAM for a shorter period of time while still working for their home organisation. Both sabbatical leave and foreign fellow participants are provided with accommodation and a room with ICT facility to work. In addition, foreign fellows are given a small stipend, while sabbatical leave participants are paid an allowance each time they teach in the centre's courses. As at the time of writing, two university lecturers from Malaysia have participated in the sabbatical leave programme, while one from New Zealand has participated as a foreign fellow. Attempts will continue to be made to attract expertise from Singapore. Experts from friendly universities from the region and beyond had supported the centre in the past and will continue to support RECSAM by making use of their formal visit to this part of the region to contribute their expertise in RECSAM either in the form of workshops or colloquiums. Currently, they include experts from Universiti Sains Malaysia and Teacher Education Institute of Malaysia, Penang branch, as well as from the University of Western Sydney and Tsukuba University.

7.5.3 *Addressing Differences Among Member Countries*

Two significant differences among the member countries are a challenge to the improvement of science and mathematics education through regional level courses like the regular course. As explained earlier, one is the proficiency of English disparity, and the other is that different countries are at different stages of economic development. It is not only that each country has its own constraint, but it may be different for different regions. To ensure that the right candidates participate in RECSAM's scholarship programmes so that they reap maximum benefits, RECSAM places certain expectations on candidates to be nominated by the various countries. They include a maximum limit of 50-year-old candidates, being medically fit, with potential maximum impact like master teachers and teacher educators and a reasonable level of English Language proficiency apart from being directly involved in science or mathematics education. However, in reality this is not strictly possible due to issues raised above.

The current solution to low English language proficiency is to conduct extra classes for the participants while participating in RECSAM's regular courses. It still does not address the needs of the three categories of countries in terms of development – the one size not able to fit all problems. Perhaps one way to address this is to have separate courses for the CLMVT countries (Cambodia, Lao PDR, Myanmar, Vietnam, Timor Leste) and the BIMPT countries (Brunei, Indonesia, Malaysia, Philippines, Thailand) and Singapore.

7.5.4 *Accounting for RECSAM's Role in Improving Science and Mathematics in the Region*

The ultimate goal of the existence of RECSAM is to contribute towards the improvement of science and mathematics education in the region. As was described at the beginning of the chapter, starting with the Ministry of Education as the main player, many other players contribute towards the improvement of science and mathematics in a country. As such it is difficult and not fair to attribute improvements to science and mathematics of a country directly to one party like RECSAM. A less direct argument will thus be made. For a start, the main focus of RECSAM is on teachers and teacher educators. This is in line with Hattie's (2003) identification of teachers as the main source of variance that can be controlled in improving student performance. Mention has already been made of the numerous training programmes in terms of length, distance and cost to impact as many teachers as possible with the limited amount of funding, not to mention the regional seminars and conferences.

For the regular courses, the multiplier effects (ME) programme has allowed more systematic and measureable ways for participants to impact clients under their care. The giving out of an achievement certificate for participants who could provide evidence of successfully conducting the ME at the home country is another attempt to ensure and measure extent of spread beyond the class in RECSAM.

At another level is the impact study conducted by RECSAM 6 months after the regular courses participants have returned to their home. A survey form is used to gather data among which include how far they could apply what they had learnt in RECSAM to their place of work or beyond.

As far as the regular courses are concerned, perhaps the comment by one of the past Centre Directors would provide a more realistic expectation of accountability:

> *…it is felt that the "multiplier" element that forms part of the course would produce a ripple-effect among the teachers in each member country as well as the student population. A "tsunami" effect would be too far-fetched, but it would make a positive contribution towards curricular reform in the region*". (Tan 2002)

For research and development work done by the centre, research results are published and disseminated to stakeholders either directly or through the Internet. However, in recent years, the SEAMEO Secretariat which coordinated the development of key result areas for the centres has more overt requirements to be used as indicators. For instance, instead of stopping at how many research that have been done, it goes further to state how many research have been adopted by the member countries. Thus, recent directions in research and development of the centre take into account the need to make the possibility of the results of such research being accepted by the member countries greater. This can be seen in the increase of regional level projects like SEARA, SEARS-MT and SEARS-ST.

One of the important areas that may be of great help to aid providers as well as great impact to countries of the region towards improvement of science and mathematics education in the region is through some form of common assessment system as well as the development of standards. Once these systems have been established, a form of best practices from countries that have managed to successfully follow desired standards as well as show improvement in the assessment system may be shared among member countries. The common assessment system touted by RECSAM is for every country to participate in existing international assessment systems like Trends in International Mathematics and Science Study (TIMSS) and the Programme for International Student Assessment (PISA). Currently, Singapore, Malaysia, Thailand and Indonesia participate in both TIMSS and PISA. Studies like TIMSS and PISA allow for more incisive technical aid to be given by the various stakeholders responsible for improvement of science and mathematics including RECSAM. Together with the promotion of a common international assessment system, the formulation of benchmarks in science and mathematics will also help as a powerful guide for countries to improve. Standards like those of science and mathematics curriculum as well as standards for science and mathematics teachers are being developed.

As discussed earlier, although the SEAMEO countries are similar in many ways, they are also different in other aspects. It has been proposed that a focus on prescription under these circumstances may not be as effective as the sharing of best practices.

A newer opportunity and challenge has risen through the formation in 2009 of SEAMEO Quality Improvement of Teachers and Education Personnel (QITEP) in Science and SEAMEO QITEP in Mathematics, both of which are located in Indonesia. The premise for their formation was that the region was large enough for the need for these additional regional centres. The centres had taken great care to avoid duplication of efforts but rather to complement each other. For instance, in the development of standards for science and mathematics, RECSAM compromised by attempting the SEARS-MT, while QITEP in Mathematics compromised by developing the curriculum standards for mathematics.

7.6 Conclusions

This chapter began with the challenge of sustaining a regional centre within the context of multiple players contributing towards promotion and improvement of basic science and mathematics education in a country. A wide variety of activities are conducted by RECSAM with support and collaboration from various stakeholders including from the host country Malaysia, SEAMEO member countries, affiliate countries and associate members as well as organisations and countries beyond that sphere. It is always a challenge for a centre like RECSAM to satisfy the needs of countries which are similar in some aspects and yet different in many other aspects. This is so true especially when the SEAMEO countries are undergoing purposeful dynamic transformation. RECSAM will continue to adapt to new challenges especially with strong support of the Ministry of Education, Malaysia, and the SEAMEO collaborative spirit.

References

Akpan, B. B. (2009, February). *The role of science teacher associations in enhancing science, technology, and mathematics (STM) education*. Invited address to the joint RECSAM/ICASE regional seminar on "The way forward for science and technology education: Implications for policy makers", held at SEAMEO RECSAM, Penang.

Aligaen, J. C. (2013). *Climate change integrated model: Building adaptive capacity for the next generation (Malaysia, Indonesia, Thailand, Philippines and Lao PDR), Julito C. Aligaen, RECSAM-APN collaborative educational project*. Penang: SEAMEO RECSAM.

Azian, A. (Ed.). (2010). *Integrating climate change issues in Southeast Asian schools. A teacher's guidebook*. Penang: SEAMEO RECSAM.

Azian, A., Devadason, R. P., Ng, K. T., & Wahyudi. (2010). Building networks for knowledge-exchange and peer learning in science and mathematics education within SEAMEO member countries and beyond – the role of SEAMEO RECSAM. *Journal of International Cooperation in Education, 13*(2), 139–153.

Cheah, U. H. (2012). *Implementing teaching and learning innovations through lesson study: Mathematical thinking and communication*. Penang: SEAMEO RECSAM.

Clements, M. A. K., & Ellerton, N. F. (1996). *Mathematics education research: Past, present & future*. Bangkok: UNESCO.
Cruz, P. C. (1980). The role of RECSAM in the eighties. *Journal of Science and Mathematics Education in Southeast Asia, 3*(2), 29–34.
de Sa e Silva, M. M. (2009). South-south cooperation: Past and present conceptualization and practice. In C. Linda & S. Gita (Eds.), *South-south cooperation in education & development*. New York: Teachers College, Columbia University.
Devadason, R. P. (2010). Malaysia's cooperation to developing countries as an emerging donor. *A Program Report by Visiting Foreign Research Fellows, 22*, 89–132.
Devadason, R. P., Wahyudi, Cheah, U. H., & Ng, K. T. (Eds.). (2010). *The state of use of ICT in the teaching and learning of science & mathematics among schools in SEAMEO member countries*. Penang: SEAMEO RECSAM.
Devadason, R. P., Zurida, I., & Ng, K. T. (Eds.) (in press). *Empowering the future generations through science education*. Penang: SEAMEO RECSAM.
Dominador, D. M., Hazura, A. B., Foo, L. K., & Devadason, R. P. (Eds.). (2014). *Making sense of science through inquiry: Problem-based learning at work. Improving science and mathematics learning module series*. Penang: SEAMEO RECSAM.
Foo, L. K. (Ed.) (2007). *Teaching science in school. La main a la pate resource materials for the primary classroom*. Edition translated from French. Penang: SEAMEO RECSAM.
Fredriksen, B. (2011, December 6–11). *Using aid strategically to promote education development: The need for a more effective global education aid architecture*. Paper presented at the Salzburg Global seminar, Salzburg.
Hattie, J. A. (2003). *Teachers make a difference: What is the research evidence?* Paper presented at the Australian Council for Educational Research Annual Conference on Building Teacher Quality, Melbourne.
Hazura, A. B., Ahmad, N. J., Ng, K. T., & Corrienna, A. T. (2013). *Children's connectedness to nature: A research report*. Penang: SEAMEO RECSAM.
Lee, S. M. (2008). *The training of trainers manual for promoting scientific and technological literacy (STL) for all* (2nd ed.). Penang: SEAMEO RECSAM.
MOE. (2012). *Laporan Strategi Mencapai Dasar 60:40 Aliran Sains/Teknikal:Sastera [Report on the strategy to cchieve the 60:40 policy. Science/technical:Arts stream]*. Ministry of Education, Malaysia.
MOE. (2013). *Malaysia Education Blueprint 2013-2015*. Ministry of Education, Malaysia. Retrieved from http://www.padu.edu.my/index.php/en/component/content/article?id=36
Ng, K. T. (Ed.). (2012). *Journey to investigative research: Selected student research projects of the 5th search for SEAMEO Young Scientists (SSYS 2006) congress*. Penang: SEAMEO RECSAM.
Ong, S. L., & Gonzalez, E. J. (2012). *TIMSS 2007. What can we learn?* Penang: SEAMEO RECSAM.
Ong, S. L., Gonzalez, E. J., & Kanageswari, S. S. (2014). *TIMSS 2011. What can we learn together?* Penang: SEAMEO RECSAM.
Robottom, I. (2013). Community-based learning: From local community to critical open society. In C. Campbell & I. Robottom (Eds.), *Learning science beyond the classroom*. Penang: SEAMEO RECSAM.
SEAMEO. (1968). *Charter of the Southeast Asian Ministers of Education Organization*. Bangkok: SEAMEO.
SEAMEO RECSAM. (in press). *Annual report 2013/2014*. Penang: SEAMEO RECSAM.
SEAMEO RECSAM. (1976). *RECSAM the first ten years 1967–1976*. Penang: SEAMEO RECSAM.
SEAMEO RECSAM. (2010). *SEAMEO RECSAM 9th five-year development plan (for fiscal years 2010/2011 to 2014/2015)*. Penang: SEAMEO RECSAM.
SEAMEO RECSAM. (2012a). *Enabling instrument for the Southeast Asian Ministers of Education Organisation Regional Centre for Education in Science and Mathematics (SEAMEO RECSAM)*. Penang: SEAMEO RECSAM.

SEAMEO RECSAM. (2012b). *Report of the 43rd governing board meeting 2012*. Penang: SEAMEO RECSAM.

Singh, P., & Ellerton, N. F. (2013). International collaborative studies in mathematics education. In M. A. (Ken) Clements, A. J. Bishop, C. Keitel, J. Kilpatrick & F. K. S. Leung (Eds.), *Third international handbook of mathematics education. Springer International Handbooks of Education, 27*. doi:10.1007/978-1-4614-4684-2_26.

Tan, K. (2002). Regional cooperation towards effective curricular reforms: The SEAMEO mission in capacity building and innovations for the 21st century. *Journal of Science and Mathematics Education in Southeast Asia, 25*(1), 1–10.

Warabhorn, P., Devadason, R. P., Wahyudi, Cheah, U. H., & Teoh, B. T. (Eds.). (2012). *Teachers' use of the geometer's sketchpad in Malaysian secondary schools: A survey report*. Penang: SEAMEO RECSAM.

White, A. L., Suhaidah, T., & Cheah, U. H. (Eds.). (2014). *Empowering the future generations through mathematics education*. Penang: SEAMEO RECSAM.

Chapter 8
Science and Technology Education Initiatives in Nigeria: The Case of STEP-B

Michael Umale Adikwu, Mercy Oluwaseyi Agunbiade, and John Monday Ubada Abah

8.1 Introduction

Over the years, many countries have thought of how to create wealth, increase youth employment and increase quality of life in various frontiers. To do this, many low-income countries have depended on international development partners, such as the World Bank, to help in assuaging the monetary needs of these nations.

Over the course of the past few years, too, the world has become increasingly complex, with continued economic malaise in the Euro zone, more modest growth in the countries of *Brazil, Russia, India and China* (BRIC) and increasing scepticism over the future prosperity in other markets. While the forces of globalisation serve to further reduce the gaps between nations, the challenges facing policymakers and development practitioners, rather than being simplified, have become even more daunting (The World Bank 2012).

On a global scale, there are lots of stressors that are confronting nations. There are floods, diseases, earthquakes and an avalanche of other problems that distort the budgets of nations. The ability and capacity to readjust is often absent in many developing countries in Africa, and as such, these nations depend more increasingly on donor agencies and countries (Inyang 2009).

M.U. Adikwu (✉)
University of Nigeria, Nsukka, Nigeria

University of Abuja, Abuja, Nigeria
e-mail: adikwum@yahoo.com

M.O. Agunbiade
Federal Ministry of Education, Abuja, Nigeria

J.M.U. Abah
NPS, Science and Technology Education, Post-Basic, Abuja, Nigeria

B. Akpan (ed.), *Science Education: A Global Perspective*,
DOI 10.1007/978-3-319-32351-0_8

Indeed, funds from donor nations can catapult poor nations to higher level of development which they crave for. A good example is the Science and Technology Education Post-Basic (STEP-B) Project, in Nigeria. This project improved the educational sector in Nigeria raising both the intake and quality of students studying science courses at all levels of the education ladder in Nigeria. The results from this project are presented in this chapter as well as the focus of future projects in the education sector.

8.2 Background of STEP-B

The Federal Government of Nigeria, in pursuance of its desire to diversify sources of revenue, create more jobs and competed more favourably at the international market and had invited the World Bank group in 2005 to conduct a study on the competitiveness of the Nigerian economy and the roles and contributions of science and technology in Nigeria's development, particularly with respect to the non-oil sector (The World Bank 2007a, b, c).

This study established several disconnects notably between graduate output from the post-basic educational institutions and the needs of the Nigerian private sector. In the same vein, the research outputs from the educational and research subsectors were also adjudged to be largely nonresponsive to the needs of the Nigerian production sector (The World Bank 2007a, b, c).

The above conclusions set the pace for the negotiation of a USD180 million World Bank credit facility to implement a Science and Technology Education Post-Basic (STEP-B) Project. The STEP-B Project essentially aims at improving the quality, equitable access to and relevance of science and technology education at the federal post-basic level so as to increase the country's competitiveness in a globalising world and to promote improved responsiveness of the post-basic institutions to the needs of the labour market (The World Bank 2007a, b, c).

8.3 The Project Development Objective (PDO)

The project development objective (PDO) of the STEP-B Project is to support all institutions under the Nigerian federal post-basic education and research subsectors to produce more and better qualified science and technology graduates and higher quality and more relevant research.

The project is designed to catalyse innovation and to encourage partnership between the eligible institutions on the one hand and the Nigerian private sector, professional associations and international academic and donor partners, on the other (The World Bank 2007a, b, c).

8.4 Project Structure

The project was formally approved by the World Bank on May 22, 2007, and was declared effective on February 11, 2008. The project, which initially had a life span of 4 years, was extended for 18 months in order to allow time for the implementation of some critical activities. The project life spanned from October 1, 2007, to June 30, 2013, extended from the earlier closing date of September 30, 2011, based on the following considerations:

- The project design involved competitive funding and wide-ranging systemic reforms, and it is too ambitious to be completed in the originally planned period of 4 years.
- The project required extensive capacity building for a range of activities at institutions at different levels (from schools to research laboratories) spread throughout Nigeria.
- The take off date was delayed by 6 months due to National Elections in 2007.
- Component 2 – Centres of Excellence needed identification of national priorities in science and technology (S&T) followed by revision in-grant proposal guidelines before grants could be disbursed.
- Most activities actually commenced during late 2009 – leaving less than 24 months for completion.

At least 3 more years were required for the impact of the project (in terms of better quality graduates and research output) to be noticeable. Some time was also required for the post-basic educational institutions (PBEIs) and the Federal Government of Nigeria (FGN) to develop strategies to ensure sustainability of the expected project gains.

Based on the extensive discussions held with the FGN during the midterm review (February 12–25, 2010) of the project and a request dated August 31, 2009, for reallocation from FGN, a two-level project restructuring was approved to facilitate (The World Bank 2010a, b).

- Streamlining/refinement of development outcome indicators and intermediate outcomes and indicators to better match the project design and improve monitoring
- Enhancement of scope of a subcomponent of the project to facilitate FGN's programme of strengthening science teaching and learning in federal secondary schools so as to get better prepared students to join S&T programmes at tertiary level
- Support for specific studies and information technology (IT) connectivity to institutions for systemic reforms and improved collaboration for research and learning
- Reallocation of the credit between subcomponents/unallocated funds and categories of expenditure for better utilisation
- Extension of closing date from September 30, 2011, to June 30, 2013

Table 8.1 Reallocation of STEP-B funds

Components	Original allocation (US$ m)	Revised allocation (US$ m)
Component 1: competitive fund for quality enhancement and innovation	81.0	81.0
1A Window A: institutional grants	32.0	40.0
1A Window B: partnership grants	18.0	18.0
1A Window C: innovator grants	4.0	4.0
1B: grants for improving S&T teaching and learning	27.0	19.0
Component 2: support for emergence of centres of excellence	54.0	54.0
Component 3: strengthening strategic planning, management and M&E in post- basic S&T education	27.0	36.0
3A: subsector-wide improvements of quality, relevance and access	18.0	25.0
3B: support for project implementation, management and M&E	9.0	11.0
Unallocated	18.0	9.0
Total	180.0	180.0

Adapted from NPS Report 2012

The STEP-B Project was initially structured to be disbursed along three main components (The World Bank 2007a, b, c):

Component 1: Competitive fund for quality enhancement and innovation. Under this window, a total sum of USD81 million was meant to be disbursed.

Component 2: Support for emerging centres of excellence (CoEx), where the sum of USD54 million was meant to be disbursed.

Component 3: Support for strengthening strategic planning, management and monitoring and evaluation (M&E) in the post-basic educational institutions was allocated the sum of USD27 million.

The restructuring also affected the funds allocation as indicated in the Table 8.1.

8.5 Brief Description of Tasks Under the Project

The national project secretariat (NPS) was in charge of the overall coordination of the STEP-B Project. In carrying out its responsibilities under the project, the NPS liaises with the World Bank and all the stakeholders regularly, chief among which are the federal ministries of education (FME) and federal ministries of science and technology (FMST), National Universities Commission (NUC), National Board for Technical Education (NBTE), National Commission for Colleges of Education (NCCE), National Project Steering Committee (NPSC) and Tertiary Education Trust Fund (TETF) among others (The World Bank 2007a, b, c).

8.6 Institutional Arrangements

The project had multiple management structures to ensure institutional ownership of grant administration and subsequently project sustainability. The project continued to enjoy the benefits of a participatory-driven process. All the regulatory agencies were actively involved in the implementation processes. This helped tremendously in reaching the beneficiary PBEIs (NPS Report 2011).

The National Project Steering Committee (NPSC) met regularly during the period, and this helped in no small measure to facilitate implementation processes (NPS Report 2012).

The national project secretariat carried out the day-to-day management of the operations of the STEP-B Project through the six operational units, namely, technical, procurement, financial management, internal audit, communications unit and monitoring and evaluation.

The World Bank also provided necessary support to move the project forward. Project supervision missions were jointly conducted by the FGN and the World Bank twice a year.

8.7 Key Focus of the STEP-B Project

With a focus on post-basic S&T education (senior secondary and tertiary education), the project aimed at catalysing an increased production of more and better quality medium- and higher-level skilled workers in Nigeria and strengthening the capacity of federal post-basic S&T institutions to carry out this task. The map (Fig. 8.1) gives a fair idea of how funds were disbursed during the project's life span.

Through a merit-based competitive approach, the STEP-B Project piloted a new, demand-driven approach to financing S&T education and training which greatly improved teaching and research within institutions and also enhanced effectiveness in resource utilisation. By providing support towards the emergence of world class centres of excellence in select disciplines of priority to Nigeria's economy, the project directly contributed to increasing Nigeria's R&D output and to strengthening the link between Nigeria's post-basic institutions and the country's growth and competitiveness agenda (NPS Report 2013). On the whole, 11 centres of excellence were created. At the end of the project, 60 research papers were published at these centres (The World Bank 2013). There were 132,001 science and technology graduates at the first degree, masters and PhD levels at the end of the project (World Bank 2013). This represented a 32 % increase over the baseline. The number of equipped and refurbished laboratories and workshops in all beneficiary institutions were 158 (The World Bank 2013). The project design also encouraged increased engagement between public post-basic S&T institutions and the private sector with a view to strengthening the relevance of teaching and learning within institutions to the needs of the Nigerian labour market.

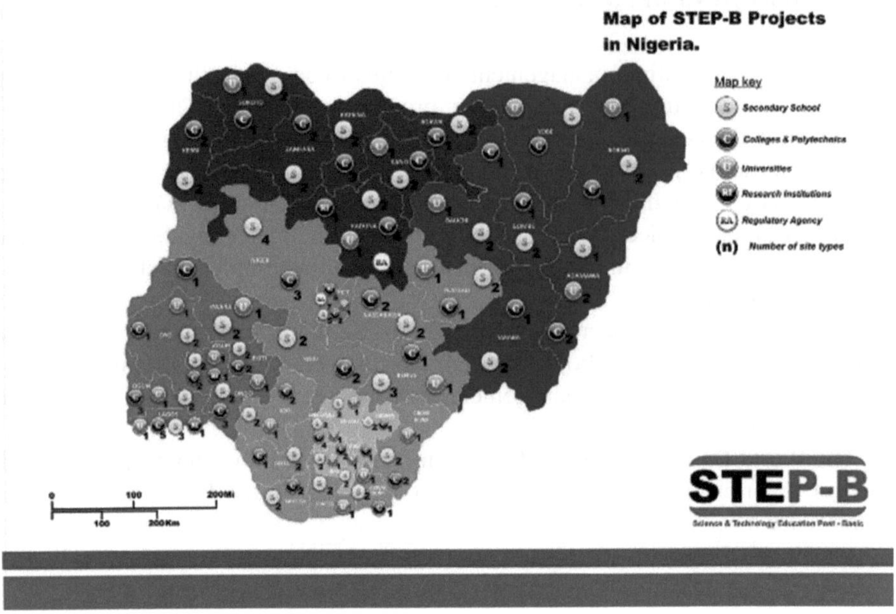

Fig. 8.1 Map of Nigeria showing STEP-B beneficiary sites (Adapted from NPS Report 2013)

8.8 Innovative Aspects of the STEP-B Project

The STEP-B Project is one of very few competitive demand-driven funding agencies in Nigeria. The awards were driven by the needs of the beneficiary institutions as conceptualised in the individual proposals. This is highly innovative in the Nigerian education system.

Funds Management Funds were disbursed directly to the institutions and World Bank guidelines were employed in the implementation process.

Result-Based Monitoring Process Regular (quarterly) reporting system from the beneficiary institutions was put in place. Also, zonal monitoring and evaluation (M&E) consultants were engaged at regular intervals to validate reports from the PBEIs. This was closely followed by the implementation support and M&E visits to address issues highlighted by the PBEIs and consultants. All disbursements, procurements and other activities were closely tied to verifiable indicators.

Capacity Building The systematic and continuous training of the project teams in all project areas helped the implementation process especially in the area of procurement and finance management (NPS Report 2013).

8.9 Distinct Aspects of STEP-B

- Award of grants to 11 institutions to emerge as centres of excellence. The concept of centres of excellence became a common educational parlance under the STEP-B Project.
- Increase in enrolment in S&T subjects at all levels: 33 % at school level, 17 % in colleges, 25 % in polytechnics and 19 % in universities (from 2007/2008 to 2010/2011); the percentage of female students is increasing from baseline of 43,232 to actual 57,006, which is a percentage increase of 32 (The World Bank 2013).
- The number of senior secondary students that obtained a minimum of five credits in S&T subjects at national examinations increased by 53 % (The World Bank 2013).
- Reporting of increased interest in research of direct relevance to Nigeria; a sharp rise in postgraduate students and research scholars in S&T.
- Increasing cooperation and partnership among groups of Nigerian institutions; collaboration with foreign institutions.
- Increasing access and exposure of faculty and students to current knowledge through visits, participation in international conferences and training and through the Internet.
- Commencement of education and research in interdisciplinary areas of critical need to Nigeria.
- Signs of leadership at national and institutional level to handle complex, multidimensional projects.
- Some institutions have achieved remarkable developments; for example, the University of Abuja is commercialising its honey produced under the STEP-B apiculture project.
- Greater transparency in public funding and procurement.

8.10 Progress Made by Components and Subsectors

8.10.1 Components 1 and 3

The STEP-B Project was largely demand driven, based on agreed criteria as contained in the project documents. All proposals to be funded under the project must meet the eligibility criteria and score a minimum of 70 % in the assessment of the Technical Review Committee (TRC). When the project started, there were three rounds of calls for proposals covering all the components (Blueprint 2008).

Although the STEP-B Project became effective in February 2008, proposals received from many eligible institutions dated back to 2006. Round 1 of the call for proposals was prolonged because of the quality of proposals received which were adjudged to be poor by the TRC. This necessitated the convening of a workshop in

April 2008 to train proponents on the art of writing award-winning proposals. International experts were invited to interact with proponents. At the end of the exercise, over 100 proposals from 60 institutions were received. These proposals were subjected to TRC review, and at the end of the review exercise, a total of 66 proposals from 56 institutions were recommended for the NPSC's approval, while 14 were referred back to the proponents for reworking. The approval was granted and funds disbursed in October, 2008. Furthermore, the 14 proposals referred back to the proponents were later approved for awards based on improvement in the proposals submitted earlier.

The Round 2 of the call for proposals opened on May 1 and officially closed on June 30, 2009. At the close of the call, a total of 265 proposals were received. These were subjected to a technical review. The TRC recommended 167 proposals, while 98 proposals found inadequate in one way or the other were referred back to the proponents with a list highlighting areas of shortcomings to be addressed before re-submission. The recommended proposals were presented to the NPSC at its meeting. The meeting approved all recommended proposals. After the mandatory no objection from the World Bank, funds were disbursed to 120 successful PBEIs on the submission of approved work and procurement plans. The third round saw a few awards of just seven institutions.

8.10.2 Innovators of Tomorrow

The Innovators of Tomorrow (IOT) Award falls under Window C, subcomponent 1A of the STEP-B Project. It aimed specifically to encourage Nigerian students to aspire to be innovators. Beneficiaries were drawn from students in the final year of study at the National Diploma (ND), Higher National Diploma (HND), National Certificate in Education (NCE), bachelors, masters or doctoral programmes from any Nigerian federal tertiary education institution with innovative projects.

During the project's life, the national project secretariat received over 1,000 applications. For the Round 1, 51 were adjudged innovative. These 51 awardees comprising of 28 males and 23 female students from 17 institutions were funded.

Another set of 615 applications were received in 2009/2010. These were subjected to internal due process in the NPS and later reviewed by the Technical Review Committee. Five hundred and thirty-six applications comprising 380 males and 156 females from 31 institutions were awarded in 2010.

On the whole, a remarkable progress was recorded in the implementation of the IOT (Innovators of Tomorrow) during the project's life span. The awardees were invited to showcase their projects and that resulted in a wide range of innovative presentations. Efforts were also made to market those that had distinct value-addedness in terms of innovation to the Nigerian private sector (NPS Report 2012). The NPS also collaborated with National Office for Technology Acquisition and Promotion (NOTAP) and Raw Materials Research and Development Council

(RMRDC) to actualise the smooth transition of research products to the industry. This could not be brought to fruitful level during the life span of the project due to shortage of time. An overview of the performance of the institutions showed that the grants made remarkable impact on the performance of the students in these institutions.

8.10.3 Component 2: Support for the Emergence of Centres of Excellence

Support for the emergence of centres of excellence comes under component 2 of the STEP-B grant. It was another area where tremendous effort was made (NPS Report 2012). Consideration of proposals under this component did not commence in time due to the absence of the strategy document for science and technology education. The initial key activity undertaken was, therefore, the development and adoption of the national strategy for science and technology education at the post-basic level. This document (NPS 2008), among other things highlighted 11 priority areas for science and technology development. These areas were:

- Biotechnology and genetic engineering
- Infectious and zoonotic diseases
- Food security studies
- Renewable energy
- Environmental protection and preservation
- Solid minerals research and development
- Advanced materials science and manufacturing
- Software engineering
- Chemical technology research and development
- Multimedia technology and cinematography
- Science, technology engineering and mathematics (STEM) education

Subsequently, a call for submission of proposals under the component was made in June 2009 with a deadline of July 31, 2009. After the deadline for the submission of proposals, a total of 93 grant proposals were received. A detailed screening exercise was carried out by the national project secretariat (NPS) out of which 43 proposals were adjudged to have met extant eligibility criteria. These 43 proposals were presented to the International Advisory Board (IAB) for detailed technical evaluation. The methodology adopted included oral defence of proposals by the proponents. At the end of the exercise, 16 proposals were shortlisted for presentation to the TRC. These 16 proposals were further reviewed by the TRC. At the end of the review, 11 centres were approved by the NPSC and got the required no objection from the World Bank. Additional support in finance management and procurement were provided by the NPS in order to fast track the implementation process.

8.11 NPSC Meetings

The NPSC was the highest decision making organ of the STEP-B Project. The committee which was chaired by the Honourable Minister of Education met periodically to provide policy guidance during implementation and facilitate coordination between the different federal ministries and their parastatals. The committee met ten times during the reporting period.

8.12 Relationship with Stakeholders

The NPS worked consistently to improve on the relationship with all the major stakeholders, namely, the Federal Ministry of Education (FME), Federal Ministry of Science and Technology (FMST), National Universities Commission (NUC), National Board for Technical Education (NBTE) and National Commission for Colleges of Education (NCCE). The NPS also continued to enlist their cooperation and support in the implementation of the STEP-B Project. This effort yielded very positive results as the stakeholders were well mobilised for NPSC meetings and level of awareness of their respective management on the project considerably increased. This aided the efficient realisation of the project development objectives (PDO) (The World Bank 2007a, b, c).

8.13 Capacity Building (Workshops/Conferences)

Capacity building for NPS staff and project officers in the PBEIs was given required attention during the period of the project. The NPS continued intensive capacity building for its staff as well as the project teams of the PBEIs. This included national, regional and international training in all project implementation areas of project management, procurement, finance, M&E and communications for all the 1,500 project team members.

Training was also carried out in the areas of team building, project communications, ICT, procurement, financial management, environmental safeguard, M&E and general project management when there were changes in project team composition due to transfer, retirement, etc.

8.13.1 End of Year 1 Review

The project became effective in February, 2008. After 1 year of operation, a team of international consultants was engaged by the World Bank to review the status of the project. This exercise took place from March 16 to April 3, 2009. Recommendations proffered at the end of the exercise helped to facilitate project implementation (NPS Report 2010).

8.13.2 Midterm Review

Going by the initial project life of 4 years, after 2 years of project operations in February 2010, a midterm review of the project was conducted from February 12 to 25, 2010. The exercise involved a comprehensive review of the project process and implementation at the end of which it was adjudged to be moderately satisfactory. Areas for improvement were highlighted and timelines were set for achievement. These areas were given attention for improvement (NPS Report 2011).

8.13.3 Approved Monitoring Framework

The monitoring framework and key performance indicators are essential documents for assessing the progress of a project and tracking achievements. There was the need to revise the document conceived at the planning stage of the project based on prevailing situations in the benefitting PBEIs. Revisions were made in order to take into account changes suggested by stakeholders based on implementation experience, comments provided by the quality assurance group (QAG) during quality enhancement reviews (QERs) of the project and to define more precisely what exactly would be measured for some indicators. The objective was also to increase reality (measurability of indicators during project lifetime, ease to collect data, etc.) of some of the indicators. The document which served as basis for the baseline survey and monitoring of implementation progress in all the institutions was revised by the NPS and PBEIs. The approved monitoring framework and arrangement for monitoring for the STEP-B Project is given in Table 8.2.

8.13.4 Monitoring of Funds Released/Implementation Support

Implementation support visits were undertaken at various times to some STEP-B beneficiary institutions spread across all the geopolitical zones from December 2008 to the end of the project. The purpose of the visits was to provide support where needed in order to ensure proper implementation of their various institutional projects in accordance with World Bank procedures and in conformity with the project development objective. During the period of implementation, all the institutions constituting the beneficiaries were visited by the NPS, RAs, World Bank and zonal consultants. All the institutions visited established functioning STEP-B teams and set up dedicated offices for STEP-B Project, and the majority of these institutions carried out procurement processes for key project equipment, materials and

Table 8.2 Monitoring framework

PDO	Outcome indicators	Use of outcome information
The project development objective (PDO) is for Nigerian federal post-basic education and research subsector to produce more and better qualified science and technology graduates and higher quality and more relevant S&T research	Percentage increase in the number of S&T students enroled, disaggregated by gender in all the beneficiary PBEIs	Assess progress towards achievement of the project development objective
	Percentage increase in the number of senior secondary school students that obtain a minimum of five credits in S&T subjects including English language and mathematics at National Examinations Council (NECO) in beneficiary unity schools	Provide feedback to the FGN, employers and the public on the performance of federal post-basic S&T institutions in implementing the national education and S&T policies; including progress towards increasing women participation in S&T
	Percentage increase in the number of S&T publications from beneficiary PBEIs in peer reviewed journals	(Inform: Database, Policy and Research)
Intermediate outcome: Component 1 A competitive, demand-driven mechanism for allocation of resources to strengthen S&T research capacity is established; quality of teaching and learning improved; and greater access to S&T programmes at tertiary level, particularly for women achieved	Percentage of fully accredited S&T programmes in beneficiary tertiary institutions	Assess contribution of the component towards overall project objective
	Number of re-equipped/ refurbished laboratories and workshops in all beneficiary institutions	Assess performance and determine need for adjustments to the procedures for the competitive fund
	Percentage increase in the number of post graduate students enroled for S&T programmes in the beneficiary tertiary institutions, disaggregated by gender	Provide feedback to the FGN and STEP-B institutions on the potential for a demand-driven, merit-based way of financing S&T education and research
	Number of students granted Innovators of Tomorrow awards	

(continued)

Table 8.2 (continued)

PDO	Outcome indicators	Use of outcome information
Intermediate outcome: Component 2 A national network of centres of excellence (CoEs) is established; high-quality R&D work in science and technology promoted; international networks and partnership with the private sector established	Number of designated centres of excellence established and assessed to be functional	Provide feedback to institutions and FGN on feasibility of achieving centres of excellence status and on necessary adjustments
	Number of joint (collaborative) projects between the centres of excellence and the private sector	Provide feedback to NPS, NPSC, the FGN and the World Bank on performance of the designated institutions for decision-making at annual reviews and the midterm review
	Number of national, regional and international teaching and research partnerships established and operational involving beneficiary institutions as assessed by	Provide feedback to the FGN on the potential for sustainability of the designated centres
	1. Number of memorandum of understanding signed by heads of collaborating institutions	
	2. Establishment of a pool of funds to support joint research activities	
	3. Number of staff involved in exchange programme between collaborating institutions	
	4. Number of conferences organised/attended in collaboration with the partner institutions	
	5. Number of joint publication or technical reports initiated or produced by collaborating institutions	

(continued)

Table 8.2 (continued)

PDO	Outcome indicators	Use of outcome information
Intermediate outcome: Component 3 national policy, planning inputs and educational services for post-basic S&T education, research and technology developed and strengthened	Annual S&T statistical report produced on regular basis	To strengthen availability of key data for performance monitoring of S&T education and R&D
	National secondary S&T education curricular review completed and implementation on course	To facilitate harmonisation of S&T qualifications and skills for teachers and graduates of S&T post-basic institutions in the country
	National framework and standards for S&T education and training in place	To assess progress towards improved quality of teaching and learning of S&T in federal post-basic institutions
	i. S&T education policy developed	
	ii. National vocational qualification framework developed and in use	
	iii. National performance standards for S&T teacher education completed and in use	
	iv. S&T laboratory standards developed and in use	
	v. National S&T indicators developed	
	vi. Labour market observatory in place	
	vii. Tracer study for graduates in place and regularly reviewed	
	National platform for developing and sharing educational resources among PBEIs established	
	i. Number of e-learning programmes (content development) implemented by beneficiary institutions	
	ii. University system portal developed and in effective use	
	iii. National Research and Educational Network (NREN) developed	
	iv. An integrated national framework for demand driven research funding developed and adopted by stakeholders	

Adapted from the Project Implementation Manual 2010

services. There was also strong support by institutional management in most institutions. Reports of the visits were incorporated into the institutional reports. During the life span of the project, 195 institutions received support with 220 subprojects.

In their bid to expand teaching, learning and research activities in science and technology-related courses, various forms of partnership/collaboration were attained by the PBEIs with several corporate bodies and individuals. This enhanced efficiency and effectiveness in operations of the PBEIs. All the PBEIs were encouraged to forge more partnership arrangements that would be beneficial to their projects even after the project has ended (The NPS 2011).

8.13.5 Enrolment and Graduate Output Pattern by Sector

8.13.5.1 Federal Universities (Figs. 8.2, 8.3, 8.4, 8.5, 8.6 and 8.7)

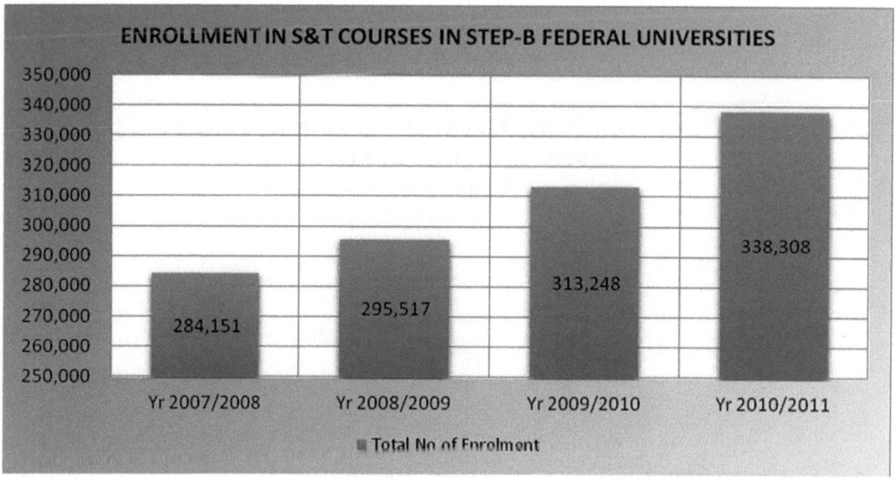

Fig. 8.2 Enrolment in S&T courses in STEP-B federal universities (Adapted from NPS Report 2013)

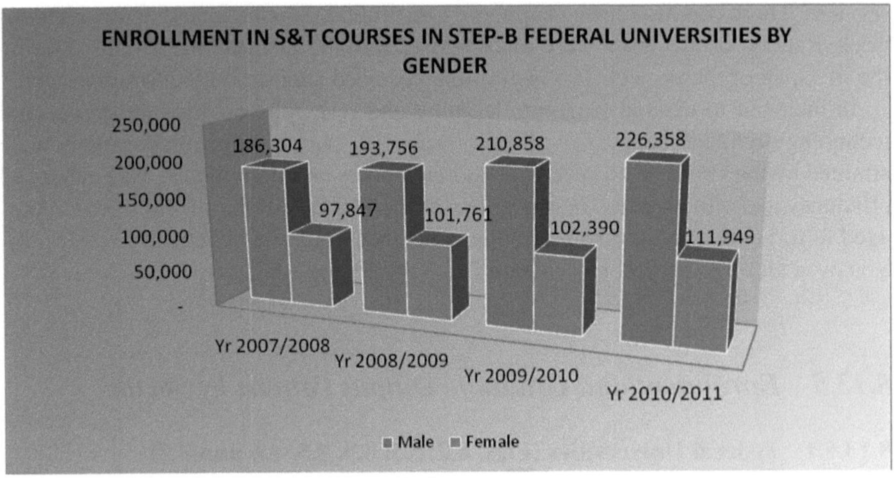

Fig. 8.3 Enrolment by gender (Adapted from NPS Report 2013)

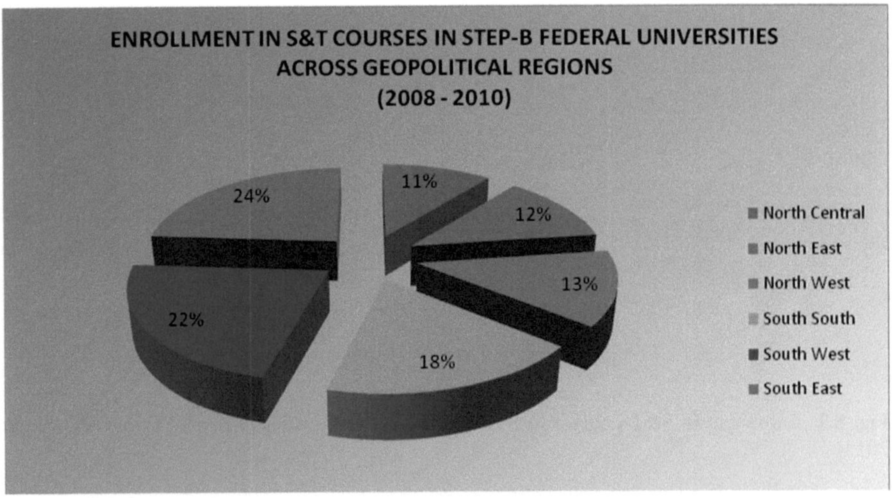

Fig. 8.4 Enrolment by regions (Adapted from NPS Report 2013)

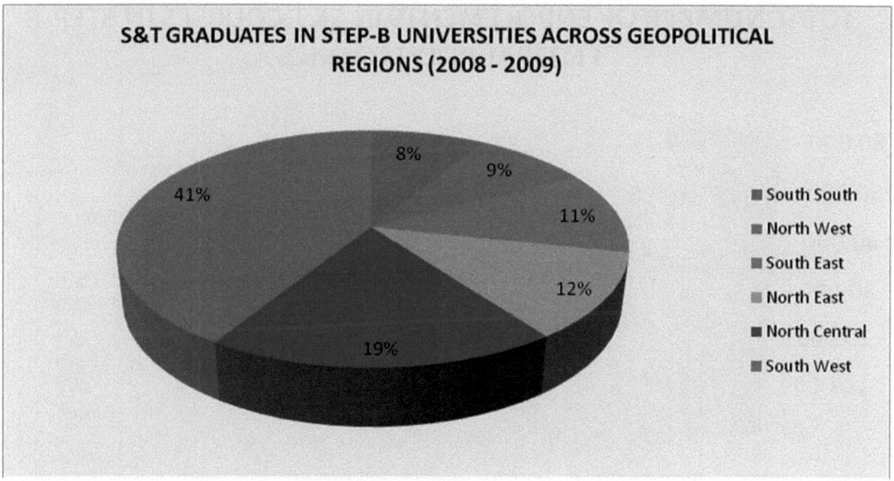

Fig. 8.5 S&T graduates by regions (Adapted from NPS Report 2013)

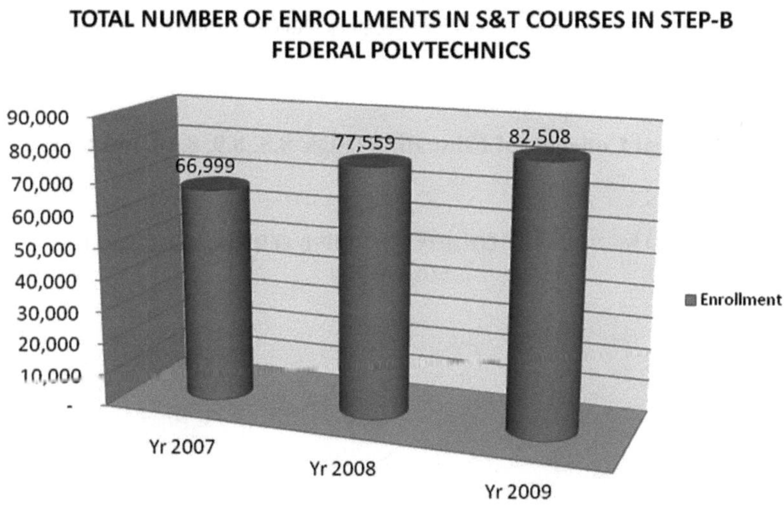

Fig. 8.6 Enrolments in federal polytechnics (Adapted from NPS Report 2013)

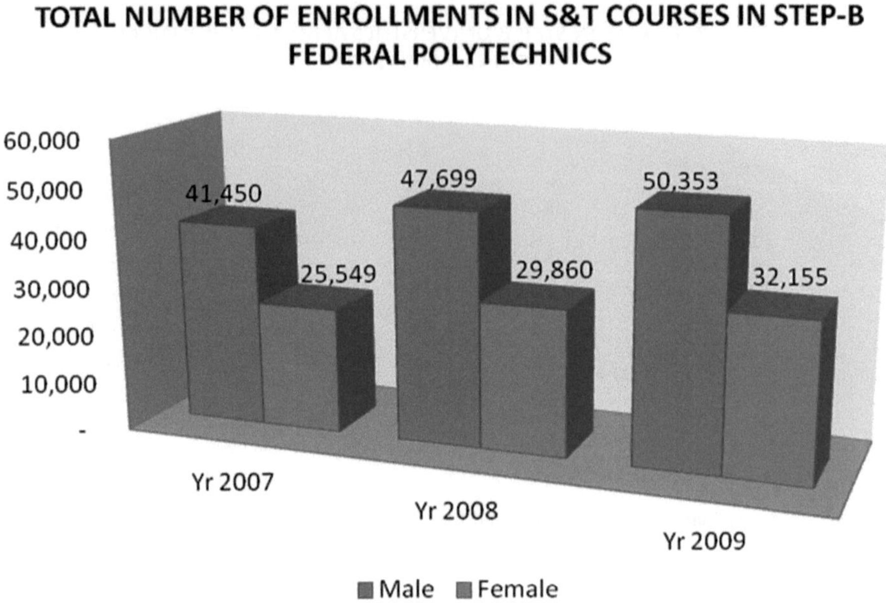

Fig. 8.7 Enrolments by gender in Federal Polytechnics (Adapted from NPS Report 2013)

8.13.5.2 Federal Colleges Of Education (Figs. 8.8, 8.9, 8.10 and 8.11)

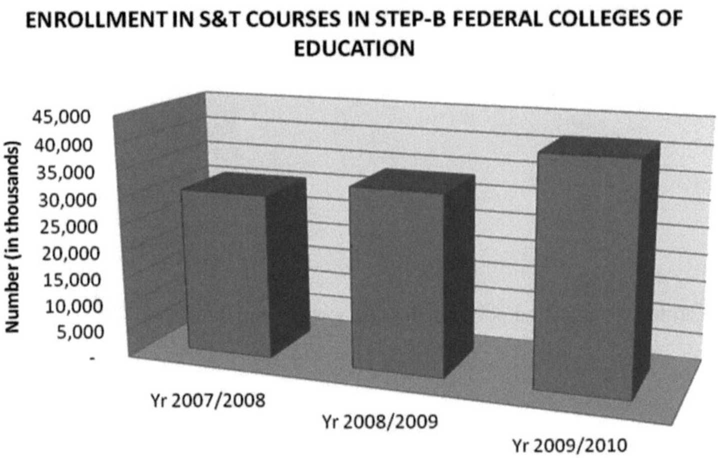

	Yr 2007/2008	Yr 2008/2009	Yr 2009/2010
Total No of Enrollment	30,605	33,415	41,479

Fig. 8.8 Enrolment in federal colleges of education (Adapted from NPS Report 2013)

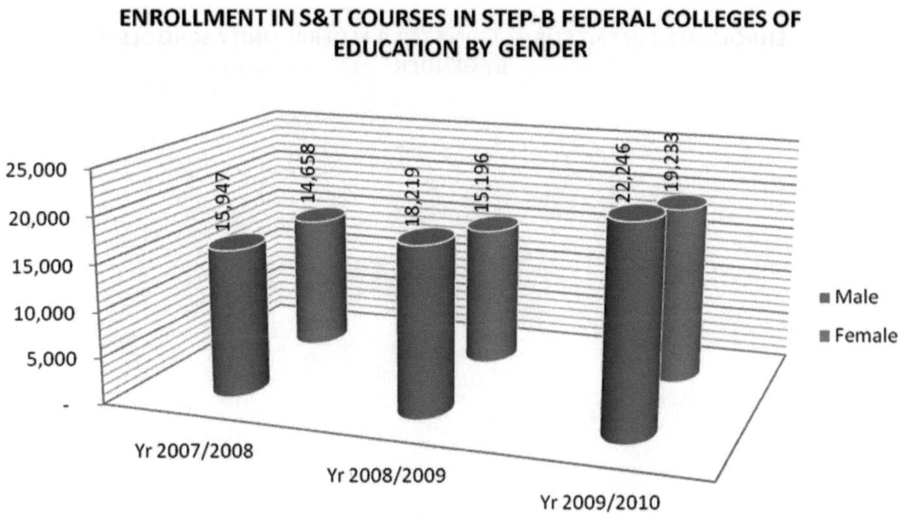

Fig. 8.9 Enrolment in federal colleges of education by gender (Adapted from NPS Report 2013)

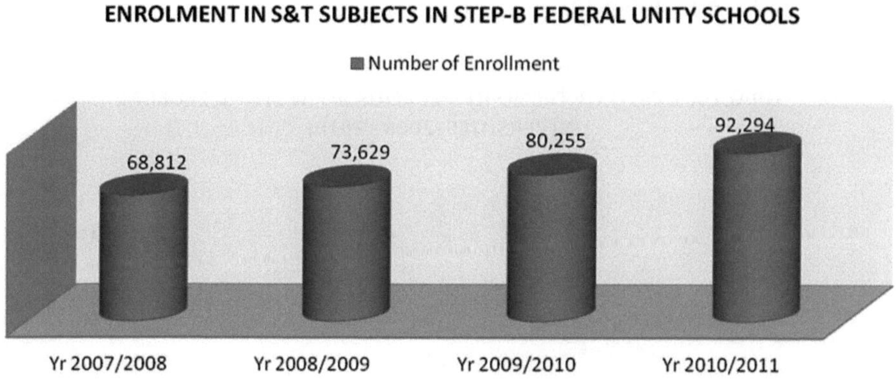

Fig. 8.10 Enrolment in Federal Unity Schools (Adapted from NPS Report 2013)

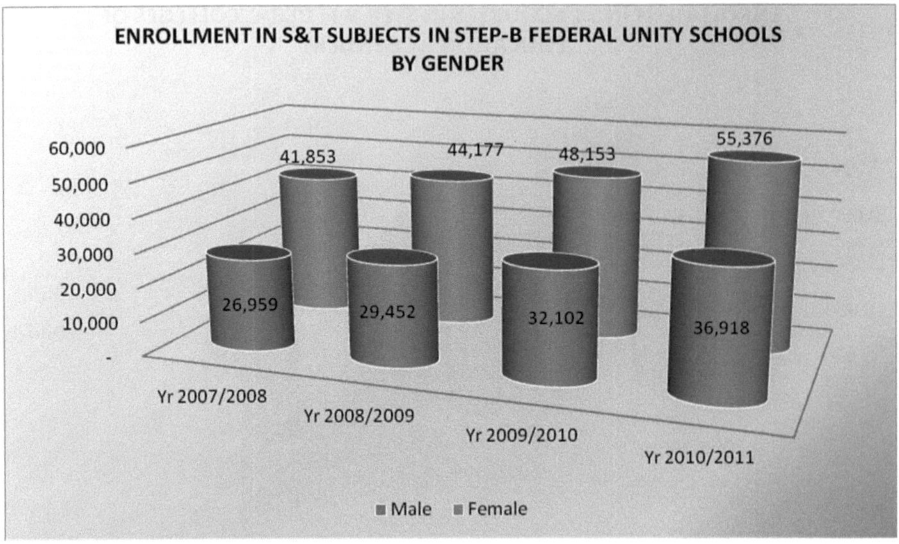

Fig. 8.11 Enrolment in Federal Unity Schools by gender (Adapted from NPS Report 2013)

8.13.5.3 Graduate Output Trend Analysis: Summary Graph Showing Graduate Output for All STEP-B Institution Types (Figs. 8.12, 8.13, 8.14 and 8.15)

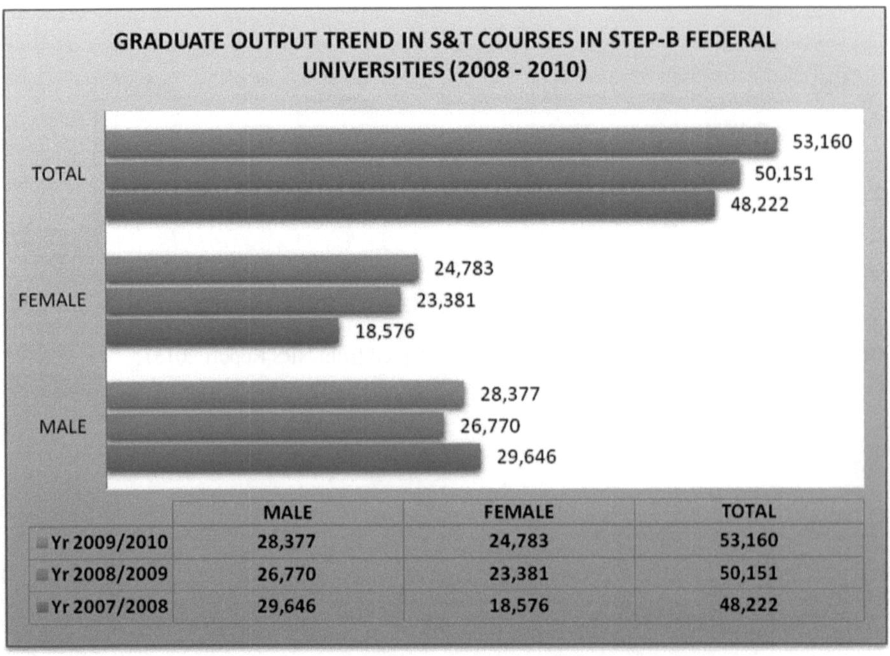

	MALE	FEMALE	TOTAL
Yr 2009/2010	28,377	24,783	53,160
Yr 2008/2009	26,770	23,381	50,151
Yr 2007/2008	29,646	18,576	48,222

Fig. 8.12 Graduate output in federal universities (Adapted from NPS Report 2013)

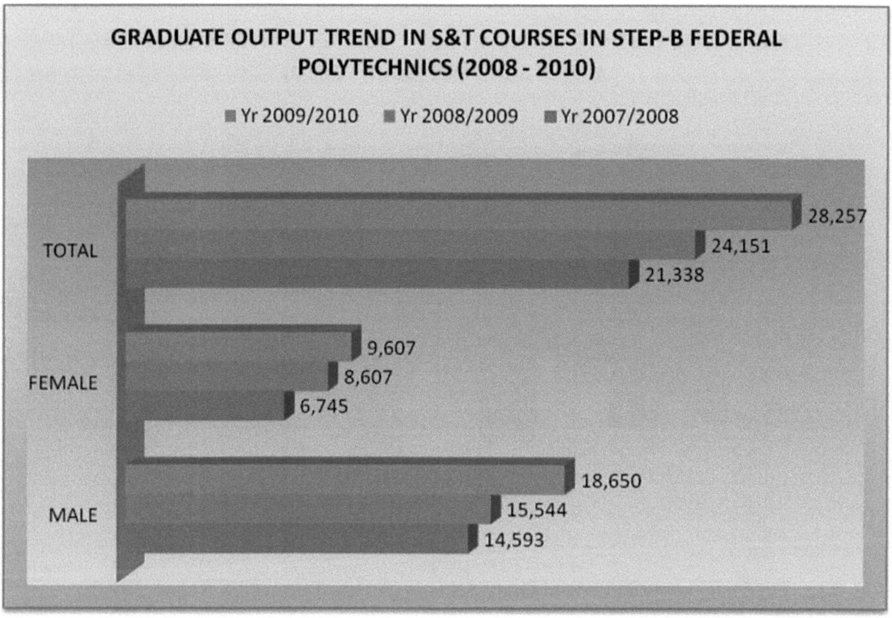

Fig. 8.13 Graduate output in federal polytechnics (Adapted from NPS Report 2013)

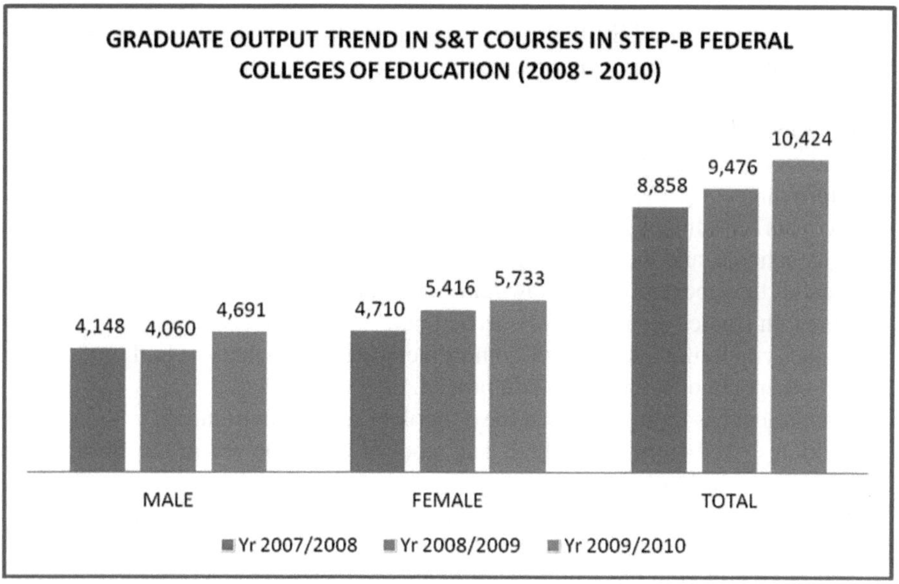

Fig. 8.14 Graduate output in federal colleges of education (Adapted from NPS Report 2013)

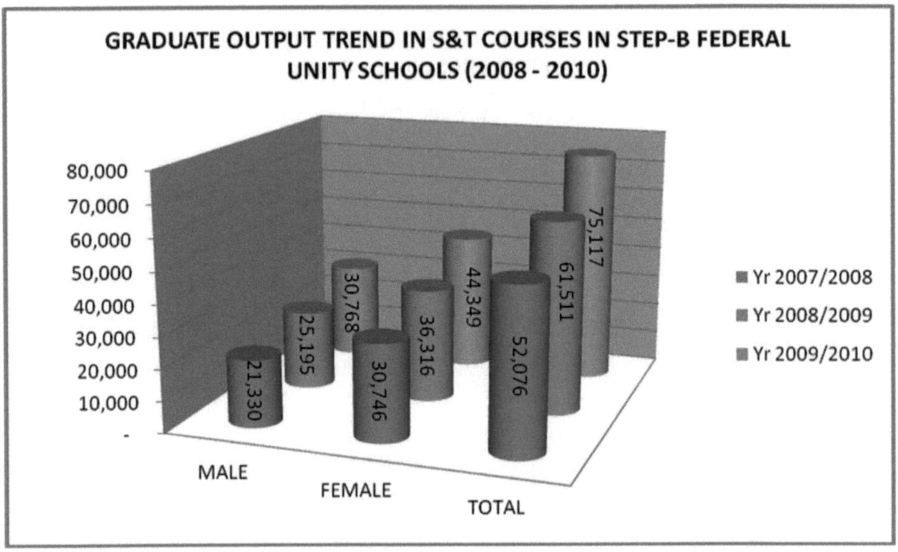

Fig. 8.15 Graduate output in Federal Unity Schools (Adapted from NPS Report 2013)

8.13.6 Key Issues During Project Implementation

- The awareness of STEP-B Project was felt more in unity schools and environment than in higher institutions. Unity schools are secondary schools owned by the Federal Government of Nigeria and were established with the purpose of fostering unity among children of Nigeria.
- Students generally now access Internet facilities, and knowledge about ICT is improving.
- Students' enrolment into science and technology subjects is on the increase in unity schools, and the percentage of female is increasing.
- Most of the laboratories are now renovated and well equipped.
- Research studies are now easily carried out.
- Capacity building and level of commitment among teachers in both the science and other school subjects have improved.
- The project has totally changed the outlook of school environment with inspiring buildings, well-furnished offices and stocked book shelves in the library with science and technology books.
- The goods procured and/or works constructed are generally of high quality.
- It was observed with satisfaction the level of technical assistance provided to PBEIs by NPS and FME in areas of finance and procurement management. It is recommended that other regulatory agencies (RA) should give similar support to their constituent PBEIs at all times whether a project is being implemented or not.

- The ME system in the PBEIs is not fully developed and responsive to project needs, because most PBEIs progress reports were on outputs and expenditure activities rather than outcomes. However, baseline data for students' enrolment and graduation were generated for all the PBEIs. ME should be institutionalised so as to all give account of institutional activities.
- There is weak and ineffective management information system (MIS) in most of the institutions funded; this makes information generation and retrieval very cumbersome. Robust MIS should be developed to capture relevant project and institutional data for future development work.

8.14 Sensitisation/Grant Proposal Writing

Many PBEIs carried out sensitisation activities in their various institutions in order to create awareness in their local environment and ensure institutional support. More sensitisation activities carried out by the NPS to ensure proper implementation and sustainability of the project after the World Bank funding. It is imperative that sustainability is key to success of any project hence there was need to ensure institutional ownership. Arrangements were continuously being made by the NPS in conjunction with the regulatory agencies to organise sensitisation campaigns and grant proposal writing workshops for the PBEIs. This was with the view to enhance awareness and also improve the quality of proposals submitted by institutions subsequently. This exercise resulted in improved awareness of the STEP-B Project and better quality of proposals received.

8.15 Communication Activities

The world has entered an age of knowledge, and the key to accessing and harnessing that knowledge lies in the ability to communicate. When the undertaking is a research project that has achieved good results, it becomes imperative not only to disseminate those results but also to get to the targeted stakeholders; otherwise the work would have been in vain. It was in recognition of this fact that the communication unit became operative at the project secretariat although this was not captured in the project appraisal document (PAD). After the project launch in October 2008, all PBEIs were directed to appoint a communication officer in addition to the four project team members. However, they did not become functional until November 2009 when they were trained and a communication strategy adopted. The unit, with the able support of the World Bank communications expert, thereafter identified ways of communicating effectively. One of the ways to successfully do this is by organising a fair for implementers to showcase their success stories.

8.16 Future Prospect in Project Implementation

With a focus on post-basic S&T education (senior secondary and tertiary education), the project aims at catalysing an increased production of more and better quality medium- and higher-level skilled workers in Nigeria and strengthening the capacity of post-basic S&T institutions to carry out this task.

Through the merit-based competitive approach, the STEP-B Project is piloting a new, demand-driven approach to financing S&T education and training which would greatly improve teaching and research within institutions and also enhance effectiveness in resource utilisation. By providing support towards emergence of world class centres of excellence in select disciplines of priority to Nigeria's economy, the project directly contributed to increasing Nigeria's R&D output and to strengthening the link between Nigeria's post-basic institutions and the country's growth and competitiveness agenda.

The project design also encouraged increased engagement between public post-basic S&T institutions and the private sector with a view to strengthening the relevance of teaching and learning within institutions to the needs of the Nigerian labour market.

8.17 Key Achievements and Results of IDA's Support

Since the date of effectiveness in February 2008, the project has recorded a number of remarkable achievements:

(i) The project was the first of its kind with a lot of capacity to produce many innovations being demand driven and competitive in nature. This design engendered the spirit of competitiveness among the post-basic educational and research institutions.

(ii) The *institutional implementation arrangements* also built capacity in management and project implementation. As a result of this, the Federal Ministry of Education proposed to henceforth allocate resources directly to the Federal Unity Schools for the implementation of their capital projects. This is novel because hitherto, funds meant for these schools were centrally managed by the ministry.

(iii) *Increased use of ICT for academic instructions*: Many of the institutions deployed ICT as a tool for teaching and learning. The University of Jos, University of Lagos and Obafemi Awolowo University, Ile Ife, for instance, have recorded remarkable success in the deployment of ICT tool to learning.

(iv) *Capacity building for project officers*: All the 1,200 project officers attended various trainings on specific project areas. This has resulted in improved performance in project management and implementation.

(v) *Provision of computers and ICT/Internet connectivity*: With the STEP-B intervention, many of the institutions now have well-equipped ICT centres and also with installed broadband radio equipment on a mast that broadcasts Internet signals that covers about 5 km radius. The impacts of this achievement include:

- Ability of teachers to download topics on various subject matters from the Internet and use the same to teach the students, making the work a lot easier and also improving curricular delivery.
- Ownership of laptops/desktops by the staff of the institutions due to the availability of Internet services.
- Easy administrative communications between the colleges and headquarters as well as other places. The STEP-B secretariat reached all the beneficiary institutions by e-mail, and reports are submitted online.
- Members of the host communities living within the transmission range of this network, who also have access to computers, are also benefiting from this intervention.

(vi) *Rehabilitation of S&T laboratories*: The rehabilitation of laboratories in some of the beneficiary institutions started yielding results before the closure of the programme. For example, in some of the Federal Unity Colleges, this has created an enabling environment for effective teaching and learning of the science subjects. Before now, most of the laboratories had dilapidated infrastructure. With the intervention, the furniture has been refurbished and equipment procured for effective teaching and learning. The lesson delivery techniques of the teachers of these subjects have been greatly enhanced/improved. Teachers now use synthetic boards with special markers to teach instead of chalkboards. Moreover, more students, especially girls now offer science subjects which were hitherto considered difficult because of the abstract nature of teaching. Similarly, in the research institutes, major equipment and rehabilitations undertaken have changed the face of research, and the institutes' capacity to take on more scientists has increased.

(vii) *Entrepreneurial skills development:*

- Key among the objectives of the STEP-B Project was the design to address the issue of unemployment and entrepreneurship. Some of the subprojects are focused on this area. Notable is the University of Abuja capacity building in honey bee production for students' entrepreneurship. The project was designed to introduce apiculture enterprise into the practical training programme of the undergraduate students of the faculty of agriculture in the university. Apiculture is the practice of rearing bees for its economic gains. Lately, it involves the use of sophisticated and artificial techniques to keep honey bees for highly desirable products such as honey, wax, pollen, propolis, bee venom and royal jelly. Similarly, the University of Agriculture, Makurdi, has a marked success with honey bee and fish farming.
- Global demand for fish and its products, especially in Nigeria as in other sub-Saharan countries, outstrips its supply. Aquaculture as the fastest

growing fish production system in the world is yet to meet up with the demands. This challenge informed the research into genetically modified-monosex fish production and its potentials for catfish aquaculture in Nigeria by an innovator of tomorrow grantee at the Obafemi Awolowo University, Ile Ife. In the study, monosex larva of African giant catfish (*Clarias gariepinus*, Burchell, 1822) was produced using simple and safe GM techniques. It is expected that its potential applications in enhancing breeding strategies for research and commercial aquaculture would continue to be fully explored.

(viii) *Learning from others*: A number of the post-basic educational institutions had undertaken study tours with the aim of learning from global best practices. The benefits of these activities are currently being harnessed to inform better decision making and productivity. The study tour undertaken by the National Commission for Colleges of Colleges (NCCE) culminated into signing a memorandum of understanding on its behalf and all federal colleges of education with Virginia Polytechnic Institute and State University (Virginia Tech) on June 1, 2009. This has also greatly enriched the content of the teachers' standards being articulated by the commission. University of Agriculture, Abeokuta, and University of Lagos have also signed memoranda of understanding with foreign universities in order to aid collaborations.

(ix) *Partnership and collaboration with the private sector/entrepreneurship development*: The national project secretariat (NPS) established a partnership and collaboration unit in May 2008. This is in line with the cardinal objectives of the STEP-B Project which seeks to mitigate the wide gap between the academia and the private sector. One of the specific objectives of the unit is to strategically position STEP-B in national development programmes aimed at the creation of employment opportunities for the teeming Nigerian youths. In a bid to realise this objective, the STEP-B seeks to promote partnership between the PBEIs and the private sector, international organisations and professional bodies. So far, stakeholders in the private sector and development partners have been sensitised through media programmes and workshops. Stakeholders in the PBEIs and private sector have called for closer linkage and partnership between them for sustainability and effective attainment of the goals of national development. Partnership and collaboration activities have since then begun to be institutionalised both at the NPS and the PBEIs. Under the guidance of the NPS, a number of partnership arrangements were meant to have been fostered with HP Nigeria, Zinox, AfriHub, STEYR Nigeria, the Bauchi Meat Company, Galaxy Backbone and many more. Discussions were also on-going with the Nigeria National Petroleum Corporation, British Council, UNESCO Paris and the Nigerian Television Authority among others in the area of promoting S&T education and research at the post-basic level. The Sheda Science and Technology Complex (SHESTCO) fostered collaborations with the African University of Science and Technology (AUST) that has led to several other collaborations with various local oil industries and Princeton University in the United States of America.

The University of Jos is equally enjoying enhanced partnership support with a grant from the HP Technology for teaching higher education. By this partnership, annotated materials are made available to students. Also, the Federal Science and Technical College, Yaba, Lagos has established working relationships with some manufacturers for enhanced on-the-job training.

(x) Biometric signalling studies undertaken by the University of Ilorin has good prospect to capture:

- Finger-printing to help in checking election and examination malpractice
- Iris and eye pictures for veiled women in Purdah for the purpose of census
- Iris and eye pictures for detection of veiled robbers for crime detection
- Handwriting processes of students to check examination malpractices
- Finger-printing to combat crime
- Toe-printing to be able to include those that have no fingers, such as lepers, in voting, census, etc.

8.18 Directions for Future Projects in Nigeria

There are arguments for and against donor and IDP projects in Nigeria. The major problems revolve around project design that makes the donor agency or country to take control of the project during the implementation phase. This leads to a situation where there are conflicts of interest between the donors and the recipients of such grants. At the end of such projects, the donor may be happy at what has been achieved, but the economies or the particular system in the acceptor country is often not stimulated to any meaningful level so as to bring about the change desired (Adikwu 2007).

For instance, in the case of the STEP-B, there was overemphasis on research without making provision for how the research materials can reach the market. Similarly there was no provision in terms of time for capacity building before project take off. The project was expected to build the capacity of its own staff. There was also no consideration for the fact that most of the equipment to be procured were not made in Nigeria and had to be imported. The project design was such that the equipment could be purchased from the equipment store next to the project office (Adikwu 2010).

The on-going turmoil in the global economic system has prompted many policymakers to revisit the efficiency of their economic development strategies in spurring economic growth, in improving resilience to shocks and in ensuring the well-being of their citizens (The World Bank 2012). While there are clearly no "one-size-fits-all" development models, policymakers continue seeking insights and inspiration from models with a demonstrated track record of success. The lingering poverty in acceptor countries shows that existing models may not be the best.

8.19 Conclusion

The aim of the STEP-B Project is for the Nigerian federal post-basic education and research subsector to produce more and better qualified science and technology graduates and higher quality and more relevant research.

The project introduced a competitive, demand-driven mechanism for allocation of resources to strengthen the teaching of science and technology as well as research. It has helped to build capacity and establish platforms that improve quality of teaching and promotes greater access to science and technology education programmes at the post-basic level of education, particularly for women. The project has helped to groom the establishment of a national network of centres of excellence that collectively promote quality research and development work in science and technology with international and private sector partnerships.

The project has made significant inputs into various national policies in science and technology education ranging from curriculum development to teacher standards, laboratory standards as well as skill standards in the area of national vocational qualifications framework. The provision of equipment, consumables and capacity enhancement resulted in remarkable increase in enrolment figures as well as performance in examinations at all levels of the education strata.

The procurement of laboratory and scientific equipment in schools ignited the interest of students and teachers, and this has enhanced teaching and learning. It has also brought about improvement in performance of students in the sciences.

References

Adikwu, M. U. (2007). *Options for financing the change management system*. Paper presented at the Stakeholders Conference on Engineering and Technology Academic Programmes. Transcorp Hilton, Abuja, November 10.

Adikwu, M. U. (2010). *Setting up of centres of excellence in STI in African institutions*. Paper presented at the Annual Conference of the African Technology Policy Studies (ATPS), Cairo Egypt.

Inyang, H. I. (2009). *Building science, technology and innovation (STI) partnerships*. World Bank's Global Forum on STI Capacity Washington DC, December 10–11.

The National Project Secretariat. (2010). *Report on the STEP-B project to the stakeholders*.

The National Project Secretariat. (2011). *Report on the STEP-B project to the stakeholders*.

The National Project Secretariat. (2012). *Report on the STEP-B Project to the stakeholders*.

The National Project Secretariat. (2013). *Report on the STEP-B Project to the stakeholders*.

The National Project Secretariat (NPS). (2008). The Blueprint for Setting up Centres of Excellence under the STEP-B Project.

The World Bank. (2007a). *Project appraisal document of the STEP-B project*. Washington, DC: World Bank Group.

The World Bank. (2007b). *Synthesis report on the STEP-B project*. Washington, DC: World Bank Group.

The World Bank. (2007c). *The financial manual of the STEP-B project*. Washington, DC: World Bank Group.

The World Bank. (2010a). *The project implementation manual of the STEP-B project*. Abuja: World Bank Group.

The World Bank. (2010b). *The procurement manual of the STEP-B project*. Washington, DC: World Bank Group.

The World Bank. (2012). *World Bank Institute's Frontiers in Development Policy*. High-Level Knowledge Forum on Pushing the Policy Frontier, October 15–16, 2012, Seoul, The Republic of Korea.

The World Bank. (2013). *Implementation completion report*. http://www-wds.worldbank.org/external/default/WDSContentServer/WDSP/IB/2014/01/08/000461832_20140108122830/Rendered/PDF/ICR28020P074130IC0disclosed01060140.pdf

Chapter 9
Harnessing Endogenous Research and Innovation in Nigeria for National Transformation

Paul A. Eniayeju

9.1 Introduction

The phrase 'the World Needs Better Economic BRICs' was coined by O'Neill in 2001 in a paper written for Goldman Sachs. The expression was referring to the four emerging "BRIC" economies: Brazil, Russia, India and China. On December 12, 2005, Nigeria was listed by Goldman Sachs investment bank among the next 11 (N-11) countries after BRICS. The 11 countries are Bangladesh, Egypt, Indonesia, Iran, Mexico, Nigeria, Pakistan, the Philippines, Turkey, South Korea and Vietnam. The N-11 countries were identified as having a high probability of becoming, along with the BRICs, the world's largest economies in the twenty-first century. The bank chose these states, all with promising outlooks for investment and future growth. The criteria that Goldman Sachs used were macroeconomic stability, political maturity, openness of trade and investment policies and the quality of education. The N-11 paper was a follow-up to the bank's 2003 'How Solid Are the BRICs?' (Goldman Sachs research paper No: 134 2005).

Nigeria's economic potential was rightly recognised in 2005 by Goldman Sachs Group, Inc. She is currently the biggest economy in Africa. Given the country's considerable resource endowment and coastal location, there is potential for strong growth. Yet Nigeria has a lot of work to do to realise this potential. One area is about harnessing endogenous research and innovation for the much-needed national transformation. The objectives of this chapter are therefore to focus attention on the:

- Compelling need for a much stronger partnership between the universities, research institutes and industries in Nigeria as a critical success factor in meeting the key elements of the transformation agenda

P.A. Eniayeju (✉)
Nasarawa State University, Keffi, Nigeria
e-mail: eniayeju1@gmail.com

B. Akpan (ed.), *Science Education: A Global Perspective*,
DOI 10.1007/978-3-319-32351-0_9

- Imperatives of harnessing endogenous research and development for rapid national development

9.1.1 National Transformation Agenda

In recent years, Nigeria has been experiencing a growth turnaround, and conditions seem right for launching onto a path of sustained and rapid growth, justifying its ranking among the N-11 countries. The government of Nigeria believes that societal transformation starts as an internal process in which each individual produces actions that through steadiness and diligence eventually translate into a significant change for good in the polity. Wikipedia (2012) quoted the secretary to the federal government as follows:

> *National Transformation is holistic overhaul of segments of the national life i.e., political, economic and social life. It places emphasis on attitudinal change along the path of patriotic zeal and commitment; it involves a resolute determination of government/citizen to systematically upgrade their national development index. It connotes the participation of all in the process that enhances the general well-being of all by improving the economic, political and social environment. Transformation as a process is brought about through intrinsic and/or extrinsic activities by either natural progression or deliberate intervention.*

Transformation of a country connotes rapid incremental changes in the development indices of a country, i.e. comparatively high gross domestic product (GDP) and high human development index (HDI), achieved through deliberate intervention policies, appropriately and faithfully implemented. The cardinal elements of the transformation agenda in Nigeria include constitutional and electoral reforms and transformation of the budgeting process, employment generation, poverty alleviation and job creation, developing critical infrastructure, human capital development and value re-orientation. Policies that seek to strengthen Nigeria's education infrastructure, the institutions that carry out research and development are crucial for the achievement of these cardinal elements of transformation (Eniayeju 2012).

Past efforts at transforming the Nigerian economy include the National Economic Empowerment and Development Strategy (NEEDS) and the New Partnership for Africa's Development (NEPAD) programmes. These programmes derive from, or are consistent with, the MDGs focus on, among others, four key issues: employment generation, wealth creation, poverty reduction and value re-orientation. One of the aims of NEEDS is to forge a stronger link between educational institutions and industries so as to stimulate rapid industrial growth and explore the use of Nigeria's abundant resources for development. But in the Nigerian private sector, many firms still depend on RD results carried out outside the country which bear little or no relevance to Nigeria's natural resource endowments and environmental characteristics (Ndiokho 1999). Undoubtedly, the incremental changes in the development indices of a country are science and technology (S&T) driven. The developed countries that have been able to harness science itself and direct it to their ends have gained considerable advantage. For them, according to Emovon (1999), the competitive task of anticipating the future has become easier since they have acquired the means to determine the future. The largest and fastest-growing actor in this sector is China, whose government has declared education and S&T to be the strategic engines of sustainable economic development. China is followed by the Asian Tigers.

9.1.2 Asian Tigers

The major development over the past decade or more has been the rapid emergence of Asian economies outside Japan as increasingly strong players in the world's S&T system, with South Korea and Taiwan being joined by Singapore, Malaysia, Thailand and others. 'Asian Tigers' is a term used in reference to the highly developed economies of Hong Kong, Singapore, South Korea and Taiwan (NSF 2007). These nations and areas were notable for maintaining exceptionally high growth rates (in excess of 7 % a year) and rapid industrialisation between the early 1960s and 1990s. By the twenty-first century, all five have developed into advanced and high-income economies, specialising in areas of competitive advantage. For example, Hong Kong and Singapore are among world-leading international financial centres, whereas South Korea and Taiwan are among world leaders in manufacturing information technology.

During the 1950s, the four Asian Tigers were in a situation of extreme poverty. The determination of not living in misery and the exponential growth of China, India and Japan motivated each of them to reorganise their economic activities. As a result, between 1960 and 1990, they all experienced rapid economic growth, mainly based in the manufacturing sector, which gave them the opportunity to become exporters. Since then, the Tigers have been experiencing high and continuous growth. These countries have adopted a liberal economy, seeking to exploit their comparative advantages of low-cost manufacturing to produce and export goods and services while also dealing with working conditions associated with low wages, long hours of work, minimum holidays and scarce social services (Wikipedia 2012).

Table 9.1 shows the human development indicators (human development index rank, human development value, life expectancy human development value and gross national income per capita) for Nigeria, Asian Tigers and BRICS. The figures in the first column indicate the relative ranking of the countries in the world. Countries that rank very high lie between 1 and 94; those between 95 and 141 have medium ranking, while those recording between 142 and 187, where Nigeria falls, are considered low.

Although the world has experienced omnipresent market- and policy-driven expansion of S&T capabilities, nowhere has this been as rapid and dramatic as in Asia. R&D is key to the rapid economic growth. What lessons can Nigeria learn from the Tigers that had once experienced low-income economy? What stops

Table 9.1 Human development indicators for Nigeria, Asian Tigers and BRICS

Country	HDI rank	Life expectancy (years)	HDI value	GNI per capita (PPP $)
Nigeria	**152**	**52.51**	**0.504**	**5353.38**
Singapore	9	82.32	0.901	72,731.23
South Korea	15	81.54	0.891	30,345.35
Brazil	79	73.94	0.744	14,274.77
China	91	75.33	0.719	11,477.15
India	135	66.41	0.586	5149.81
South Africa	118	56.92	0.658	11,787.91
Malaysia	62	75.02	0.773	21,823.93

Source: UNDP Human Development Index: Trend 2014

developing countries in Africa from using these countries as role models? Research and innovation have vital roles to play in this regard.

9.1.3 Research and Innovation

Research is quest for new knowledge. Basic research refers to study and research on pure and applied science that is intended to increase our scientific knowledge base. It is often purely theoretical with the intent of acquiring new knowledge without a specific application in view. Applied research on the other hand is investigation of the findings of 'pure' or basic research, to determine if they could be used to develop new products or technologies. Applied research is designed to solve *practical problems* of the modern world, rather than to acquire knowledge for knowledge's sake. The goal of the applied scientist is to *improve the human condition.* According to the OECD (1993), R&D refers to 'creative work undertaken on a systematic basis in order to increase the stock of knowledge, including knowledge of humans, culture and society, and the use of this stock of knowledge to devise new applications'. Technology is central to many of the changes now taking place in the manufacturing and service sectors of countries across the world.

Innovation according to Wikipedia is the making of better or more efficient products, processes, services, technologies or ideas that are accepted by markets, governments and society. It refers to the use of a new idea or method or to the notion of doing something different rather than doing the same thing better (Rogers 1995). A vital distinction is normally made between invention and innovation because the two terms are closely linked. Invention is the first occurrence of an idea for a new product or process, while innovation is the first attempt to carry it out into practice (Fagerberg 2004). In many cases, however, there is a considerable time lag between the two. In fact, a lag of numerous decades is not uncommon (Rogers 1995). Such lags according to Fagerberg (2004) reflect the different requirements for working out ideas and implanting them. While invention may be carried out anywhere, for example, in research institutes or universities, innovations occur mostly in firms.

To be able to turn an invention into an innovation, a firm normally needs to combine several different types of knowledge, skills and facilities, market knowledge, a well-functioning distribution system and sufficient financial resources. This implies that the role of the innovator (what the innovation theorist Joseph Schumpeter called the 'entrepreneur') may be quite different from that of the inventor. In fact, history is replete with cases in which the originators of major technological advances failed to reap the fruits of their breakthroughs. Originating major technological invention and advances requires endogenous research.

9.1.3.1 Endogenous Research

An emerging tendency since mid-1990s is endogenous research and development (ER&D). ER&D is defined as development from inside or that is based on local initiative, knowledge, institutions and resources (Gefu 2004). The main objective of ER&D is the attainment of technological advancement, whereas in basic and applied research, the objective is the advancement of scientific knowledge. ER&D therefore emphasises or builds on people's knowledge, local resources and innovativeness for value addition especially to items with comparative advantage (Eniayeju 2012; Rebelo 1991; Rogers 1990). This is particularly relevant for the growth of small and medium enterprises (SMEs), the main engine of economic growth of the Asian Tigers – South Korea, Taiwan, Singapore, Hong Kong, Malaysia and Indonesia. Countries are no longer rated developmentally based on their endowment with material and human resources but rather on the pool of knowledge and their application for deliverables in terms of goods/packages, processes or services. Rapid economic advances and overall human development index (HDI) are now knowledge-based private sector led, S&T driven and mainly government facilitated via appropriate policy instruments (Adeniyi and Aletor 2005).

The phrase 'endogenous growth' embraces a varied body of theoretical and empirical work that emerged in the 1980s. This work distinguishes itself from neoclassical growth by emphasising that economic growth is an endogenous outcome of an economic system, not the results of exogenous forces that impinge from outside (Romer 1994). For this reason, the theoretical work does not involve exogenous technological change to explain why income per capita has increased by an order of magnitude since the Industrial Revolution. The empirical work does not settle for measuring a growth accounting residual that grows at different rates in different countries. It attempts instead to uncover the private and public sector choices that cause the rate of growth of the residual to vary across countries. As in neoclassical growth theory, the focus in endogenous growth is on the behaviour of the economy as a whole. The endogenous growth theory also holds that policy measures can have an impact on the long-run growth rate of an economy. For example, subsidies for research and development or education increase the growth rate in some endogenous growth models by increasing the incentive for innovation.

An endogenous growth theory implication is that policies which embrace openness, competition, change and innovation will promote growth (Romer 1994).

Conversely, policies which have the effect of restricting or slowing change by protecting or favouring particular industries or firms are likely over time to slow growth to the disadvantage of the community. Howitt (2007) has also written:

> *Sustained economic growth is everywhere and always a process of continual transformation. The sort of economic progress that has been enjoyed by the richest nations since the Industrial Revolution would not have been possible if people had not undergone wrench changes. Economies that cease to transform themselves are destined to fall off the path of economic growth. The countries that most deserve the title of "developing" are not the poorest countries of the world, but the richest.* [They] *need to engage in the never-ending process of economic development if they are to enjoy continued prosperity* (p.40)

Many developing countries including Nigeria, however, remain primarily involved with incremental technology-generating efforts, concentrating on the adoption, use, assimilation and modification of imported technologies. Shifting from this state to becoming a creator of new technologies demands a deliberate investment in technological learning, that is, in accumulating the capability to introduce increasingly complex changes in technologies. This is important because even to assimilate imported technologies effectively, firms in developing countries need to invest in ER&D and other local technology-generating efforts. And to encourage them to do so, governments must design a variety of financial and nonfinancial instruments.

The universities, polytechnics and research institutes have been and remain the centres of knowledge generation worldwide. When knowledge, jobs and wealth are created, poverty is mitigated and global competitiveness and overall development are enhanced. While the more advanced countries of the West and Southeast Asia aggressively engage the synergies of their universities, R&D institutes and industries in knowledge generation, skills acquisition and intellectual property pursuits, most African countries are yet to fully embrace these imperatives as the major driving force of global national development and wealth creation.

9.1.4 The R&D Journey in Nigeria So Far

Technology is central to many of the changes now taking place in the manufacturing and service sectors of countries across the world. In Nigeria, however, much creation of new knowledge is still confined to government research institutes and universities, and these tend to be isolated from the production system. As a result, businesses remain peripheral to the production of knowledge, limiting both their competitiveness and economic development:

- Over the years, several Nigerian universities and R&D institutions in the North and South have made modest strides in development of a variety of S&T products and processes. Most of these products are yet to be patented or put into meaningful commercial usage due to lack of patronage or linkage of the institutions with industry and commerce.

- Poor funding has limited the development of S&T, ICT, biotechnology, nuclear and space research.
- The reluctance of the private industries and commerce to partner in the development of these technologies has denied the Nigerian universities and research institutes of the much-needed human and financial resources. Such partnerships are known to have aided the rapid development of the Tigers.
- One of the aims of NEEDS is to forge a stronger link between educational institutions and industries so as to stimulate rapid industrial growth and explore the use of Nigeria's abundant resources for development, but in the Nigerian private sector, many firms still depend on R&D results from abroad, which bear little or no relevance to our natural resource endowments and environmental characteristics (Ndiokho 1999). The consequences of these are twofold: (1) lack of significant absorption of local R&D/ER&D results into their operations with negative effects on value addition to our natural endowments including human capacity and (2) over-dependence of most of our indigenous firms on R&D conducted overseas.

9.1.5 Arguments by the Indigenous Firms Against ER&D

Foreign-based firms in the country have not been eager to engage in ER&D on the following grounds:

- More ready availability of "imported" R&D results than local ones, i.e. ER&D
- Higher cost and long gestation periods of R&D results and returns thereof
- The absence of government support for companies engaged in local R&D/ER&D
- The absence of suitable research facilities in our universities and other R&D institutions
- Fear of losing to competitors
- Unstable socioeconomic policies of government

9.1.6 High-Technology Manufacturing Exports

Science and technology are changing the world in profound ways. High-technology exports are products with high R&D intensity, such as in aerospace, computers, pharmaceuticals, scientific instruments and electrical machinery. S&T help to support competitiveness in international trade and to generate revenues needed for further investments in innovation. High-technology manufacturing exports provide an indication of a country's ability to produce high-technology goods that can compete in the international marketplace. The OECD defines five industrial sectors as high-technology: aerospace, pharmaceuticals, computers and office machinery, communication equipment and scientific (medical, precision and optical) instruments.

Service sector industries account for about two-thirds of world economic activity, and high-technology or knowledge-based service industries are a large and growing part of the sector. OECD identifies five broad service industries that report high R&D intensities: communication, finance and insurance, business, education and health services. High-technology industry value-added activity has also grown rapidly in Taiwan, Singapore and Malaysia since the late 1990s.

According to Aletor (2006), low participation ER&D has the following implications:

(i) Universities and research institutes are denied the much-needed funds for R&D.
(ii) Low value addition to raw materials such that 30 % of our manufacturing industries have shut down; 60 % are ailing and only about 10 % operating at installed capacity!
(iii) The private sector contributes less than 0.5 % of local R&D budget!

The question then is, 'can Nigeria's high-technology export be boosted?' The answer is yes, if the South Commission prescriptions are followed. The initiative of the South Commission had offered a way out of the situation. This initiative was taken by an Asian think-tank and nonaligned movement by the prime minister of Malaysia and had Dr. Julius Nyerere of Tanzania as the commission's chairman. The commission met for 3 years (1987–1990) and produced the third world's strategies for economic development. The main thrust of their immutable recommendations is shown in Table 9.2. You may wish to score Nigeria in the second column of the table against the parameters listed.

9.1.7 Patterns of R&D Practice in Advanced Countries

Before proposing the way forward in form of recommendations, it is instructive to take a glimpse at obvious R&D practice in advanced countries. The European Union's (EU) innovation leaders are Sweden, Denmark, Germany and Finland. The four countries tend to have the following ingredients for innovation:

(i) Above average R&D expenditure, especially in the business sector
(ii) Higher investment in skills and finance
(iii) Strong national research and innovation systems with a key role for partnerships between public and private sectors
(iv) Better results in turning technological knowledge into products and services

Almost all EU countries have become better at fostering innovation, according to an annual research and innovation scoreboard. The scoreboard makes clear that the EU will have to increase efforts to stimulate and speed up innovation if it is to boost – let alone maintain – its competitiveness. The scoreboard compares countries

Table 9.2 South commission primers for sustainable economic development

S/N	Score	Parameter
1		A strong educational infrastructure which emphasises science at all levels, adult literacy and universal primary education
2		Developing domestic capacity in technology through a network of training in modern manufacturing skills
3		The establishment of centres for technology information with access to scientific information
4		Application of S&T efforts to agriculture to ensure food security
5		A vigorous export drive of surplus food product
6		Domestic savings schemes to accumulate investment capital to reduce foreign borrowing with strong emphasis on fiscal and monetary discipline
7		A fight against three identified negative tendencies – lack of accountability, reckless/flamboyant consumption pattern and militarisation
8		A caution that as development proceeds, the states must shift their pattern of production and export from raw materials to manufacturing of products with high and medium local R&D intensity. This is to avoid decline in prices of raw materials in the commodity market and to encourage local industries to add value to their resource endowments via local R&D efforts, i.e. ER&D. It opined that technology transfer and absorption are only possible when the recipient country has developed in the basic and applied sciences
9		An urgent review of incentives and reward systems for S&T practitioners to ensure that at the secondary and tertiary levels, an increasing proportion of students will opt for science, technological and entrepreneurial courses

Source: (South Commission 1990)

based on R&D investment levels – a key performance indicator – along with 23 other factors. Research and innovation – and removing bottlenecks that prevent good ideas from reaching the market – is at the heart of all EU countries. The approach adopted for achieving this includes promoting partnerships between the public and private sectors, facilitating access to funding and skilled workers, reducing red tape and lowering the cost of patenting new ideas.

Attracting international companies to locate R&D activities in their home country is top priority for all European investment and promotion agencies. In particular attracting new R&D facilities is of interest since these investments are seen as key to generate growth and high skilled job creation.

The observable practice of Asian Tiger countries as regards R&D involves:

(i) Training of more graduates to support the development of an increasing knowledge-based economy
(ii) A higher and further education system fashioned towards lifelong learning
(iii) Pursuits of cutting-edge research which target commercial applications, i.e. aggressive R&D/ER&D

(iv) An expanding pool of educated people who can participate creatively in policy discourses and respond effectively to pressing socioeconomic and political challenges (Aletor 2006)

The Asian Tigers have based their transformation on the economic model of export-led growth, employed first by Japan. To achieve this, government policies were directed towards:

- Economic stability which is very important for private investment and attraction of foreign capital inflows.
- Protection of domestic markets in order to make them competitive against the international markets. The first goal was to strengthen the domestic market and then compete better internationally.
- Imposing specifically the types of industries that were needed for development. For example, at certain times in Japan and South Korea, the automotive, shipbuilding and manufacturing of chemicals sectors illustrate this.
- Targeting the most attractive companies and specific industries.
- Creating agencies for the commercialisation of exports.
- Increasing the capacity of the workforce. This established improvements in educational attainment through universal primary education and wider access to secondary education. With regard to university education, they prioritised science and technology. Some countries brought foreign teachers or sent their students to be perfected abroad;
- Openness to ideas and technologies from abroad.
- Providing subsidies to the declining industries.
- Investment of public funds in research and industrial activities and exports.
- The extensive exchange of information between public and private sectors.
- Adoption of low to not a bit of corruption rate, commitment of the states and governments to improve quality of life and national economy and profit. The state's presence was very important in this process, especially to manage the economy into the transition to globalisation, to increase exports and find a strategic direction for insertion into the global economy.

Other observable patterns of innovations include:

(i) Training of more graduates to support the development of an increasing knowledge-based economy
(ii) A higher and further education system fashioned towards lifelong learning
(iii) Pursuits of cutting-edge or top flight research which target commercial applications, i.e. aggressive R&D/ER
(iv) An expanding pool of educated people who can participate creatively in policy and discourses and respond effectively to pressing socioeconomic and political challenges.

9.2 Conclusion and Recommendations

The universities, polytechnics and research institutes worldwide have been and remain the centres where knowledge, jobs and wealth are created, poverty is mitigated and global competitiveness and overall development are enhanced. While the more advanced countries of the West and Southeast Asia aggressively engage the synergies of their universities, R&D institutes and industries in knowledge generation, skills acquisition and intellectual property pursuits, most African countries are yet to fully embrace these imperatives as the major driving force of global national development and wealth creation. Consequently, universities and research institutes are denied the much-needed funds for R&D. As rapid economic advances and overall national development have become knowledge based, private sector led and S&T driven via appropriate R&D, universities, polytechnics and research institute-industry synergy remain a critical determinant of the extent to which Nigeria and indeed the developing countries of Africa will meet the key elements of the MDGs.

The 'Asian Tigers', some of whom were at the same level of technological and economic development with Nigeria about 20–30 years ago, owe their current technological advancement and relative economic prosperity to the productive and synergistic R&D partnerships between their universities, research institutes and the private sector with government providing the enabling environment. Developing countries of Africa need to embrace this! The desire of Nigeria to be technologically and economically self-reliant will remain elusive if our universities, polytechnics and R&D outfits fail to follow global trends or if they are denied the means to do so by governments and the society, including the private sector.

To transform the situation in Nigeria, policy-makers and researchers in both the private and public sectors must be encouraged to commit more resources to generating new technologies through R&D. And to optimise the synergies of university-industry R&D partnership for economic prosperity, wealth creation and overall national development, Nigeria would require a paradigm shift involving a careful integration of the following:

(i) The need to adopt internationally accepted research practice and an evaluation of current research practice.
(ii) A revisit and a faithful adoption of the key elements of the South Commission prescriptions.
(iii) The establishment of a well-funded national R&D foundation or council to harness the university-industry synergies to generate local R&D products and services. This body should be empowered to fund competitive, market-oriented or need-driven research in which the industries, especially the SMEs, universities, polytechnics and research institutes collaborate.
(iv) Active promotion of local content policy for the manufacture of strategic basic equipment, foods, goods and supplies with incentives: subsidies, grants, tax holidays, etc.

(v) Strict enforcement of the patent laws with stiff penalties for intellectual property breaches.
(vi) Re-evaluation of our value and reward systems to retain our best brains as is done in the more advanced countries.

References

Adeniyi, P. O., & Aletor, V. A. (2005, November). Harnessing the synergies of University-Industry Partnership for economic growth and wealth creation: The Nigerian perspective. *Proceedings of the 1st African regional conference of vice chancellors, provosts and deans of engineering and technology*, Accra, Ghana.

Aletor, V. O. (2006, November). *Leveraging endogenous (local) research and development for national development: The Nigerian perspective*. Power point presentation at the first international conference on Research and Innovation Management in West Africa (WARIMA), Ikeja, Lagos

Emovon, E. U. (1999). National research policy for sustainable development. In P. O. Adeniyi (Ed.), *Research capacity for sustainable development in Nigeria*. Lagos: Unilag Consult.

Eniayeju, P. A. (2012, June). *Transformation agenda: What can Nigeria learn from endogenous research and innovation of Asian tigers*? Paper presented at the 1st national conference of the faculty of education, Nasarawa State University, Keffi.

Fagerberg, J. (2004). Innovation: A guide to the literature. *The oxford book of innovation* (pp. 1-26). Oxford: Oxford University Press.

Gefu, J. O. (2004). *Research for development and cooperation*. Abuja: Inaugural Conf. of AvH Foundation Club (Nig. Chapter).

Goldman Sachs. (2005). Global economics paper No: 134. *How solid are the BRICs*.

Howitt, P. W. (2007). *Innovation, competition and growth: A Schumpeterian perspective on Canada's economy* (p. 40). Montreal: C. D. Howe Research Institute.

Ndiokho, B. U. (1999). The State of, and the imperatives of private sector funding of, and investment in research and development capacity building. In P. O. Adeniyi (Ed.), *Research capacity building for sustainable development in Nigeria*. Lagos: Unilag Consult.

NSF. (2007, August). Special report. *Asia's rising science and technology strength: Comparative indicators for Asia, the European Union and the United States*. http://www.nsf.gov/statistics/nsf07319

OECD. (1993). *Frascati manual* (5th ed.). p. 13. New York: OECD Publishing.

Rebelo, S. (1991). Long-run policy analysis and long-run growth. *Journal of Political Economy, 99*(3), 500.

Rogers, E. (1995). *Diffusion of innovation* (4th ed.). New York: The Free Press.

Romer, P. M. (1990). Endogenous technological change. *Journal of Political Economy, 98*, S71–S102.

Romer, P. M. (1994). The origins of endogenous growth". *The Journal of Economic Perspectives, 8*(1), 3–22.

Wikipedia. (2012). http://en.wikipedia.org/wiki/Nigeria—Vision 202020=cite note-1. Retrieved July 12, 2012.

Part III
Curriculum and Pedagogical Considerations in Science Education

Chapter 10
Motivational Science Teaching Using a Context-Based Approach

Jack Holbrook and Miia Rannikmae

10.1 Introduction

In most countries, science subjects are part of the core school curriculum, taught from the earliest grades until preuniversity level. The reason often put forward is because science and technology are an integral part of our lives and that an understanding of science is important in the ever-changing developments. Yet the foundation for teaching science in school, the earliest rationale, was to provide a science background for students wishing to study science subjects at university level and hence the need to provide a conceptual background for further studies (Fensham 2008).

This earlier approach has heavily influenced science curricula, concentrating on the so-called fundamental ideas linked, in some cases, to applications that take place in society. Thus, an understanding of the various biological systems, whether in botany or zoology, has been a curriculum mainstay, as has a study of a range of natural phenomena from forces, energy changes to heat, light, sound and properties of matter such as magnetic, or semiconducting, as well as considerations of motion such as waves or circular acceleration. And for chemistry, curricula have tended to focus on fundamental particles and their interaction and combinations in new structures, plus a systematic look at the properties of elements and compounds. All of this is very much subject content centred and identifying with an academic science, rather than a student relevant approach.

Alas, major changes in the field of science and technology have been and are currently taking place, especially advances through biochemistry, nanotechnology and digital devices such as mobile phones or through pharmaceutical developments, satellite transmission or advances in the technological developments of polymers. It

J. Holbrook (✉) • M. Rannikmae
Centre for Science Education, University of Tartu, Tartu, Estonia
e-mail: jack.holbrook@ut.ee; miia@ut.ee

B. Akpan (ed.), *Science Education: A Global Perspective*,
DOI 10.1007/978-3-319-32351-0_10

is not surprising that science and technology has become much more complex and all too often far removed from the fundamental science studied at school. This has led to a growing dissatisfaction among students, especially teenagers who are shortly to be faced with subject and career choice. With questions of relevance and value of studying such seemingly unrelated science, seen as far removed from society and abstract in nature, is there little wonder that the popularity of school science is low in many areas of the world, such as Europe (Osborne et al. 2003; EC 2007)? Typically, the emphasis is, taking one example, on a study of electricity in school, while electronic devices are far more attractive and desirable within our growing consumer society. And on top of it all, we live in the age of the Internet offering sources of entertainment in today's global society and often referred to as an increasingly knowledge-based society.

The modern world is more and more associated with the importance of skills of accessing rather than remembering and entrepreneurial or creative approaches based on this rather than knowledge acquisition. Education is beginning to be associated with an education enabling students to be capable of meeting challenges, solving problems and making decisions (Partnership for 21st Century Skills 2008). Such attributes are far from a fundamental approach to the teaching of science, often stressed by teachers as 'the backbone' of science teaching.

Inevitably, things need to change, and change is very much behind today's refocusing of science curricula and the need for reinterpretations associated with this by science teachers. But central is the recognition that science subjects cannot be separated from our technological world; there is an essential need to promote creativeness and the ability to meet a challenge utilising processes provided by the use of technology. Additionally, increasingly being recognised as integral to science study is the process of engineering and engineering design. Whereas technological products find their involvement in science study more as tools to allow more sophisticated study of the complex phenomena in our world and as aids in supporting conceptualisations of this complexity, engineering is associated with the design and creating of innovative artefacts and structures.

It is little wonder then that a complex introductory consideration, leading to a subsequent breakdown to levels of scientific conceptualisation, is taking over as a learning approach from the more traditional building blocks of science – the so-called fundamentals. This change is sometimes referred to as the context-based approach, often stemming from a social context. Under such an approach, atomic structure is no longer the accepted place to begin studies in chemistry. Rather approaches are entertained that begin from the familiar materials around us and the phenomena we experience, even if this familiarity differs at the local level and thus leads to contextual study options.

So what then is the goal of science teaching today? The European Commission (Eurydice 2011) very much sees this as the development of competences, while in the USA, standards of attainment of students within a twenty-first-century skill frame are the greater focus (NRC 2010). Other expressions seek the development of capabilities (MCEETYA 2008), but all with the goal of seeing the study of science subjects for the purpose of attaining a certain degree of scientific literacy (SL) or, as identified by some (Holbrook 2008), of STL (scientific and technological literacy).

So why the change? And what is the change and how does it affect the teaching of science subjects?

10.1.1 The Why?

Simply put, the change is in response to a realisation that acquisition of knowledge is not enough and that education is also about the skills, attitudes and values that support societal development. Education is perceived as being more than learning to read and write, even though still recognising the crucial nature of these in our digital, pictorial age. Today, the role of education is towards an expectation that it provides students with an intellectual base and attributes to seek suitable and rewarding employment and a capability to play a meaningful role in the development of society as a responsible citizen (Holbrook and Rannikmae 2007).

10.1.2 What Is the Change?

Education encompasses the total school provision. However, this is subdivided within the school system and almost universally is a breakdown by subject domains (still, it seems, a hangover from past structures), and thus each domain is an education provider. This, therefore, applies to the teaching of science subjects. Actually, science teaching is within the domain of science education and as such must be expected to play a role in developing the goals of education as put forward by the system (Holbrook and Rannikmae 2007). From this perspective, it is clear that, once again, knowledge is not enough, and skills, attitudes and values are also components to be taught. From this stance, a major realisation is that science education is not the same as science, no matter how much teachers would wish it to be and no matter how science education is actually expressed in educational circles, such as schools. Students do not study science; rather they are being educated 'through a context of science' (Holbrook and Rannikmae 2007). This is a key message and allows contrast with the old paradigm of 'science through education'. The older focus was solely on intellectual studies, even if supported by hands-on, student-centred activities. Furthermore, 'science through education' is not perceived as a focus on 'education for all (EFA)', a major development heavily promoted around the world (UNESCO 2012). Actually 'science though education' focuses on an elitist role, favouring a narrow concentration on conceptual developments and seeing content and scientific concepts as the overriding target. It separates school science from society and inevitably leads to the image of scientists as isolated persons in white coats, often male and elderly. Table 10.1 attempts a comparison of the new 'education through science' vision with a vision of 'science through education'.

Table 10.1 A comparison of similarities and differences in philosophical emphases between 'science through education' and the alternative 'education through science'

Science through education	Education through science
Learn fundamental science knowledge, concepts, theories and laws	Learn the science knowledge and concepts important for understanding and handling socio-scientific issues within society
Undertake the processes of science through inquiry learning as part of the development of learning to be a scientist	Undertake investigatory scientific problem-solving to better understand the science background related to socio-scientific issues within society
Gain an appreciation of the nature of science from a scientist's point of view	Gain an appreciation of the nature of science from a societal point of view
Undertake practical work and appreciate the work of scientists	Develop personal skills related to creativity, initiative, safe working, etc.
Develop positive attitudes towards science and scientists	Develop positive attitudes towards science as a major factor in the development of society and scientific endeavours
Acquire communicative skills related to oral, written and symbolic/tabular/graphical formats as part of systematic science learning	Acquire communicative skills related to oral, written and symbolic/tabular/graphical formats to better express scientific ideas in a social context
Undertake decision-making in tackling scientific issues	Undertake socio-scientific decision-making related to issues arising from the society
Apply the uses of science to society and appreciate ethical issues faced by scientists	Develop social values related to becoming a responsible citizen and undertaking science-related careers

Holbrook and Rannikmae (2007)

It is suggested that society development needs more from science education. Science education needs to cater for all students and meet the needs of society, where change is a major focus. Science education needs to make its contribution to the education of students and hence contribute to the development of all educational goals put forward by the society, even if these encompass general attributes such as communication skills, collaborative working and promoting values for the common good as a key attribute for sustainable development (Holbrook 2009).

The 'education through science' movement thus sets out to provide a more balanced portrayal of the educational provision labelled science education. It seeks to provide, within a more relevant contextual frame, the cognitive developments needed for reasoning and critical thinking to flourish the skills needed for problem-solving and interacting with others, as well as the means to formulate decisions all of which provide a foundation for being a responsible citizen or for the world of work. And in the face of students' disinterest in science studies, develop positive attitudes towards scientific and technological developments.

10.1.3 The How?

Inevitably, the change advocated leads to the way science subjects are portrayed and taught in schools. While the curriculum emphasis needs to change, so also does the teaching approach. Students can no longer be expected to be 'spoon-fed' – given the knowledge and thinking by others and asked to assimilate. Clearly this has little purpose unless it can be seen as meaningful, especially with respect to the development of employment skills, not only for science and technology careers, but for careers in general. And in developing responsible citizens, students' self-direction skills of organising their learning, students' self-directing ability to monitor their learning and students' directed self-evaluation of their learning are all very important. And for this, student motivation is essential, not simply derived from the extrinsic motivation provided by the classroom environment set up by the teacher, important though this is. The motivation needs to be provided by the learning itself. For that, students need to perceive the relevance of science education to their lives and to possess some familiarisation with the learning situation. In other words, the motivation component necessitates a more context-based approach coming from society and not from a scientists' perception of the discipline of science. The approach can no longer be abstract science learning leading to applications in society, where such applications are becoming increasingly rare. Rather approaches are needed which initiate learning from contextual phenomena within society and lead, in a motivational manner, to the unknown science required for better appreciation of the familiar, but not understood, context put forward (Holbrook and Rannikmae 2010). The move to a motivational context-based approach is seen as an important paradigm shift. Of course, this has advantages and disadvantages:

Advantage

Motivation for students and hence favourable for student learning
Provides a situation from which to determine students' prior science knowledge

Disadvantage

A systematic approach to scientific content is not possible.
Teachers are unfamiliar with the approach and thus need strong professional development.

10.1.4 Preparing for the Change of Paradigm

10.1.4.1 Reconsidering the Role of the Textbook

The old teaching paradigm, based on the identification of scientific topics, presented in a well-designed sequence, was closely aligned to textbooks. The books provided the knowledge source and put this in a conceptual learning context. Inevitably, in the face of scientific and technological developments, the textbook content and examples became out of date, and replacement books were required. The scientific approach was seen as aligning students with scientists (with a focus on being 'little scientists'). Clearly, while textbooks could be a useful source of references, they were ill-placed to be the driving force for teaching in a context-based situation, where student motivation was paramount. For this, teaching materials, referred to here as teaching modules, became important, whether downloaded from suitable sources such as the Internet or created by the teacher themselves.

10.1.4.2 A Change of Approach

The view of the nature of science education (NSE) being advocated is seen as a challenge to the prevalent content acquisition approach to science education (Holbrook and Rannikmae 2007). It necessitates a change of approach, based on a view that sees enhancing scientific literacy as a step towards developing, as the science teaching contribution, capabilities for life through education (life skills). It sees utilising student motivation (intrinsic motivation) as a very important step, with the approach based on familiar situations that promote relevance and interest. The change of approach recognises the importance of relating to the society and thus supporting the goal within science education as the enhancement of scientific literacy (SL) or perhaps more appropriately (Holbrook and Rannikmae 2009) scientific and technological literacy (STL).

10.1.4.3 Operationalising the STL Approach in the Classroom

There is no suggestion that the various education learning components are taught in isolation or that the following descriptors are unique and clearly reflect only one attribute. The descriptors merely try to point out there are different aspects through which science education teaching materials should be recognised and hence they give some direction for tackling the attributes involved.

An intrinsically motivational approach to science teaching is taken to be based on three key components:

- Familiarity to the student
- Intimate involvement of the student in terms of need, metacognition and action
- Relates to conceptual science

In this approach, the frame of reference is familiarity and relevance to the student and thus indicates a society beginning. Science conceptualisation is not the organiser of the teaching at this stage. The starting point is considered as a relevant socio-scientific aspect in the society (Marks and Eilks 2009). In this case, relevance is seen as being associated with a familiar issue or concern in which the students are likely to be involved or through which the lives of the students are affected in some way. Thus, at a beginning stage, science lessons focus on a socio-scientific relevant issue, concern and situation, engaging students in a desire for self-actualisation

(Maslow's triangle; see Holbrook and Rannikmae 2010). The learning thus begins in the context of the society in which the students function. It is context-based teaching and learning.

10.1.4.4 Contextualised Teaching and Learning

The previous paragraph points to a social circumstance view in forming the contextual beginning to the teaching (Gilbert 2006). But unlike most context-based approaches, the learning focus is in an educational learning context, rather than the learning of science ideas that pertain to the context (Parchmann et al. 2006). The learning is thus involving educational goals in general with science education seen as merely one component. This viewpoint has been described as 'education through science', as opposed to 'science through education' (Holbrook and Rannikmae 2007). And unlike other approaches to context-based science learning, there is no attempt to organise the learning from a conceptual science perspective (Bennett and Lubben 2006) nor to provide a 'ladder' approach to science learning (Schwartz 2006). The learning, stemming from relevant society scenarios, includes the determination of prior science learning while paying much attention to personal attitudinal attributes, deemed part of the overall goals of education. The scientific conceptual learning that is the eventual target is not, in itself, recognised as a key learning focus at this stage (there is no attempt to perceive specific scientific concepts as 'basic'), but setting this up as a need-to-have (to the depth needed and also building from the student's current starting point) for enabling students to put forward scientific attributes for reasoned decision-making within the social context. Nevertheless, the context to be studied is carefully chosen. Not only is it deemed to be of relevance to at least the majority of the students and hence designed to trigger intrinsic motivation, but the conceptual science learning that evolves is seen by the teacher as within the zone of proximal development (Vygotsky 1978) for the students involved.

The teaching approach focuses on:

- How to identify relevance at least for a sufficient number of students within the class.
- Determining students' views on the aspect of relevance in a manner which has educational value.
- Learning more about students' scientific understanding or prior conceptual knowledge.
- Establishing, in the minds of students, the need that by gaining a stronger conceptual science background, they would be better placed to make decisions within society which are needed in relation to an aspect of relevance.
- Identifying the way forward – that is, the approach to the gaining of science knowledge and skills, needed in order to begin to develop a wider view of the nature of science, greater intellectual skills and the competencies for the ultimate discussion, or consensus, on decision-making.

This approach to teaching also means that the sequence is no longer 'science driven' (i.e. the sequencing is not necessarily that seen as logical by scientists). The initial sequence is not necessarily in line with any recognisable concept map, as the teaching is not conceptual. Rather, the teaching progresses from an issue or concern to, eventually (after gaining the needed conceptual science), the interpretation and subsequent action. It involves tackling the science component from a society level of complexity in the context in which it is met. The science learning, identified as an important need so as to be able to further consider the issue or concern, is sandwiched between the initial approach and the subsequent societal interpretation and broken down to the needed level of conceptual complexity for comprehension by the students involved.

10.1.4.5 Beyond Initial Motivation

But providing an intrinsic motivational start is obviously not enough. Cognition and metacognition by students are expected to be important components of learning, if the issue or concern is to be given more meaning. Hence the sandwiched, science conceptual learning by the students can be expected to occupy the majority of learning time in science lessons. However, this follows after the issue or concern has been identified, and students have expressed their desire to become involved in activities that can play a role in building up their science background. In this, students need to be suitably guided by the teacher to operate within their zone of proximal development (Vygotsky 1978) and determine scientific conceptual needs which meet their expectations. Acquiring suitable science is first associated with identifying the scientific component from the society, contextual setting. Then, from this contextualised beginning, the teaching moves the learning into one or more conceptual science components. The teacher does this by moving to a decontextualised mode so that the conceptual science learning takes place in a sequence that enables the students to bridge the gap between their prior knowledge and the learning needed to appreciate the science component, or components in the conceptual issue or concern.

10.1.4.6 Decontextualisation of the Learning

As indicated in the previous section, this phase is driven by 'need to know' the science, which provides a scientific bearing on the social concern/issue. The decontextualised learning focuses on the scientific ideas, solving scientific problems and the seeking and evaluating of relevant scientific information. It builds on the students' prior learning (as determined by the teacher in the previous section) and with appropriate scaffolding (guidance, support and extrinsic motivation) by the teacher to promote the development of intellectual self-actualisation and self-efficacy. This phase, by necessity, is decontextualised from the society and builds from a scientific conceptual perspective (recognising, nevertheless, the importance of generic – goals of education – skills, e.g. cooperation, communication, positive social values). In this phase, the intrinsic motivation of the students is heavily reinforced by extrinsic motivation from the teacher and other attributes recognised by the teacher as adding to the motivational aspect.

In the decontextualised mode, the teaching is no longer context-based learning (CBL). Instead, it moves to a more inquiry-based science education (IBSE) approach. Thus, the teaching, within this phase, is driven by:

- Structuring the scientific learning so that it, ultimately, is able to support the societal decision-making process, related to the initial issue or concern that was taken to be motivational learning for students
- Providing the needed scientific knowledge so as to give a background for the students' subsequent conceptual acquisition process related to the decision-making
- Providing the needed scientific skills (process skills) or additional practice in such skills which provide a platform towards developing a competency in 'scientific problem-solving'
- Focusing on inquiry-based science education (IBSE) as a component of self-actualisation by students

The teaching approach can relate to the appropriate level of IBSE as indicated below. Where intrinsic motivation is established and the need for student involvement strong, an 'open' inquiry mode for learning at the secondary education level is preferred. Nevertheless, the teachers will need to determine whether the students are ready to operate in this mode, having had sufficient prior involvement in inquiry learning at a structured and guided inquiry level (Zion 2007) (Table 10.2).

Once the scientific conceptual learning phase has been conducted to the satisfaction of the teacher, the context can be revisited, and the new-found science learning acquired by the students can be applied to the original issue or concern.

Table 10.2 Extent of teacher involvement in different structures of scientific inquiry

Levels of IBSE	Teacher provides to students (√)		
Level	Inquiry question	Method of investigation	Interpretation and analysis
Structured	√	√	–
Guided	√ (probably)	√ (possibly)	–
Open	–	(Fully student developed project)	–

10.1.4.7 Recontextualised Teaching and Learning

Recontextualisation is put forward as a further important, not to be omitted, phase. This phase allows the newly gained science learning to be consolidated by guiding the students to transfer their learning to a relevant society context and in so doing lead students towards enhancing their scientific literacy in a wider society context. This phase includes revisiting the initial issue or concern and allowing the students to undertake reasoned decision-making within the complexity of the social environment in which the issue or concern was first addressed. However, this time the students are expected to draw on their newly gained science ideas and are taught to transfer these to relate to the context of the society issue or concern. This recontextualised phase can also encompass the need for consensus decision-making in a social environment and also the promotion of the students' presentation skills, both oral and written.

In the recontextualised teaching approach, the teaching is driven by:

- Consolidating the scientific learning in terms of the knowledge and process skills gained and their transference to a social situation
- Developing generic skills identified with the goals of education (e.g. argumentation, debate, role playing), enabling the value of the science learning to be included into a relevant situation/concern/issue coming from society
- Determining a justified, collective decision which illustrates the value of enhancing scientific literacy for all, the value of scientific careers and, as appropriate, the role scientists play within society

The teaching approach therefore needs to focus on:

- Reflection and student consolidation of the learning in step 2 (*the decontextualised phase*)
- Enabling students' to express their views on aspects of relevance in a manner which encompasses a scientific component (*developing capacities and hence enhancing scientific literacy*)
- Establishing, in the minds of students, the value of the scientific component when making specific decisions (*developing capacities and hence enhancing scientific literacy*)
- Deriving justified and hence well-reasoned decisions, expressed both orally and in written (including posters, models, newspaper) formats (*developing capacities and hence enhancing scientific literacy*)

The three phases that constitute the teaching approach are illustrated in Fig. 10.1.

Science learning initiated by a **familiar contextual frame** of reference, linked to a relevant need in the eyes of students.	Meeting the cognitive and process skills science learning needs by **decontextualised** learning giving due attention to NOS.	Consolidation of the conceptual scientific learning, through transference to a new contextual frame and promoting socio-scientific decision making.
Begins in a social context which involves science	Takes place in non-social context (gain new science)	Enhancing scientific literacy in a socio-scientific context

Fig. 10.1 A contextualisation-decontextualisation-recontextualisation model of STL teaching (Holbrook and Rannikmae 2010)

10.1.5 A Philosophical Look at Science Education

It is thus postulated that science education be seen as:

(i) Enhancing scientific literacy towards enabling a wider career choice and developing responsible citizens, able to play a full role in society (whatever career path is chosen), depending on their status, position and orientation. The science knowledge and competences acquired, attributes developed and judgemental skills gained prepare students for future innovations, meaningful careers and becoming citizens able to appreciate science (and through this technology) in society and take appropriate actions with regard to issues and concerns in their future lives.

(ii) Promoting an interest in science and scientific problem, able to seek meaningful information, evaluate its relevance and reflect on solutions or concerns about socio-scientific aspects considered relevant by them. Science education helps students make use of science knowledge and ideas introduced on a 'need-to-know' basis and to inter-linking this with other pertinent thinking from other discipline areas.

(iii) Portraying a balanced view of science, one which recognises that science does not have all the answers (it is not the absolute truth and certainly unable to answer ethical or spiritual questions). Gaining an insight into the nature of science, as a way of appreciating the importance of science in our lives, recognising it is an important component of learning for all and illustrates the importance of logic, creative thinking, the need for reproducibility of data and conducting careful interpretation of observations.

10.2 Conclusion

The STL (scientific and technological literacy) teaching approach proposed differs from an un-contextualised emphasis on scientific principles and concepts used as the focus of learning in most textbooks. It considers the textbook approach as providing insufficient width or relevance of learning, especially with regard to enhancing the intrinsic motivation of students. It sees a socio-scientific approach as essential, but as the science and technology in use within society is often very complicated and demanding in conceptual understanding, the STL science taught in schools needs to engage in intrinsically motivational challenges and also enable science education to play its role in promoting the goals of education. Students are definitely required to think (minds-on), but the depth of treatment reflects the 'need to know' required for the learning being promoted (there is no requirement that the whole of a topic, as expressed by the subject curriculum, must be followed at any particular time nor that it is approached in any given sequence). The inclusion of scientific principles and scientific concepts, as advocated in phase 2, ensures a strong demarcation between social studies (science) and natural science teaching and allows students to develop their social capabilities and interaction skills with a strong conceptual science background. Intrinsic motivation is stressed as important for meaningful, long-lasting science learning. The transference of conceptual science from the decontextualised setting, as is the case for learning in science classrooms, is promoted as a further essential capability to enhance scientific literacy.

Appendices

Example of a Module Following This STL Approach

Is Home-Made Soap Appropriate and Viable in Today's World?
[*A grade 10–11 science (chemistry) module on the preparation and cleaning action of soap*]

Abstract
This set of activities introduces soap as a common chemical substance which is familiar to students. It examines the making of soap and guides students to put forward an explanation for the cleaning action of soap. The activity reinforces the meaning of the saponification process and compares this in terms of reversibility to the process with the hydrolysis of fats and oils in the body.

Overall Competences: The Students Are Expected to Learn to Be Capable to:

- Recognise factors that affect the choice of soap to buy
- Put forward and carry out procedures for making home-made soap
- Put forward and carry out a procedure for testing the cleansing ability of soaps
- Cooperate and meaningfully communicate orally as member of a group in planning, carrying out the preparation and interpreting the testing of soap so as to undertake this in an effective manner
- Explain saponification, emulsifying power and the action of soap as a cleansing agent
- Explain how the energy stored in fats can be transported around the body by changing fats into water-soluble substances
- Decide, with justification, how to determine whether home-made soap is appropriate and viable

Curriculum Content Soap as a chemical substance; Saponification (consolidation of concept); making soap; cleaning action of soap; reversible reaction

Kind of Activity Group discussion; planning experiments, conducting experiments (to make soap) and carrying out self-devised tests (related to soap). Putting forward justified decisions (related to viability/suitability of home-made soap)

Anticipated time: six lessons

About the Module

This unique teaching-learning material is intended to guide the teacher towards promoting students' scientific literacy by recognising learning in four domains – intellectual development, the process and nature of science, personal development and social development.

Its uniqueness extends to an approach to science lessons which is designed to follow a three-stage model. For this the approach is intentionally from society to science and attempts to specifically meet student learning needs.

This uniqueness is specifically exhibited by:

1. A motivational, society-related and issue-based title (supported in the student guide by a motivational, socio-scientific, real-life scenario)
2. Forming a bridge from the scenario to the scientific learning to be undertaken
3. Student-centred emphasis on scientific problem-solving, encompassing the learning of a range of educational and scientific goals
4. Utilising the new science by including in socio-scientific decision-making to relate the science acquired to societal needs for responsible citizenship

Part 1: Student Activities (To Be Seen by Students)

Scenario

There are many brands of soap on the supermarket shelves. Yet there is a growing interest in making home-made soap, especially for festive occasions when many people are looking for suitable gifts for relatives and friends.

But how to make home-made soap and how do you decide whether buying home-made soap is safe, suitable for use and is a viable proposition? And is it important that such soap also be able to clean well? Should such soaps have a pleasant smell, look good and have a good feel on the hands? Is price a factor in determining what might be the most appropriate commercial soap? And does this mean home-made soaps are only viable to sell on festive occasions?

Your Tasks

1. Hold a group discussion on the value and advantages of home-made soap, including how home-made soap is made by considering the following questions. Be prepared to share your comments with the rest of the class.
2. Undertake a library search on (a) How soap is made? What are the essential ingredients? (b) How are soaps useful for the body? (c) Why is soap a valuable substance. Record your findings in a report.
3. Put forward a procedure for making soap in the laboratory. Use the guiding questions below to help you in this task. After discussion with the teacher, carry out the preparation of soap using your plan or the teacher's modification of the experimental approach. Separate out your soap and obtain a solid sample which can be used in subsequent tests.

Guiding questions

1. What is the main ingredient that is used as the base?
2. Is the main ingredient soluble in water?
3. The main chemical reaction to make the soap 'breaks down' the main ingredient. What substance can be used to do this?
4. What suggestions do you have to speed up the reaction?

4. Making comparisons – testing the soap you made and commercial soaps.
5. Undertake the following:

   (i) Discuss, as a group, home-made soap and its appropriateness. Reflect on whether commercial soaps are better? And on what does 'better' mean?
   (ii) After weighing samples of the soap (as given), soak them in water for 1 h. Do not move them during this time. After one hour remove them and allow the soaps to dry for 1 day. Reweigh the bar. Calculate the loss in mass.
   (iii) Design further experiments to compare the different soaps based on (a) pH, (b) feel on the hands (be careful with the home-made soap and only carry this out if the pH is very similar to that of commercial soap), (c) cleaning ability with cloth and (d) lathering ability.
   (iv) Answer the following questions in writing:

      (a) What useful comparative information can you get from the test 2 above?
      (b) Do you think perfume is added to soap? Is this likely to be a factor in determining the appropriateness of the soap in tests 2 and 3?
      (c) Imagine you are a counsellor working for the consumer society. What factors, based on tests 2 and 3, would you suggest were important when choosing which soap to buy?
      (d) In test 3c, I will clean (type of cloth to clean)..
      ....
      In this, I will compare.................................
      (For this comparison, I will control the following .. ..
      ...)

(e) Did you think you can find any soap that is black in colour? (Very unlikely) Can you suggest a reason for this?
(f) The chemical reaction to make soap is known as saponification. Explain the meaning of the term saponification and indicate how the product differs from the initial main ingredient.
(g) Soap has the ability to form links with oils and fats. Is soap soluble in water (check results of test 2)? How do you suggest soap acts in the way it does (it cleans)?

6. Comparing the home-made soap with commercial soap and determining the appropriateness and viability of home-made soap:

   I suggest the best soap is ....
   I based my choice on the following ....
   I feel that home-made soap is appropriate to use if
   ....
   I feel home-made soap is not good because
   ....

Part 2: Teacher's Guide (For the Teacher)

This project enables students to learn about

(a) Why the body needs fats and oils and how it converts them to transport needed substances around the body
(b) The preparation of soap
(c) The functions of soap
(d) Procedures for testing the cleansing ability of soaps
(e) Controlling variables to arrive at a valid conclusion about the cleaning ability of soaps
(f) How home-made soap compares to commercially available soaps

Learning Outcomes by Lesson

At the end of lesson 1, students are expected to be able to:

- Be aware that soap can be easily made at home
- Be aware that soap making is a very old process
- Put forward an opinion with suitable supporting arguments whether there are advantages in making and using home-made soap
- Suggest approaches to the making of soap in the school laboratory

At the end of lesson 2, students are expected to be able to:

- Recognise the body needs substances that can be insoluble in water (fats, oils), but can be soluble for mobility by changing into an ionic, water-soluble format (hydrolysis)
- Recognise that the hydrolysis process is reversible
- Explain the reversible process using water as the reactant with (a) single esters and (b) fats/oils
- Recognise the structure of fats/oils as complex trihydric substances

At the end of lesson 3, students are expected to be able to:

- Explain that although soap making involves hydrolysis of fats, the process is not reversible
- Explain why the soap making process is not reversible
- Explain soap making symbolically
- Put forward a practical plan to make home-made soap
- Make soap based on a suitable procedure

At the end of lesson 4, students are expected to be able to:

- Separate out the soap from the lye
- Explain substances that are suitable for colouring and adding perfume to soap
- Develop a plan for testing the suitability and cleaning ability of a range of soaps
- Carry out an experiment to compare the solubility of soaps

At the end of lesson 5, students are expected to be able to:

- Carry out a set of tests to determine the cleansing ability of soaps
- Compare the actions of different soaps
- Explain the manner in which soaps act as cleansing agents
- Argue whether industrially made soaps are more appropriate for cleaning

At the end of lessons 6, students are expected to be able to:

- Determine whether home-made soaps have advantages and disadvantages
- Decide as a group whether home-made soap is appropriate and viable
- Be able to make a presentation to the rest of the class on the suitability of home-made soap

Suggested Teaching Strategy (*For the Teacher*)

1. This project is initiated by drawing on the students' prior knowledge about home-made soap and from what soap is actually made. It allows students to discuss whether home-made soap has any advantages in the eyes of the public.
2. This can lead to the questions why some soaps are more effective than others or why we now use detergents rather than soaps to clean clothes. It leads to a consideration of what soap is or how it is made.

3. Students determine from a library search the need for fats in the body as energy stores and the need to move this energy around the body. This leads to the need to change fats into water-soluble substances and hence the concept of hydrolysis whereby fats and esters are changed to acids and alcohols, both more polar than the fats and capable of dissolving in aqueous solutions.
4. Finding out through the library search can also include the fact that the hydrolysis of fats is a reversible process and hence esterification reoccurs between the acids and alcohols. By removal of the acid however in the hydrolysis process means that the reaction is no longer reversible. This can come about by using an alkali instead of water for the hydrolysis process. Students need to understand that this is the essence of soap making.
5. Students, in groups, are now ready to plan how to make soap for themselves using an oil (rather than a fat for convenience) and making use of an alkali such as sodium hydroxide (a substance with which they are already familiar and can write the formula in a molecular and/or ionic manner).
6. The soap making process will take time as the hydrolysis (saponification) process is slow and involves much stirring to bring the non-polar and polar substances together. The reaction needs heat and it is appropriate to allow the mixture to simmer rather than boil. Great care needs to be taken not to spill the mixture or get any on the skin as the mixture is very hot and the alkali is corrosive to both skin and clothes.
7. To compare commercial soaps and the home-made soaps, the teacher may ask students themselves to bring in samples of soap. If this is the case, the teachers should instruct students to ensure that the soap is clearly marked so that the brand name is clearly visible.
8. The experiments to determine the solubility of the soap need some planning on the part of the teacher. As it is not important, the experiment is undertaken for exactly 1 h; probably the best procedure here is to initiate this experiment at the beginning of the lesson and to remove the soap at the end of the lesson. The soap then dries until the students again meet in class which may be 1 day or longer.
9. During the rest of the lesson, students can design a further experiment and careful explain what they hope this experiment will contribute to a comparison of different soaps. Students can also be guided to discuss the questions given for this part of the project. But most important of all, students can develop their ideas on how they might study the cleaning power of the soaps in the main part of the project which will follow in the next lesson. The teacher can collect these at the end of the lesson.
10. The teacher will be able to study the suggestions of the students for their project and prepare the required apparatus for this. Largely this will mean making available 'dirty' samples of cloth that can be compared easily using the facilities available in the laboratory.
11. After comparing the suitability (especially pH) and the cleaning action of the home-made soap, the students are now in a position to initiate a group discussion on the appropriateness and viability of using home-made soap in today's

world. Is it simply a gimmick for festive occasions or does home-made soap have real advantages that need to be considered further? Students need to be able to put forward convincing arguments in support of their position and to reach a group consensus. This can be recorded by all students to give a written record.

12. The topic can conclude by students, based on their written records, presenting to the rest of the class their understanding of the saponification process, the cleansing action of soap and their case for determining whether home-made soap is appropriate and viable.

Achieving the Competences

1. *Put forward reasons for deciding on whether home-made soap is appropriate and viable.*

   This is an integral component of the project, and the teacher will be able to determine how far this is being achieved by seeing the written records of students.
2. *Recognising factors that affect the choice of soap and the appropriateness and viable of home-made soap.*

   This aspect is achieved by undertaking a group discussion. It should be further consolidated by the reasoning given by the students for their particular stance.
3. *Put forward and carry out a procedure for making soap.*

   The teacher will determine the students' ability to achieve this objective by marking written record of their suggested procedures and then observing and guiding their actions during the following practical session.
4. *Put forward and carry out a procedure for testing the cleaning ability of soaps.*

   The teacher will determine the students' ability to achieve this objective by marking written record of their suggested procedures and findings and then observing and guiding their actions during the following practical session.
5. *Ability of students to work as a member of a group.*

   This is achieved by students working as a group. Special attention to cooperation can be placed in the designing of the testing experiments where the results across groups will most probably be required.
6. *Understand saponification and the manner in which soap is able to clean.*

   This is determined by the teacher interacting with the students, asking questions orally and guiding students towards conceptualisation of saponification and an emulsifying agent.

Part 3: Suggested Assessment of Student Learning (For the Teacher)

Different assessment methods (and possible criteria for each) are put forward for the teacher to consider. It is expected that only part A, or part B, or part C is likely to be used (and even then only to the extent the teacher feels it is viable). As different assessment approaches can be used, all are put forward as examples of possible formative assessment approaches, i.e. *takes placed during the learning. This means that the marking needs to be kept simple and a 3-point scheme is suggested (any 3-point marking scheme can be used). It is left to the teacher to divide whether marks are given to individual students or whether some or all occasions marks are given for the work of the group as a whole, thus encouraging teamwork and student-student assistance where additional understanding is required.*

Assessment Based on Skills

This guide to assessment strategies is put forward from different perspectives. In part A the assessment is based on the skill to be developed in the student. Part B is based on the assessment strategies to use in each lesson, whereas part C illustrates the assessment by the three different approaches which a teacher may use for formative assessment – observation, by oral communication or by marking of written work. Summative assessment strategies are not shown, but these could relate to viva type oral communication and/or to the marking of written tests/examination questions:

Award of Social Value Grade (Learning Outcomes 1 and 2)

Teacher listens to the debate on whether home-made soap is appropriate and viable

x Not able to take part meaningfully in the discussion and suggest a justifiable decision

√ Actually takes part in the discussion but is not able to justify any meaningful decision to be taken on whether home-made soap is appropriate and viable

√√ Is able to play a major role in the discussion and in making meaningful and justified decisions related to home-made soap

Award Scientific Method Grade (Learning Outcomes 3 and 4)

Teacher observes the students and notes the observations recorded

x Carries out the experiments, but the observations are either not accurate or inappropriate

√ Able to carry out the experiments in a careful and safe manner and make meaningful observations and obtains a good product in the case of home-made soap

√√ Able to carry out the experiments, obtain a good yield in the case of home-made soap and undertake sufficient repeat observations to make the comparison experiments meaningful and reliable

Award of a Personal Skill Grade (Learning Outcome 5)

Teacher observes the students and notes the observations recorded

x Carries out the experiment, but without enthusiasm and the observations are either not accurate or inappropriate

√ Able to carry out the experiment in a component and safe manner and make and record meaningful observations

√√ Able to carry out the experiment in an efficient and safe manner, taking sufficient repeat observations so as to make the experiment meaningful and reliable. Able to record the experimentation and discussions on whether home-made soap is appropriate and viable, in a suitable manner

Award of a Cognitive Skill Grade (Learning Outcome 6)

Teacher reads the student reports

x Not able to explain the saponification process and the cleansing action of soap

√ Able to explain saponification and how this differs from hydrolysis. Able to explain the cleansing action of soap. Able to compare the cleansing action of soap and indicate the appropriateness and viability of home-made soap in a written record

√√ Able to explain saponification and how this differs from hydrolysis using suitable models and equations. Able to explain the cleansing action of soap in a diagrammatical as well as written format. Able to compare the cleansing action of soap and indicate the appropriateness and viability of home-made soap in a written record with strong justification

Assessment by Lesson

Lesson 1

	Dimension	Criteria for evaluation The student	Mark/grade given (x, √, √√)
1.	Socio or socio-scientific reasoning	Puts forward ideas on whether making soap at home is appropriate and viable	
2.	Developing experimental procedure	Suggests ideas related to ways to make soap	
3.	Willingness to respond to questions	Willing to attempt to provide answers to the purpose of soap and historical aspects related to soap making	

Lesson 2

	Dimension	Criteria for evaluation The student	Mark/grade given (x, √, √√)
1.	Seek information	Able to gather information from suitable sources on the reason for fats in the body and the type of substances they represent	
		Able to seek information on the process of hydrolysis and how this can be applied to the reaction of water with (a) esters and (b) fats	
		Able to gather information on reversible reactions and the structure of fats/oils	

	Dimension	Criteria for evaluation The student	Mark/grade given (x,√,√√)
2.	Answers questions	Provides correct verbal or written answers to questions on fats in the body and how they can be made soluble through hydrolysis	

Lesson 3

	Dimension	Criteria for evaluation The student	Mark/grade given (x,√,√√)
1.	Writes a plan for a procedure to make soap	Develops an appropriate procedure to make soap starting from an oil and sodium hydroxide	
2.	Carries out the experimental procedure	Undertake experiments on the preparation of soap when the teacher approves the plan or provides specific instructions	
3.	Answers questions	Provides correct written answers to questions given orally or in written format to question explaining saponification and why the process is not reversible	

Lesson 4

	Dimension	Criteria for evaluation The student	Mark/grade given (x,√,√√)
1.	Writes a plan to compare soaps	Develops an appropriate procedure to compare soaps	
2.	Collect experimental data collected	Separates out the home-made soap in a suitable manner	
		Undertake experiments on the solubility of soaps and record observations	
3.	Interpret or calculate from data collected and making conclusions	Correctly carries out calculations on the solubility of soaps	
		Draws appropriate conclusions	
4.	Answers questions	Provides correct written answers to questions given orally or in written format related to which substances are suitable to use to colour soap and to add a smell	
5.	Cooperate as a group	Cooperates with others in a group and fully participates in the work of the group	
		Illustrates leadership skills – guiding the group by thinking creatively and helping those needing assistance and summarising outcomes	

Lesson 5

	Dimension	Criteria for evaluation The student	Mark/grade given (x,√,√√)
1.	Record experimental data collected	Carries out tests on the cleansing action of soaps and records observations/data collected appropriately (in terms of numbers of observations deemed acceptable/accuracy recorded/errors given)	
2.	Interpret or calculate from data collected and making conclusions	Interprets data collected on the solubility of soaps in a justifiable manner	
		Draws appropriate conclusions	
3.	Answers questions	Provides correct written answers to questions given orally or in written format on whether industrial soaps are more appropriate	
		Provides answers in sufficient detail especially on pH factors and other differences between home-made soap and commercial soap when called upon to give an opinion or decision	

Lesson 6

	Dimension	Criteria for evaluation The student	Mark/grade given (x,√,√√)
1.	Answers questions	Able to explain the manner in which soaps act as cleansing agents	
		Able to explain whether home-made soap has any advantages and disadvantage over commercial soap	
2.	Scientific or socio-scientific reasoning	Gives a justified decision on whether home-made soap is appropriate and viable	

Assessment by Teacher's Strategy

Assessment Tool Based on the Teacher's Marking of Written Material

	Dimension	Criteria for evaluation The student	Mark/grade given (x,√,√√)
1.	Writes a plan or report of an investigation	Puts forward an appropriate research/scientific question and/or knows the purpose of the investigation/experiment	
		Creates an appropriate investigation or experimental plan to the level of detail required by the teacher	
		Puts forward an appropriate prediction	
		Develops an appropriate procedure (including	
		apparatus/chemicals required and safety procedures required)	
2.	Record experimental data collected	Makes and records observations/data collected appropriately (in terms of numbers of observations deemed acceptable/accuracy recorded/errors given)	
3.	Interpret or calculate from data collected and make conclusions	Interprets data collected in a justifiable manner	
		Draws appropriate conclusions	
4.	Answers questions	Provides correct written answers to questions given orally or in written format	
		Provides answers in sufficient detail especially when called upon to give an opinion or decision	
5.	Scientific or socio-scientific reasoning	Gives a justified decision on the 'best' choice of soap	

Student Assessment Tool Based on the Teacher's Observations

	Dimension	Criteria for evaluation The student:	Mark/grade given (x,√,√√)
1.	Performing the experiment (at the pre-inquiry and inquiry phases)	Performs the experiment according to the instructions/plan created	
		Maintains an orderly and clean worktable	
		Understands the objectives of the experimental work and knows which tests and measurements to perform	
		Uses lab tools and the measurement equipment in a safe and appropriate manner	
		Behaves in a safe manner with respect to himself or herself and to others	

	Dimension	Criteria for evaluation The student:	Mark/grade given (x,√,√√)
2.	Functioning in the group during experimentation or discussion	Contributes to the group discussion during the theoretical inquiry phases (raises questions and hypotheses, designs the experiment, draws conclusions, makes justified decisions)	
		Shows tolerance with, and gives encouragement to, the group members	
		Cooperates with others in a group and fully participates in the work of the group	
		Illustrates leadership skills – guiding the group by thinking creatively and helping those needing assistance (cognitive or psychomotor) and summarising outcomes	
3.	Presenting the experiment orally	Presents the activity in a clear and practical manner with justified decisions	
		Presents by illustrating knowledge and understanding of the subject	
		Uses precise and appropriate scientific terms and language	

Part 4: Additional Notes for the Teacher

For these lessons it is assumed that students are already familiar with the saponification reaction and the type of substances which form the raw materials for making soaps. These lessons reinforce this learning and relate the reactions to the properties of actual soap in the marketplace. It is expected the lessons illustrate that it is difficult to make a decision on which soap is best and factors such as price and advertising may well play a greater role in enabling society to choose than does the cleansing properties of the soap itself. The lessons thus cover the cost of the soap and the solubility of the soap if left to stand in water for some time and allow students to introduce other factors that they feel may have influence, such as packaging, size of the soap bar and colour.

The actual testing of the cleansing properties reinforces the need for a control if comparisons are to be meaningful. Students are required to suggest suitable experiments, give appropriate apparatus for this (based on their experiences if prior working in the school laboratory) and, importantly, how the control experiment will be set up in each case.

Expected experiments are:

1. pH of the soap bar (a measurement) and controls are the fixed amount of soap, fixed quantity of water used and that the water must come from the same source, i.e. all tap water or all distilled (deionised, if water is truly pH=7) water. This

experiment is probably not meaningful without the use of a pH metre as the differences are likely to be small.

2. Ability to remove stains from a piece of cloth (controls are same cloth; same size of cloth; same type of stain; same intensity of stain; same temperature; same water; same quantity of water; same type of container for undertaking the experiment; same additional aids, e.g. stirring; same length of time for experiment; same post-experiment check).
3. Variations in these factors may affect the cleansing ability of the soap and hence experiments could vary one variable at a time using different soaps and then checks made on the effectiveness of the various soaps under each condition.
4. Ability to lather (controls here are same quantity of soap, same water, same quantity of water, same time, same additional aids such as shaking, same type of container, same instrument for measuring depth of lather.

Experimental Details for the Saponification of Fats

The breakdown of sunflower oil by sodium hydroxide

Apparatus

Each pupil or pair of pupils will need:
Page from *Laboratory Investigations* (experiment 21.5)
Teat pipette
Two beakers, 250 cm^3
Tripod and gauze
Bunsen burner and asbestos square
Glass rod
Filter flask, funnel, filter papers and filter pump (if possible)
Spatula
Sunflower oil
5M sodium hydroxide solution
Common salt

Procedure

1. Place about 10 cm^3 of sunflower oil in a 250 cm^3 beaker, with about 50 cm^3 of 5 M sodium hydroxide solution. (The pupils should be warned not to get this on their skins or on the bench.) Warm the beaker and stir the contents with a glass rod until it is boiling. Boil it gently for 10–15 min, stirring throughout.
2. Add about 50 cm^3 of distilled water and spatula measures of salt (sodium chloride) to saturate the solution; boil gently and stir for 3–5 min. Let the mixture cool, stir to break up any large pieces of solid and filter this off. Wash the solid residue in the funnel with a little distilled water and then allow the product to dry.
3. Shake a small quantity of the solid with water in a test tube, when it should lather, showing the formation of soap (colouring matter and perfume can be added and the solid put in a press to form a solid bar of soap).

What Is Soap?

Soap is a cleansing agent made from fats and oils with alkali.

Ingredients

Oils and fats for soap are esters of fatty acids which react with alkalis such as sodium hydroxide to form glycerol and the sodium salt of the fatty acid. The fatty acids required for soap making can come from animal fats, grease, fish oils and vegetable oils. The hardness, lathering qualities and transparency of soap vary according to the combinations of fats and alkalis used as ingredients. An experienced soap crafter uses many combinations of oils.

How Does Soap Clean?

Most soaps remove grease and dirt because they (or some of their components if we consider the colouring and perfumes added) are surfactants (surface-active agents). Surfactants have a molecular structure that acts as a link between water and the dirt particles. This loosens the particles from the underlying fibres or surfaces to be cleaned. One end of the soap molecule is hydrophilic (attracted to water), and the other is hydrophobic (attracted to substances that are not water soluble). This peculiar structure allows soap to adhere to substances that are otherwise insoluble in water. The dirt is then washed away with the soap.

A Scientific Explanation

Water molecules consist of two hydrogen atoms and an oxygen atom. The oxygen atom is linked to the two hydrogen atoms at a bond angle of about 104°. Oxygen is far more electronegative than hydrogen and so it tends to have a higher electron density. Consequently the water molecule is *polar – it has a* positive charge at one end of the molecule (the hydrogen end) and a negative charge at the other (the oxygen end).

The positive end of one water molecule will be strongly attracted to the negative end of another water molecule. When an ionic compound, like sodium chloride, dissolves in water, the oxygen (negative) end of the water is attracted to the cations (positive ions), while the hydrogen (positive) end of the water is attracted to the anions (negative ions). The solubility of a substance in water is largely determined by the relative strength of the attraction of water to the substance compared to the strength of the attraction between water molecules.

In contrast to oxygen, carbon has almost the same electronegativity as hydrogen and the carbon-hydrogen bond is *non-polar*. For example, the octane molecule (a component of gasoline) consists of eight carbon atoms in a chain, with two hydrogen atoms attached to the interior carbons and three hydrogen atoms on the end carbons. Since the electron density is evenly spread, the molecule is electrically neutral along its entire length.

The simplest way to understand solubility is to remember the rule 'like dissolves like', that is, polar and ionic substances are soluble in polar and ionic substances, while non-polar substances are soluble in non-polar substances. Thus salt dissolves in water, but not in gasoline. Oil dissolves in gasoline, but not in water.

Living Cells and Polar/Non-polar Substances

Living cells need both polar and non-polar substances. The cell uses non-polar substances, fats and oils, to make up the cell membrane which separates the interior of the cell from the exterior. If the cell membrane were soluble in water, it would dissolve away and soon there would be nothing to divide the cell from the non-cell. But in order to get to the cell in the first place, all the parts of the cell must be water soluble because that's how materials are transported from place to place. What nature needs is a non-polar material that can be dissolved, moved around and then made non-polar again. This material is known as a *lipid (fat)*, or *triglyceride*.

A lipid is an ester and basically consists of two parts – a fatty acid and a trihydric (three OH groups) alcohol called glycerol. Both the fatty acid by itself and the glycerol by itself are water soluble, because of the polar oxygen atoms on the ends of these molecules. In a lipid, three fatty acids are bonded to the three oxygen atoms (three OH groups) on the glycerol. Although the oxygen atoms are still there, they are now buried inside the molecule and the lipid is essentially non-polar. The lipid is therefore insoluble in water.

A fatty acid (saturated) has the formula $C_nH_{2n+1}COOH$. The chemistry is dominated by the properties of the COOH group. Because this group is polar, fatty acids tend to be soluble in water. Octanoic acid, $C_8H_{17}COOH$, is just one of a very large number of fatty acids. In fact, most fatty acids are longer than octanoic acid. Two very common components of lipids are palmitic acid ($C_{15}H_{31}COOH$) and stearic acid ($C_{17}H_{35}COOH$). Solid lipids are generally called *fats*. Another class of fatty acids is the *unsaturated* fatty acids, with less than 2n+1 hydrogens for every n carbons. Oleic acid, for example, has the formula $C_{17}H_{33}COOH$ and linoleic acid has the formula $C_{17}H_{31}COOH$.

Saturated fats contain saturated fatty acids and are solids at room temperature. Lard and butter are examples of saturated fats. Soap made from these fats tends also to be solid at room temperature. Unsaturated fats contain unsaturated fatty acids and are liquids at room temperature. Generally these are called oils and examples include corn oil and sunflower oil. These oils produce liquid soap. While unsaturated fats are generally more healthy than saturated fats, a liquid is often not very convenient. Thus, margarine, which is made from unsaturated plant oils (e.g. corn oil), is hydrogenated to change it from an unsaturated oil to produce a saturated (solid) fat.

To make soap, the trihydric ester (fat) is hydrolysed (broken down) into its fatty acid and glycerol constituents. The fatty acid has a long hydrocarbon tail which is soluble in fats and a polar oxygen end which is soluble in water. Thus a fatty acid in solution acts as a soap by dissolving fats in one end of the molecule and water in the other. When a strong base, such as lye, is used to hydrolyse (saponify) the fat, the fatty acid is then present as a large cation, which is polar at one end and non-polar at the other. Just as sodium chloride and sodium carbonate which are soluble in water, sodium octanoate, the sodium salt of octanoic acid, is also soluble in water.

Saponification

Saponification is the term applied to the hydrolysis of fats using a strong alkali-like lye. If we take a fat derived from palm oil (containing palmitic acid) and hydrolyse it using sodium hydroxide, the reaction is

$$\begin{array}{llllll} [C_{15}H_{31}CO]_3[C_3H_5O_3](s) & +\ 3\ NaOH(aq) & \rightarrow & 3\ C_{15}H_{31}COONa(aq) & + & C_3H_5(OH)_3(aq) \\ \text{fat(s)} & +\ 3\ \text{lye(aq)} & \rightarrow & 3\ \text{sodium palmitate(aq)} & + & \text{glycerol(aq)} \end{array}$$

While this reaction may appear intimidating because of the long formulas, it is, in fact, quite simple. It could be written generally as

$$[RCO]_3[C_3H_5O_3](s) + 3NaOH(aq) \rightarrow 3RCOONa(aq) + CH_2OH.CHOH.CH_2OH(aq)$$

where ‘R’ is some long carbon-hydrogen chain.

If you look on a list of ingredients on a soap, you will find things like ‘sodium stearate’, or ‘sodium palmitate’. This is simply specifying the particular fatty acids present in the soap.

When fat is introduced to a soap solution, the non-polar tail of the fatty acids dissolves in the non-polar fat, leaving the water-soluble oxygen end at the surface of the fat globule. With enough soap, these fat globules become covered with a water-soluble coating and disperse throughout the solution, as in the last figure. They are not truly dissolved since individual fat molecules are not dispersed in the solution. Rather, the fat is emulsified.

References

Bennett, J., & Lubben, F. (2006). Context-based chemistry: The salters approach. *International Journal of Science Education, 28*(9), 999–1015.

European Commission (EC). (2007). *Science education now: A renewed pedagogy for the future of Europe*. Report by a High Level Group on Science Education. Brussels: Author.

Eurydice. (2011). *Science education in Europe: National policies, practices and research.* Retrieved from eacea.ec.Europa.eu/education/eurydice

Fensham, P. (2008). *Science education policy-making: Eleven emerging issues*. Paris: UNESCO.

Gilbert, J. K. (2006). On the nature of "Context" in chemical education. *International Journal of Science Education, 28*(9), 957–976.
Holbrook, J. (2008). Introduction to the special issue of science education international devoted to PARSEL. *Science Education International, 19*(3), 257–266.
Holbrook, J. (2009). Meeting challenges to sustainable development through science and technology education. *Science Education International, 20*(1/2), 44–59.
Holbrook, J., & Rannikmae, M. (2007). Nature of science education for enhancing scientific literacy. *International Journal of Science Education, 29*(11), 1347–1362.
Holbrook, J., & Rannikmae, M. (2009). The meaning of scientific literacy. *International Journal of Environmental and Science Education, 4*(3), 275–288.
Holbrook, J., & Rannikmae, M. (2010). Contextualisation, decontextualisation, recontextualisation- A science teaching approach to enhance meaningful learning for scientific literacy. In I. Eilks & B. Ralle (Eds.), *Contemporary science education* (pp. 69–82). Aachen: Shaker Verlag.
Marks, R., & Eilks, I. (2009). Promoting scientific literacy using a sociocritical and problem-oriented approach to chemistry teaching: Concept, examples, experiences. *International Journal of Environmental and Science Education, 4*(3), 231–245.
MCEETYA. (2008). *Melbourne declaration on education goals for young Australians.* Retrieved from (June 2010) www.mceecdya.edu.au/mceecdya/melbourne_declaration,25979.html
National Research Council. (2010). *Exploring the intersection of science education and 21st Century skills: A workshop summary.* Margaret Hilton, Rapporteur. Board on Science Education, Center for Education, Division of Behavioral and Social Sciences and Education. Washington, DC: The National Academies Press. Retrieved from www.nap.edu/catalog/12771.html
Osborne, J., Simon, S., & Collins, S. (2003). Attitudes towards science: A review of the literature and its implications. *International Journal of Science Education, 25*(9), 1049–1079.
Parchmann, I., Graesel, C., Baer, A., Nentwig, P., Demuth, R., Ralle, B., & ChiK Project group. (2006). "Chime im Kontext": A symbiotic implementation of a context-based teaching and learning approach. *International Journal of Science Education, 28*(9), 1041–1062.
Partnership for 21st Century Skills. (2008). *A resource and policy guide*. Retrieved from www.21stcenturyskills.org
Schwartz, A. T. (2006). Contextualized chemistry education: The American experience. *International Journal of Science Education, 28*(9), 977–998.
UNESCO. (2012). *EFA global monitoring report: Youth and skills putting education to work.* Paris: UNESCO.
Vygotsky, L. (1978). Problems of method. In: M. Cole (Trans.). *Mind in society.* Cambridge, MA: Harvard University Press.
Zion, M. (2007). Implementation model of an open inquiry curriculum. *Science Education International, 18*(2), 93–112.

Gilbert, J. K. (2006). On the nature of "context" in chemical education. *International Journal of Science Education, 28*(9), 957–976.

Chapter 11
In *The New Zealand Curriculum*: Is It Science Education or Education Through Science? One Educator's Argument

Steven S. Sexton

11.1 Introduction

In anticipation of the full implementation of *The New Zealand Curriculum* (Ministry of Education 2007a), a great deal of literature was supplied to schools to inform New Zealand teachers about its intention, meaning and purpose (see, e.g. Arcus 2009; Cowie et al. 2009; Foster 2009). This curriculum document officially replaced the 1993 documents for the start of the school year in February 2010. Prior to this, a separate document detailed the scope, essential skills, purposes and possible learning experiences students could be exposed to in each of the learning areas (e.g. in science see, Ministry of Education 1993). With the present curriculum, one document now encompasses the educational experiences of English-medium students for years 1–13 (students aged 5–18). As New Zealand is a trilingual country, it recognises English, te reo Māori (the language of the indigenous people of New Zealand) and New Zealand Sign Language as its languages. Therefore, there are two additional curriculum documents: *Te Marautanga o Aotearoa*: *He tauira hei kōrerorero* (Ministry of Education 2007c) for Māori-medium schools and *New Zealand Sign Language in the New Zealand Curriculum* (Ministry of Education 2007b). However, the majority of students attend English-medium schools.

The New Zealand Curriculum is a political document targeted to a nationwide audience. This document was written in consultation with numerous parties with vested interests. Nevertheless, it still has to be implemented in the classroom by individual teachers. Many of these teachers found themselves having to implement a new curriculum based on an integrated approach to teaching. Previously, teachers worked to develop essential skills and attitudes in students through a set of predetermined learning experiences in each of the learning areas. Now, teachers are presented

S.S. Sexton (✉)
University of Otago College of Education, Dunedin, New Zealand
e-mail: steven.sexton@otago.ac.nz

B. Akpan (ed.), *Science Education: A Global Perspective*,
DOI 10.1007/978-3-319-32351-0_11

with the task of putting into practice rich and meaningful guidelines underpinned by extensive national and international research through content determined by the local community. Teachers should be taking into consideration the content material that is seen as relevant, useful and meaningful (Sexton 2011) to their students based on their worldview. Schools, however, were offered no professional development support by the Ministry of Education in interpreting what all of this meant or how to incorporate this new curriculum into classroom practice.

As stated, this curriculum change was to be fully implemented for the beginning of the 2010 school year. It has been estimated that at least half of New Zealand teachers were unprepared for this change (Hipkins and Hodgen 2012). One of the reasons for teachers reporting a lack of readiness was the new curriculum's emphasis on the Nature of Science as the overarching strand. Here, the Nature of Science refers to the overarching strand of science for *The New Zealand Curriculum* as opposed to the lowercase nature of science (often abbreviated in the science literature as NOS) broadly referring to how science is a "knowledge-based enterprise" (Hipkins et al. 2005, p. 243).

This Nature of Science emphasis no longer requires teachers to focus on the content strands of the living world (biology), the material world (chemistry), the physical world (physics) or planet Earth and beyond. Teachers should now focus their students' learning, using whatever science content that is appropriate, through any or all of the Nature of Science's four elements:

- Understanding about science – this requires that students are able to ask questions about the science they are doing and accept that there may be more than one explanation.
- Investigating in science – it means that students are expanding their world through activities, play, questions and/or simple models.
- Communicating in science – students are able to use the terminology and vocabulary appropriately as they discuss the science they are doing.
- Participating and contributing – first, students are able to relate the science they are doing to their world; and second, they are able to make informed decisions that impact on their world based on this science (Ministry of Education 2007a; Sexton 2010).

This conceptual shift from prescribed learning activities in science to an explicit directive to incorporate student-initiated topics in their teaching has resulted in many teachers self-reporting a lack of content knowledge. This perceived science content knowledge limitation, compounded with perceived lack of resources for science, results in many students not experiencing effective learning in science (Education Review Office 2012). Duschl et al. (2007) highlighted that many of the key ideas of, and about, science may be impossible for students to grasp without instructional support, that is, the classroom teacher. They reiterated that for science learning to successfully engage students, it must be meaningful to them, and they must be supported by the teacher. To be very clear here, primary teachers are not trying to be scientists or somehow turn their students into scientists. Primary teachers need to provide effective learning opportunities, and science is one of the contexts for these learning opportunities.

What counts as quality and effective science in schools in New Zealand has been highlighted by the Education Review Office (ERO). The ERO 2012 report noted the characteristics that were evident in the classrooms in which effective education through science was taking place. These schools evidenced characteristics of students' engagement, specifically students like doing science, are motivated by their classroom science activities, think they are learning well in science and are enthusiastic about doing more science (see Education Review Office 2012). Similarly, teachers in these schools modelled some of the teacher characteristics ERO identified as good practice indicators, in particular:

- High-quality planning, including strategies for identifying and responding to students' prior knowledge and for teaching them the significant scientific concepts (or big ideas).
- Flexible approaches that take advantage of students' curiosity and are able to meet their diverse needs.
- An emphasis on the quality of students' thinking or conceptual development.
- High-quality investigations, reflection and discussions that help students develop their understanding of scientific knowledge and processes.
- Engaging practical activities that allowed students to investigate their own ideas as well as those of others – these activities were collaborative and relevant and drew on local context as well as interests of students (see Education Review Office 2010).

11.2 Science in Primary Education

Much has been written, in New Zealand and internationally, about science in education (Bolstad and Hipkins 2008; Parker 2004; Traianou 2006; Tytler 2008). Parker (2004) highlighted the fact that the need to know what you are teaching and how to teach it is central to all teachers' work. However, must teachers have a great deal of subject content knowledge to teach effectively and to be able to present the various opportunities their students require for accessing that knowledge? Shulman (1986) described this duality as pedagogical content knowledge (PCK), which means that teachers need to be able to apply more than one pedagogical approach to the subject matter and be able to break the subject matter down to suit the pedagogical tool that is being applied. In all, teacher expertise in pedagogy (P) and content (C) enables students to access knowledge (K) (see Fig. 11.1).

In relation to primary science, this means providing students with the experiences or opportunities to make sense of the world around them. Ideas get linked to the wider world and create broader concepts grounded in a deeper understanding when they emerge out of a student's own environment (Otrell-Cass et al. 2009; Parker 2004). In line with this, *The New Zealand Curriculum* empowers local communities to select their own content and, thereby, to have their interests reflected in their children's education.

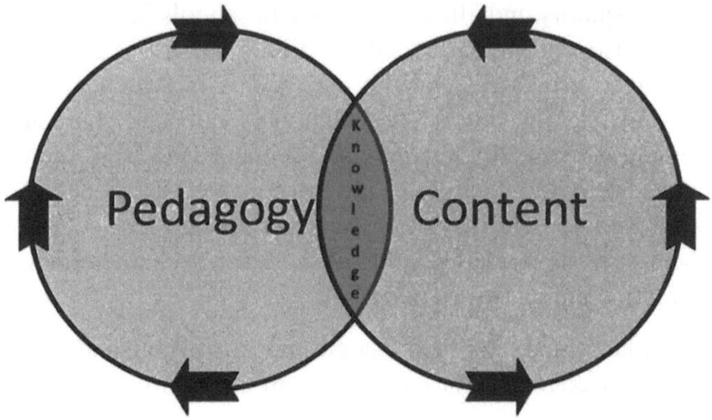

Fig. 11.1 The knowledge students are able to access due to teacher expertise

In itself, however, a meaningful context is not enough. Students also need to see the relevance and usefulness of their learning and teaching environment. Traianou (2006) used the example of one British primary teacher to describe what relevant, useful and meaningful teaching could look like in the primary classroom. This teacher's expertise enabled her to structure the class so that her students were encouraged to ask, and address, their own questions. As a result, her teaching allowed the students to engage in activities that challenged their preconceptions or misconceptions about their own world. For this teacher, the goal of primary science is that 'children find out about things that are obvious in everyday life but they haven't thought of in this way before' (p. 68). This required her students to participate in both the doing of science and discussions about the science they did. Her success in assisting her students in accessing new knowledge supports findings about what turns students 'off' science. Tytler's (2008) review of the state of Australia's Science Education reported three central characteristics of what makes science teaching unappealing to students:

- The transmissive pedagogy that characterized school science
- The decontextualised content that did not engage students' interest or commitment
- The unnecessary difficulty of school science (see p. 9)

Similar findings were reported by Bull et al. (2010). In their paper to encourage debate within New Zealand on how to better engage students in science, they noted that 'much of the science taught in schools is not especially useful in everyday life, and many students do not achieve sufficient understanding of it to be able to contribute to scientific debates' (p. 3). It seems that for many students, science is learning science knowledge rather than an education through a scientific context. More importantly for this chapter, Bolstad and Hipkins (2008) reported the progressive disengagement from science as primary-aged students advanced through their schooling.

Bolstad and Hipkins (2008) reviewed data provided by the New Zealand National Education Monitoring Programme (NEMP), which recorded trends in year 4 and year 8 students' attitudes towards science between 1995 and 2007. Less than a third of all year 8 (students aged 12) students in 2007 reported doing 'really good things' in science more often than not. For New Zealand Māori, less than a quarter reported positive experiences in science. Bolstad and Hipkins go on to highlight the fact that the emerging research provides strong evidence that students are aligning themselves to careers in or out of science fields prior to years 7 and 8. As a result, some primary-aged students are making life-long learning choices based on their primary schooling experiences.

This disengagement by primary-aged students may be because science in the 1993 New Zealand curriculum documents (New Zealand's official curricula from 1993 to 2010) focused on doing science in a scientific manner with the explicit intention of developing scientific skills and attitudes through investigation (Ministry of Education 1993). The final objective was for students to be able to carry out a complete scientific investigation, starting with focusing and planning, then information gathering and finally processing and interpreting this information to reach the reporting stage. To achieve this, classroom teachers were encouraged by the Ministry of Education to incorporate a number of predetermined activities into their programmes. In reality, this resulted in many of us simply working our way through disjointed lists of unrelated activities over the course of the school year for each science content area. This prescriptive approach to teaching led to many of my colleagues reporting a lack of confidence in their own scientific understanding, and this has resulted in a predominantly reticent attitude to teaching science (Otrell-Cass et al. 2009).

11.3 Science and the Nature of Science

In April 2011, Sir Peter Gluckman delivered a report from the Office of the Prime Minister's Chief Science Advisor (Gluckman 2011), in which he highlighted the challenges and opportunities for science in New Zealand. He recognised the changing nature of science and the role it plays in society. More importantly, he stated that 'a forward looking science education system is fundamental to our future success in an increasingly knowledge-based world' (p. vi). He then went on to note that a particular challenge was in how New Zealand primary school teachers would be provided with the skills necessary for this forward-looking science education system. But is it science education or education through science that teachers should be aiming at?

In *The New Zealand Curriculum*, students should be developing the four elements of the Nature of Science, understanding about science, investigating in science, communicating in science and participating and contributing, by working with the four content strands. This is where I believe primary education comes into conflict with how New Zealand's Ministry of Education sees the purpose of science.

Science in primary education should not be focused on getting students to understand what a 'scientist' is and 'how scientists work' but on how to get them to make sense of their world. With students being turned off by science by the time they complete primary education (Bolstad and Hipkins 2008), science in primary education needs to focus on learning through science contexts not science education. For example, one science activity that occurs in many New Zealand classrooms is the use of batteries, wires and bulbs to demonstrate how circuits operate. This type of activity lends itself more towards science education than learning through the context of science. Students may very well generate a great deal of in-depth and richly rewarding questions, but these invariably refocus attention back on the teacher. While the intention is student centred, this type of lesson will more often than not revert to a transmissive pedagogy as less-experienced or content-challenged teachers seek to address their students' questions. Accordingly, the learning environment loses relevance to most students, as what they are discussing no longer reflects their world or the content progresses to a level beyond their understanding.

It is for reasons such as this that primary education through science contexts works brilliantly with *The New Zealand Curriculum*. The strong association between science and constructivist theory has been well documented (see, e.g. Darling-Hammond 1997; Skamp 2004; Taber 2010). More importantly for this chapter concerning New Zealand, it has been one of the theories used to ground the development of New Zealand's curriculum documents since the 1980s (Bell and Baker 1997). However, the fact that students continue to report transmissive pedagogy as a deterrent to their continued engagement in science indicates that many teachers still do not practice how they are supposed to implement *The New Zealand Curriculum*.

11.4 Education Through a Science Context

The 5E approach (Skamp 2004) has been shown to be successful in teaching scientific content to primary-aged students. The 5Es are engage, explore, explain, elaborate and evaluate. This approach is very useful for teachers struggling with their own pedagogical content knowledge. Specifically for this chapter, the 5E approach will be used to complement *The New Zealand Curriculum's* overarching strand of the Nature of Science for education through a science context. It should be noted that many teachers are getting caught up in using learning theories and approaches that try to be as student centred as possible, forgetting that they also play a vital role in education. Teachers need to be reminded that no learning theory or approach is perfect. Sometimes direct acts of teaching, directing, telling, explaining, prompting, modelling, giving feedback and questioning (Ministry of Education 2006) while being very teacher centred, are necessary. Once students have the information or material that is needed, it is then appropriate to switch from teacher-centred to student-centred learning approaches. Not every lesson should follow the same routine; both students and teachers need some variety.

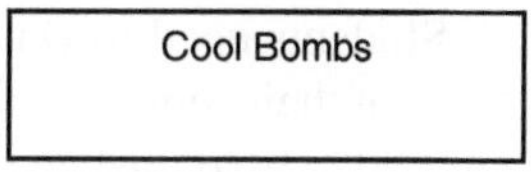

Fig. 11.2 Cool bombs' title on board

The following will be both a description and discussion of how an activity titled 'cool bombs' (see Fig. 11.2) is approached as education through a science context rather than science education. It will explicitly link the 5E approach to *The New Zealand Curriculum's* Nature of Science. The first step in 5E is to get the students engaged in the activity; the use of a title such as this has more of a positive impact than saying or writing up on the board 'chemical reactions' or 'citric acid and baking soda mixture'. Writing this title on the board is a simple step to begin the discussion about what the students already know about bombs and blowing things up. Should any question(s) arise as to why the word 'cool' is also written up, then that question can be explored as to possible meanings, but it will become clear as the activity continues. In this instance, only two words are generally needed to engage most students' interest; other ideas to introduce the topic, depending on students' age, interest or experience, could be picture books, reading to or a classification activity using pictures and/or words.

The explore step is important as it is where the students get the chance to try something out and see what happens. This is also the step that is more likely to provide those experiences that challenge what they thought they knew. This step, however, cannot be the end of the science lessons. Students need to experience the 'WOW' factor; but without the 'WHY' discussions that need to follow this activity is most likely only going to be a fun activity as there is probably no real learning involved. Students should have the time to explore long enough to challenge what they think they know, but not so long that they lose interest.

In pairs, tell the students to mix 1 'spoon' of citric acid and 1 'spoon' of baking soda in a small sealable bag. In primary education through a science context, the focus should not be on precise measurements. At the primary level, science should be seen as a verb and more specifically an action verb. Specifically, students should associate science with them doing. In fact, the science strand of *The New Zealand Curriculum* requires students to recognise, observe, explain, extend, act on, build, share, describe, compare, find and explore. Therefore, they do not need to concentrate on getting 1 level tablespoon. The measurement is not the activity. If you want a tablespoon, then use one and tell the students they need one of these 'spoons'. In addition, use small sealable bags as they are easy to handle and then pairs of students are able to do this activity instead of watching the teacher or a select few. Next, add water to the bag to wet the mixture – the amount of water is not really important. For younger students, turn on the tap to a moderate steady flow and have them line up and hold the bag under the water while they say 'cool bombs', and then the next in line comes forward. Seal the bags, if they can, and let them see and hear what happens. Using small bags means any popping will not be so messy, but if the room is carpeted or spillage is a concern, this is easy to do outside with a pitcher/bottle of water.

Students need to watch, hear and feel what is happening. This activity relies on three of their senses to gather the necessary data, so after any ‘popping’, the students are told to hold the bag in their hand while they figure out together what to say about what happened and what is happening in their hand. Teachers need to remember that learning starts with what the students know and how they are being challenged, so listen what are they saying, what can they tell you, what are they describing and what are they seeing, hearing and feeling?

As stated, the explore step is where the students get the chance to try things out and experience what happens. This step mirrors the relationship between the Nature of Science’s understanding about science and investigating in science elements. By undertaking this activity, students explore what different ingredients are able to do. This in turn, hopefully, raises further questions students will need to explore as they seek to discover the answers. This is the intent behind understanding about science and investigating in science. The questions generated about ‘how’, ‘why’, ‘what if’ and ‘what about’ as they seek deeper understanding about the science they are doing lead to further and more meaningful investigations in the science.

As stated, the focus of the explore step is to allow the students the opportunity and if necessary opportunities (some activities need to be repeated more than once to allow students the time) to see, hear, think and begin to discuss what is happening and how they are beginning to make sense of the event. Why tell students what would happen when they can see, feel, hear, taste, touch or smell it for themselves? Positive learning effects have been seen in students who use discussions as a key part of the science process (Duschl and Hamilton 2011). Students need to be explicitly asked to explain their explorations. This is a deliberate step to discuss what they did, why and what the results of what they tried were. The students need the opportunity to talk about the science. This deliberate communication in science allows the teacher to hear what students are actually thinking and how they are making sense of the science. This facilitates the transition into the explain step of the 5Es.

Explain is not so much the teacher talking but more the teacher listening to what the students are saying. Here is where the teacher gets to hear what the students are thinking and how they are making sense of it. It is also where the teacher gets a great deal of informative assessment on what they are getting out of and taking away from the activity. The explain step allows the teacher to determine what terminology is necessary for the students to discuss their experiences. For this activity, there could be appropriate usage of vocabulary such as explosion, decrease (of) temperature or expansion (of the bag). In addition, teachers will also get to hear where the gaps are in their students’ comprehension of the activity and the science behind it. If students are new to this, prompt with questions that get them describing and talking about what happened. Most students have a far greater speaking vocabulary than a writing one, and once the cognitive load of spelling and grammar are removed, they can amaze you with what they are able to talk about. This is where many teachers start to panic and say they do not have the content knowledge to do science. Teachers need to be reminded that you do not have to know it all! If teachers insist on holding to the position that they have to know everything, how much are you limiting your students by what you do not attempt in class? In addition, how are students expected to learn how to find out what they do not know if they are not given the opportunity by the teacher? The key is keeping the explanations at an age-appropriate level so the students get both WOW and WHY. Care must be taken, however, not to minimise or simplify the science the students are experiencing to the point of meaninglessness. It needs to be simple enough to make sense to them but no simpler. Primary teachers start with what the students know, are able to know and go from there. This third E is vital as it lets the students show what they know and what they got, but it also directs the teacher to what will probably be needed in the fourth E of elaborate.

The elaborate step is most likely going to follow naturally any explanations that confuse or challenge what students thought they knew. As stated, teachers do not have to know it all, but they do have to model to students what to do when you do not know the answer. Teachers must be prepared to say, 'I don't know, but let's find out'. As a group, find the answers to any questions that they do not understand or want further explanations around. This can generally be done with classroom computers and/or reference materials already gathered to support the science content learning.

The elaborate step is where the 'science' can be explicitly brought out, but this really is just where the students get the missing information they need to better discuss what they just experienced. This is where teachers shine or fail. This should be where the teacher makes sure the science that was targeted is covered and the students are exposed to what the main point was for the lesson. Teachers also need to make sure they return focus to the main point(s) of the science – tangents are great and often perfect teaching moments, but the students should have been doing this activity for a reason. When the students were explaining, what did they leave out or not quite get for the 'accepted scientific explanations' – this is in single quotation marks for a reason. Not every student needs to understand the full scientific explanation.

The teachers must know their students, where they are at and how far they can be taken before too much information or too high a cognitive level is sought. Sometimes, we do have to accept that they got close enough for now, and next year or even years later, it can be covered again in more detail. In the case of cool bombs, the main focus for primary students is the temperature of the bag decreasing, the bag gets colder in their hands, and they can feel this temperature change. For many students (and teachers), this is counterintuitive as we have learnt to associate explosions with heat not cold. This is one reason, the other being that this is actually a fun activity, why this activity is titled 'cool bombs'. Every teacher knows there is a fine line between trying to elicit the information you want and badgering students to no avail. If they do not have the correct terminology, give it and move on so they can learn to use it. You want them to be able to use the words temperature and thermometer in their discussions.

Even children as young as five can describe that the bag is getting colder; most children gain a working knowledge of temperature by their parents/caregivers very early on in life. This is then repeatedly reinforced by most parents when they start exploring the world around them. Primary-aged children know the idea of temperature just maybe not the spelling.

Modelling the use of questions, I could ask them, 'What do you mean by saying that it is getting cold?' I want them to use the word temperature. 'How can we tell if something is hot or cold?' If possible, I also want to elicit the word 'thermometer' from them as the tool we use to measure or find out what the temperature of something is (Fig. 11.3).

Thermometer – this is now the third word that has been written down so far; the first two were cool bombs. Now would be a good time to repeat the activity to back up the idea that this gets colder not hotter. Stepping back to the explore step ensures the students get the opportunity to see how the temperature changes and how the thermometer changes in this activity. This time the students could use two to three spoons of each ingredient so the reaction lasts longer. Young students do not need to be able to read a thermometer to tell you what is happening; they can see what is happening. Older students may be able to read a thermometer, and part of their learning experience could be to record temperature versus time. When the thermometer changes what is happening to the temperature? Simple directing questions still keep the students engaged, and now they have the chance to show they are able to use the vocabulary appropriately, i.e. communicating in science.

I would ask the older students, 'Who knows what it is called when something gets colder after you mix it?' Once again if eliciting does not work, give the next needed vocabulary 'endothermic' (Fig. 11.4).

Now there are four words on the board. But lining up the 'therm' in endothermic with that of 'thermometer' is a visual way to help students grasp the meaning. Review what the students should know now as to the function, purpose or usage of a thermometer which shows the temperature of something. So 'what would you think the "therm" part of thermometer would mean?' Almost every time someone is able to make the connection of 'therm' with temperature. In this case of cool bombs, 'what happened to the temperature?', 'Did it give off heat?', 'It got colder; so, who thinks they know what "endothermic" means now?' If necessary, more directed follow-up questions are: 'who thinks endothermic means the temperature goes down or temperature goes up?' and 'So, does endothermic mean heat goes into it and the temperature gets colder?'

'If a thermometer is used to tell the temperature of something and endothermic tells us the temperature is getting colder, what could the "therm" part of these words

Fig. 11.3 Teacher's writing on board

Cool Bombs

Thermometer

Fig. 11.4 Teacher's final writing on board

Cool Bombs

Thermometer

Endothermic

mean?' I want them to connect this with the idea of 'heat'. There is always room to extend and challenge those who are willing and able to take it further. Further explanations for older students might be useful here with the extension vocabulary of 'exothermic' with the description of why we put another log in a fire (want more heat to be given off, so exothermic = heat given off, heat comes out) to help them get the idea of endothermic = heat goes in, gets colder.

The final E of the 5Es is evaluate. The step provides the students the opportunity to express not only what they understand about science but also how it relates to their world and what they are able to do with the science, that is, participating and contributing. This is where the students need to be able to discuss what they did, how they did it and why this science is relevant, useful and meaningful to them. What did they do that challenged what they thought they knew? How did this impact on their understanding? In accordance with *The New Zealand Curriculum*, a teacher could ask, 'What did they do that was like a scientist?' The point, however, to these discussions is for the teacher to hear all the summative assessment that is needed for these activities: what went well, what still needs to be worked on and where to go next? All this assessment is based on what the students have been discussing. I am arguing that primary science should not require a written scientific investigation report or test/quiz; it should be them talking and discussing.

This version of the fifth E is new to most teachers as this is where the Nature of Science is explicit. Students need to be exposed to content from the sciences (living world, material world, physical world and planet Earth and beyond), but it is not the content that is the focus anymore. It is how the students address this content and why they are being exposed to this content, i.e. the Nature of Science. In this activity, one of the main points was for the students to be able to use the scientific vocabulary appropriately. Remember the main focus was to use these words, not have to spell them; a wall display or modelling book can do that for the students. This activity (and not all activities will do this) did use all four elements of the Nature of Science so the students need to have this drawn out and explained. So, now that they can use the vocabulary (not every student will get it the first time; some might need to revisit this later. There are other activities that let you use this same vocabulary in a different activity), I would explicitly ask, 'Who can describe how we investigated in science?' This question assumes your students also know what they ought to have done to investigate in science. This was both an activity and fun play. This activity does not use a simple model, so if this comes up, students need to know why this is not a model. It is an endothermic reaction, not a model of one. Also teachers need to be aware of and able to distinguish between questions that are procedural from those that are about the science itself as the latter are not part of investigating in science but understanding about science. This important distinction is addressed below.

As stated, primary science is not teaching them to be scientists. Primary science uses science as a learning context, which is why they need to be explicitly asked, 'How have we acted like a scientist?' This gets them to discuss their understanding about science and what they got from the activity. The teacher, if at all possible, should be able to recall the questions they had about the science and show them the distinctions between them and those that are not questions about the science like 'do we use one spoon or two?' This is not a question about the science; it is a procedural question to do the activity. Questions about the science could be 'What would happen if we used cold water or hot water?' and 'Would there be a difference if it was a cold, warm or hot day?' Students should never ask 'Why did we do this?' Students do need to have fun and experience the WOW factor of science, but they also must know the WHY factor, how has this impacted/affected their life and their world? If they cannot do this, then you probably only gave them a fun activity without any of the learning they could have had. They will more than likely need scaffolding through this the first few times, but then they will amaze you with the depth of learning when they realise it is all about them and not writing it up.

This activity can take anywhere from 15 min to well over an hour, students need to learn and be taught how to discuss their ideas, and science is a discussion class not a writing class. Once they get the hang of this, the discussions may take longer and longer – but the learning gets deeper and more meaningful. I would argue for quality of learning over quantity in learning every time.

Since students need to be taught how to discuss and share ideas and opinions, they may also have to be untaught what they think a science lesson is. Actually, sometimes teachers have to be untaught this first. Science is a discussion class with the students doing most of that around the activity that they are doing. This could easily be a stand-alone activity, but it also fits in nicely in a series of activities around the topic of 'bubbles'. The first activity in the bubble series uses soap and water in a cup and uses a straw to 'make' bubbles in the cup. This normally gets soapy water on the table so see if students can make bubbles on the table with the straw. Then can they make bubbles without the straw using just their hands, then arms and hands, then bubbles inside bubbles, then how large a bubble, etc. This can be around the idea of what makes a bubble, the shape of bubbles (yes, you can make an almost square bubble) and the sizes of bubbles. If we can play with size and shape of bubbles, what else could we do? How can we make them cold or hot? Lesson 2 would be cool bombs for cold bubbles, and Lesson 3 is monster foam for hot bubbles (see Appendix 11A).

11.5 Conclusions

If we want students to become interested in science in secondary (or high) school or at the tertiary/university level, then they need to be exposed to effective science learning in primary school. Primary school students need to be excited, but also they need to know why they are doing it. Science that is only fun is only filling time as students might enjoy it, but what are they learning from it? Science should let students explore their world and help them to explain how and why their world does weird and wonderful things. Sometimes this means we let them go off with wrong ideas to be fixed later, but science is like that; if it makes sense and explains the student's world, how can what they think be wrong? One of the greatest advantages of education through science is that as students (and teachers and scientists) learn more, they are allowed to change their answer and still be correct. Pluto being reclassified is just one example of how what scientists think they know is able to change as they learn more about the topic.

Duschl et al. (2007) highlighted the need for in-service teachers to continue in their own professional development. This is often easier said than done. Teachers have enormous demands placed on them for the day-to-day running of their classes. I would argue that most teachers do in fact want to provide challenging and engaging learning opportunities for their students; it is just that in some subject content areas, this is not that easy. Teachers, like their students, come to class with their own personal baggage, and a self-perceived lack of science content knowledge is one of those obstacles (Simon 2000). Simon (2000) concluded that teachers need to be both enthusiastic and knowledgeable about their subjects so that they provide 'well-ordered and stimulating science lessons' (p. 115).

But there is more to designing a teaching and learning opportunity than just addressing each learning demand in turn as students need to be first guided and then taught how to independently draw upon and link together the big ideas (Leach and Scott 2000). Linn and Eylon (2006) further elaborated on this issue, claiming that students tended to localise their learning in the specific context in which it is occurring when they need to integrate all the ideas. *The New Zealand Curriculum* was developed to facilitate this integration of science with social, cultural, educational and personal experiences (Linn and Eylon 2006). The problem, unfortunately, has been that the teachers who were expected to implement this new curriculum were expected to do so without any guidance or professional development.

As a result, in 2011, Sir Peter Gluckman noted that a particular challenge was in how New Zealand primary school teachers would be provided with the skills necessary for education through science. He went as far as to say all primary schools in New Zealand needed a science champion, 'to assist teachers less confident in providing that sense of scientific enquiry and scientific enthusiasm to young minds' (Gluckman 2011, p. 5). In 2013, a reference panel was appointed to make recommendations to the New Zealand government in three areas: enhancing the role of education, public engaging with science and science sector engaging with the public. As a member of this panel, we were charged with examining what is being done, what is working, what is not working and how funding could be better utilised. This resulted in the publication of a final report in 2014, *A nation of curious minds* (Ministry of Business Innovation and Employment, Ministry of Education, and Office of the Prime Minister's Chief Science Advisor 2014). In conjunction with the release of this report, the Ministry of Education announced funding for the Science Skills in Education Initiative (Joyce and Parata 2014). The Sir Paul Callaghan Science Academy is this initiative. We have been working with primary teachers for the past four years not only to provide the professional development that is needed in effectively implementing *The New Zealand Curriculum* but also to provide an ongoing community of practice to support our alumni teachers (Kennedy et al. 2015). Education through science is at an exciting crossroads in New Zealand and now seems to be going in the right direction.

Appendix 11A

Monster Foam

Place a drink bottle in a bucket or tray as this makes the clean up so much easier. To this bottle, add about 2 cm depth of 30 % hydrogen peroxide using a funnel. And then squeeze in some dish soap; a good squeeze will make more bubbles. If you want, add some food colouring.

In a cup that is one-half full of warm water, add a spoonful of yeast and stir still until it starts to smell and gets a little bubbly. This takes a few minutes.

Then using the same funnel, pour the yeasty water into the bottle. This will result in an immediate reaction of foam spraying out. Most students will not get even half of the yeasty water into the bottle. The bubbles are warm to the touch and demonstrate an exothermic reaction.

Caution: this uses 30 % hydrogen peroxide (if you use a lower concentration, the amount of foam is reduced), and students should wash their hands after touching the foam. Should any of the hydrogen peroxide get on any skin, wash the area with water.

If done in a bucket or tray, it is easier to clean up. This does produce a lot of foam. If poured down a drain, it will come up through the lowest point of the drainage system. This has happened to me in one school. The boys came out of the bathroom with foam all over their shoes and pants after spending several minutes playing in it before telling anyone what was happening.

References

Arcus, C. (2009, August 24th). Seizing the opportunity. *New Zealand Education Gazette 88*(15), 10–11.

Bell, B., & Baker, R. (Eds.). (1997). *Developing the science curriculum in Aotearoa New Zealand*. Auckland: Longman.

Bolstad, R., & Hipkins, R. (2008). *Seeing yourself in science: The importance of the middle school years. Report prepared for the Royal Society of New Zealand.* Wellington: New Zealand Council for Educational Research.

Bull, A., Gilbert, J., Barwick, H., Hipkins, R., & Baker, R. (2010). *Inspired by science. A paper commissioned by the Royal Society of New Zealand and the Prime Minister's Chief Science Advisor*. Wellington: New Zealand Council for Educational Research.

Cowie, B., Hipkins, R., Boyd, S., Bull, A., Keown, P., & McGee, C. (2009). *Curriculum implementation exploratory studies: Final report*. Wellington: Ministry of Education.

Darling-Hammond, L. (1997). *The right to learn: A blueprint for creating schools that work* (1st ed.). San Francisco: Jossey-Bass.

Duschl, R., & Hamilton, R. (2011). Learning science. In R. E. Mayer & P. A. Alexander (Eds.), *Handbook of research on learning and instruction* (pp. 78–107). New York: Routledge.

Duschl, R. A., Schweingruber, H. A., & Shouse, A. W. (Eds.). (2007). *Taking science to school: Learning and teaching science to grades K-8*. Washington, DC: The National Academies Press.

Education Review Office. (2010). *Science in years 5 to 8: Capable and competent teaching*. Wellington: Education Review Office.

Education Review Office. (2012). *Science in the New Zealand curriculum: Years 5 to 8*. Wellington: Education Review Office.

Foster, G. (2009). Implementing the new curriculum. *New Zealand Science Teacher, 120*, 34–35.

Gluckman, P. (2011). *Looking ahead: Science education for the twenty-first century*. Auckland: Office of the Prime Minister's Science Advisory Committee.

Hipkins, R., & Hodgen, E. (2012). *Curriculum support in science: Patterns in teachers' use of resources*. Wellington: New Zealand Council for Educational Research.

Hipkins, R., Barker, M., & Bolstad, R. (2005). Teaching the 'nature of science': Modest adaptations or radical reconceptions? *International Journal of Science Education, 27*(2), 243–254. doi:10.1080/0950069042000276758.

Joyce, S., & Parata, H. (2014). *Kiwi curiosity at heart of science engagement*. Retrieved 17 Sept 2014, from http://www.beehive.govt.nz/release/kiwi-curiosity-heart-science-engagement

Kennedy, I., Smith, P., & Sexton, S. S. (2015). Ensuring New Zealand's future prosperity: A Professional Learning Development initiative to bridge the gap between theory and practice. *Science Education International, 26*(1), 42–55.

Leach, J., & Scott, P. (2000). Children's thinking, learning, teaching and constructivism. In M. Monk & J. Osborne (Eds.), *Good practice in science teaching: What research has to say* (pp. 41–56). Buckingham: Open University Press.

Linn, M. C., & Eylon, B.-S. (2006). Science education: Integrating views of learning and instruction. In P. A. Alexander & P. H. Winne (Eds.), *Handbook of educational psychology* (2nd ed., pp. 511–544). Mahwah: Lawrence Erlbaum Associates, Publishers.

Ministry of Business Innovation and Employment, Ministry of Education, and Office of the Prime Minister's Chief Science Advisor. (2014). *A nation of curious minds, He whenua hihiri i te mahara: A national strategic plan for sciecne in society*. Wellington: New Zealand Government.

Ministry of Education. (1993). *Science in the New Zealand curriculum*. Wellington: Learning Media.

Ministry of Education. (2006). *Effective literacy practice in years 5 to 8*. Wellington: Learning Media.

Ministry of Education. (2007a). *The New Zealand curriculum*. Wellington: Learning Media, Ltd.

Ministry of Education. (2007b). *New Zealand sign language in the New Zealand curriculum*. Wellington: Learning Media.

Ministry of Education. (2007c). *Te marautanga o Aotearoa: He tauira hei kōrerorero*. Wellington: Ministry of Education.

Otrell-Cass, K., Cowie, B., & Glynn, T. (2009). Connecting science teachers with their Māori students. *Set, 2*, 34–41.

Parker, J. (2004). The synthesis of subject and pedagogy for effective learning and teaching in primary science education. *British Educational Research Journal, 30*, 819–839.

Sexton, S. S. (2010). Science education a vehicle for the key competencies in praxis. *New Zealand Science Teacher, 125*, 38–40.

Sexton, S. S. (2011). Revelations in the revolution of relevance: Learning in a meaningful context. *The International Journal of Science in Society, 2*(1), 29–40.

Shulman, L. (1986). Those who understand: Knowledge growth in teaching. *Educational Researcher, 15*(2), 4–14.

Simon, S. (2000). Students' attitudes towards science. In M. Monk & J. Osborne (Eds.), *Good practice in science teaching: What research has to say* (pp. 104–119). Buckingham: Open University Press.

Skamp, K. (Ed.). (2004). *Teaching primary science constructively* (2nd ed.). Southbank: Thomson.

Taber, K. S. (2010). Paying lip-service to research? The adoption of a constructivist perspective to inform science teaching in the English curriculum context. *Curriculum Journal, 21*, 25–45.

Traianou, A. (2006). Understanding teacher expertise in primary science: A sociocultural approach. *Research Papers in Education, 21*(1), 63–78.

Tytler, R. (2008). Re-imaging science education: Engaging students in science for Australia's future. *Australian Education Review, 51*, 1–77.

Chapter 12
Towards a Socially Responsible Science Education

Gilbert Onwu

12.1 Introduction

The world is facing almost insurmountable challenges that transcend national boundaries. The Millennium Project has identified 15 global challenges, which according to Glenn et al. (2009, p. 10) "provide a framework to assess the global and local prospects for humanity". Africa, in particular, is facing daunting challenges with respect to access to education and healthcare, averting deepening poverty and worsening environmental degradation such as erosion and loss of biodiversity, access to clean water and improved sanitation, access to energy, enhancing food security to meet the needs of growing populations, addressing income inequity, enhancing women empowerment and emancipation especially of rural women, lack of democracy and good governance, poor infrastructural development and so on. The list seems endless. How should we best address this litany of issues of human development, equity and sustainability? Our ability to provide life's essentials such as water, food, shelter, energy and good governance will require not only major advances in science and technology but also a critical mass of scientifically literate citizenry able to mobilise the intellectual resources required to meet the needs of our societies and our global obligations. What should a socially responsible science education look like within that context?

As has been argued elsewhere (e.g. Kyle 2006; Onwu and Kyle 2011), formal science education has much to offer in helping to facilitate knowledge and skills development required to take action to counteract those debilitating human conditions that persist for a significant percentage of the African population. Regrettably, our current approaches to education in science still perpetuate a way of thinking that is not in synch with preparing our (African) youth to develop the competencies

G. Onwu (✉)
Department of Science Mathematics and Technology Education, Faculty of Education,
University of Pretoria, Pretoria, South Africa
e-mail: gilbert.onwu@up.ac.za

B. Akpan (ed.), *Science Education: A Global Perspective*,
DOI 10.1007/978-3-319-32351-0_12

required for active participation in today's complex world. In many countries of Africa, and indeed worldwide, the situation is such that science education is more focused on transferring canonical knowledge than with promoting students' understanding and application of relevant scientific knowledge for the purpose of community development and social transformation (Sinnes et al. 2010). Many contemporary students feel alienated by science; they fail to see the link between science and the knowledge necessary to transform and improve the quality of life of the community in any sustainable way (Carter 2008; Lyons 2006). Science education is now perceived by many students as not being particularly relevant in addressing personal interests and related social issues.

The international community has pointed to education as the most viable means of promoting sustainability and transformation. The theme, "Education for Sustainable Development", for instance, places much emphasis on people empowerment through providing knowledge and skills to address human development and environmental issues, which ultimately bring about change in peoples' behaviours and attitudes (UNESCO 2005). The application of knowledge has become one of the key sources of growth in the global economy. Precisely for this reason, the term knowledge economy (KE) has been coined to reflect this increased importance of knowledge. This emphasis on knowledge economy, or the knowledge revolution as it is sometimes called, is manifest in different ways: there are closer links between science and technology; there is increased importance of science and culture, and so greater attention is given to the complementary frameworks of orthodox science and indigenous knowledge systems (IKS) (Onwu and Mosimege 2004; Onwu 2012a); there is increased advocacy for socio-scientific issues (as defined, these refer to social issues that have a scientific basis and are important to the society) to be considered in the course of students' formal education as a way of connecting science education to the goals of sustainable development (Kyle 2006; Onwu and Kyle 2011; Ratcliffe and Grace 2003; Sadler 2009). Given the knowledge explosion, clearly there is every indication that over the next 20 years, a different kind of community of practice in science classrooms is going to have to emerge. A business-as-usual mind-set with respect to science education can no longer be the norm nor will it be acceptable or even tolerated (Holbrook 2009).

The central issue in this chapter is about the call for a shift in focus in science education from one that has been characterised by a view of science as being disconnected from socio-scientific Discourses to one that emphasises science as a human activity that is embedded in sociocultural and political issues (Carter 2008; Mudaly 2011). I concur with Sadler (2009, p. 12–13) when he states that "an important objective for science education ought to be for learners to come to identify themselves as willing and able to engage in socio-scientific Discourses…as active contributors to society with competencies and willingness to employ scientific ideas and processes, understandings about science and social knowledge to issues and problems that affect their lives".

The link between science education and "real-world" experiences is almost always tenuous in the minds of present-day science students. The failure of science educators, to date, to develop what may be referred to as subject matter curriculum connection, notably, between science and the day-to-day lived experiences of students say, – especially in the way science is taught and what is taught – is always likely to obscure and diminish the relevance of science in their lives (Bennett and Holman 2002; Onwu 2000). And that is why I am in agreement with Sadler (2009, p. 12) when he calls "for the development of communities of practice in science classrooms that prioritise socio-scientific Discourses and development of identities reflective of engaged citizenship". Such a transformation in science education will progressively lead towards a real-world context-based approach to science teaching and learning (Holbrook 2009), which will promote the active participation of science students in ways that would inevitably help them to develop and acquire the knowledge, competencies and skills necessary to take part in sustainable human development and informed decision-making.

Thus, the main goal of science education "should be to facilitate both students' and the public's ability to identify possibilities, to seek challenges, to use their imagination to transform" (Kyle 2006, p. 14). Further, science education ought to expand and enhance students' engagement and facility in problem-solving and decision-making within the context of the community in which they live (Gray et al. 2009; Kazeni and Onwu 2013). It is against this backdrop that the call is being made to reformulate the agenda in science education, so as to make it more pertinent and empowering to students, and for science to become more relevant, socially connected and responsible.

One may very well ask: but what is a socially responsible science education? Kyle (2006) draws upon Habermas (1972) when he describes a socially responsible science education as an education that not only focuses upon development of students' technical knowledge imperatives, "where the aim is to understand nature, but also their emancipatory knowledge interests, where their experiences with science are self-involving and socially just" (Sinnes et al. 2011, p. 2).

In the Executive Summary to the 2011 State of the Future, Glenn et al. (2011) state that there is no question that the world is getting richer, healthier, better educated and better connected and so people are living longer, and yet half the world is potentially disadvantaged and unstable. They further contend that there is no question the world can really be better than it is if we do the right things. Glenn et al. (2011, p. 2) assert that "the world has the resources to address its challenges. What is not clear is whether the world will make good decisions fast enough and on the scale necessary to really address the global challenges". Clearly, what is at stake here is whether the ability and the political will to make those "good decisions" exist and are readily applicable. Those decisions will ultimately be influenced largely by the type of science education or more precisely the philosophy underpinning the science curriculum (whether formal or informal) experienced by students and citizenry of today and also of the future. Hopefully, it would be the type of education in science which is responsive to the conditions and demands of the time.

It is relevant to point out at this juncture that for the past 100 years or so, science has been progressively inserted into the school curriculum. And yet, its original purpose, as Holbrook (2009) has pointed out – that of preparing students for university studies – has more or less remained the predominant determinant of what content is taught and how it is taught at school level. Indeed, several other authors (Kyle 2006; Ratcliffe and Grace 2003) have also noted that there is still a strong conceptual tone in science content learning, which essentially characterises science preparation as deserving of a selected few or an elite group of students. Additionally, as Kyle (2006) notes, the content and teaching would seem to ignore or take little cognisance of the wider complex of sociocultural and political factors that influence the ways schooling is structured for the benefit of some students more than others.

With this perspective in mind, I discuss some remaining issues arising from much contestation about approaches to science teaching, which have failed to provide contemporary students with the opportunity to see the relevance of studying science in their lives. Young people's interest in the study of science and science-related courses is on the decline in Africa and indeed worldwide (Economic and Social Research Council (ESRC) 2008). It is not unlikely that this state of affairs would continue unless meaningful and relevant science education is put in place, particularly one that has much to offer in enhancing the emancipatory interests of students in such a way as to enable them to play a responsible role in society (Kyle 2006). A major educational issue facing all countries today is how to get the right balance between having a sufficient number of students going on to scientific and technological careers and providing all students with enough motivation and knowledge of science to appreciate its strengths and limitations (Holbrook 2009).

Sterling (2009, p. 105), in a similar vein, has also argued that rather than retain "the policy and practice of a Western and Westernized education still largely built on the assumptions and epistemology of a previous age", science educators should insist on a progressive relevant science education that is "fully responsive to the conditions and needs of our time". What should be the main focus of a relevant science education is a long-debated issue that concerns what a relevant science education really is (Onwu and Kyle 2011). As an exemplar, Holbrook in the earlier chapter (see Chap. 10) proposes a real-world, motivational science teaching at all levels of the school curricula using a context-based approach.

Education, and science education in particular, has been advanced as an imperative for development, and yet questions persist about how science education can be made more relevant to enable students deal with complex everyday issues and participate in decision-making oriented towards development. If as science educators we were to focus more on students' personal benefits and incorporate socio-scientific issues in the course of formal teaching and learning in science, it would probably be a more credible way of maximising the relevance of science for improved access and motivation. A relevant and socially responsible science education would thus tend more to what is important to the student including scientific knowledge as priority, other student needs and student motivation than what is important to the scientist (Malcolm et al. 2009; O'Donoghue 2010). Hence, we have implied that science education for relevance, social knowledge and responsibility is essentially the province of science teachers and science educators.

12.2 Calling for a Shift

In calling for a shift in emphasis in science education from one that is disconnected from social issues to one that emphasises and is embedded in social knowledge and environmental issues, I will draw upon my South African experience in terms of exemplars, convinced that the narratives would resonate with what obtains in other African countries and elsewhere. The shift is especially pertinent in the African context for meeting the requirements of "multicultural" science as opposed to a monoculture science that tends to serve the interests of a privileged few. Proponents of multiculturalism hold "a more contextual and historical view of science…and recognise that different cultures have disparate ways of understanding the natural world that might be viewed as science" (le Grange 2004, p. 208). It is beyond the scope of this chapter to discuss the notions of multiculturalism and universalism as competing worldviews from which to make sense of natural phenomena. Suffice it to say that the claim that science is important has not always been accompanied by the knowledge of what kind of science and curriculum, for instance, is responsive to students' aspirations in the communities in which they live or, indeed, what kind of science best translates into wealth and improved quality of life for people in Africa. Thus, science as learned and taught at school level, must of necessity seek to do more than just provide the foundations or groundings for success at higher levels; if not, higher-level science will continue to progressively be given a wide berth by students, as a future career option. Besides, school science is only one among many other school subjects or academic resources that students can draw from in order to succeed at that level.

In 2009, in recognition of the need for socio-scientific issues to be considered in the course of students' formal education in science, the International Council of Associations for Science Education (ICASE) sponsored a conference in Abuja, Nigeria, with the theme "Meeting Challenges to Sustainable Development in Africa through Science and Technology Education" (ICASE 2009). The conference was intended to provoke debate and discussions regarding the ways in which science education ought to be connected to the goals of sustainable development. In the course of the various discussions, it became quite clear that any initiatives proposed, or embarked upon, ought to necessarily include examples, which highlight the value of science learning relevant to the needs of a changing society and to the needs of both boys and girls as future citizens whatever their future career aspirations may be (Onwu and Kyle 2011). In other words, developing a scientifically literate citizenry, which is appropriate to the particular community, is a major goal of sustainability. One could offer the assertion that the aims of science education are intrinsically linked to one's conception of scientific literacy. In that regard, one agrees with Roberts (2007), when he proposes that the visions of scientific literacy materialise from the contexts in which science subject matter is taught. He refers to these contexts as curriculum emphasis (Roberts 1982, 1988).

Onwu and Kyle (2011) note that Roberts (2007) seems to indicate two visions of scientific literacy representing the extreme ends of a continuum which are referred to as *Vision I* and *Vision II*. For Roberts, Vision I is "rooted in the products and processes of science" (p. 730). Advocates of Vision II, the perspective from which this author emanates, stress the importance of starting "with situations then reaching into science to find what is relevant" (Roberts 2007, p. 730). Roberts elaborates further in stating

> What is important to recognize is that advocates of Vision II stress that all students in democratic societies – regardless of their career plans – need to develop SL (scientific literacy) that is appropriate to situations other than conducting scientific inquiry. Thus, for example, an understanding of scientific inquiry is not only important for potential scientists, in Vision II thinking. It is a vital component of a citizen's ability to keep scientific perspective in balance with others. Thus, students need classroom experience with situations in which different perspectives are deliberately brought to bear in socio-scientific issues. (Roberts 2007, p. 771)

Traditionally, scientific literacy has been couched in terms of past knowledge and this development has been limiting. Such a perspective, for instance, has failed to capture the dynamic emergence and/or disappearance of new literacies (Roth 2007). Furthermore, van Eijck's (2009, p. 256) notion of scientific literacy supports the viewpoint that "grounding the concept of scientific literacy in a cultural-historical perspective allows the articulation of what being *scientifically literate* means".

Given those varied perspectives on scientific literacy, clearly, it is problematic for science education to be primarily aimed at, or limited to, promoting future success in the culture of schooling only, without links beyond school science to society, communities, self and social empowerment and transformation. Sadler (2009) had this to say about such a limited view, which falls short of the aims that science education ought to assume:

> Science as it is practiced in the lived experience of engaged citizens can serve as the basis for developing a different kind of community of practice in science classrooms. The underlying premise is that we position science classroom as contexts for students and teachers to actively express issues and problems with two necessary elements: (1) conceptual and/or procedural connections to science and (2) social significance, preferably as judged by the community participants themselves. This class of issues and problems has been termed socio-scientific issues (SSI). (Sadler 2009; p. 11)

Thus, from Sadler's position, science education in a multicultural society like South Africa's for instance, should aim at providing equitable opportunities for both Black and White students of varying sociocultural backgrounds that would facilitate an understanding of the identities constructed by individual students and how these identities are retained or reconstructed to include scientific reality (Onwu 2012b). What is being implied is that science is contextual and culturally produced and that cultures have disparate ways of understanding the natural world and that the different ways of knowing should be recognised by Western modern or orthodox science as well. With such an understanding, there are areas of self-empowerment in science education that a multicultural approach could possibly enhance. Some of these areas would include (i) the incorporation in a legitimate and valid way of local or indigenous knowledge including aspects of students' culture and beliefs in science education and (ii) collaborative participation of the community in schools and science classrooms, to name just a few. The interfacing between the community and the science classrooms could serve as the basis for developing a different kind of community of practice in science classrooms.

In Africa however, the bond between formal science and society is still relatively weak. There are certain socially related misgivings and communication gaps between activities of the scientific communities and the public understanding of science. The upshot is the absence of a strong science culture among the various constituent stakeholders, who under normal circumstances ought to be the drivers of scientific and technological development. This state of affairs came about largely as a result of the failure of African governments at the dawn of political independence to invest strategically in science and technology. Throughout the 1980s and 1990s, in spite of the glaring evidence from South East Asia and other regions that investment in science and technology was really worthwhile, science and technology received scant attention by way of funding support in Africa. Almost a generation later, Africa is yet to recover from the missed opportunity.

12.3 Cultural Contexts and Science Education

A discernible consequence of Africa's low investment in science and technology is the declining quality of science and engineering education at all levels of educational systems. Student enrolment in science, engineering and technology subjects at primary, secondary and tertiary levels is also falling. In most developing countries of Africa, since science has become a major factor in national, social and economic progress, low uptake of science subjects and related careers are likely to impact negatively on the quality and quantity of scientific research and national economic development (Economic and Social Research Council 2008).

African countries, like Nigeria and South Africa, in recent times are faced with problems of low uptake and underperformance in science subjects. In South Africa a demonstrably low percentage of all post-secondary education degrees are in mathematics, science or engineering. South Africa's ratio of scientists and engineers to the population, for instance, stands at 3.3 per 1000 compared with 21.5 per 1000 and 71.1 per 1000 in the USA and Japan, respectively, as reported by the National Research Foundation (NRF) Report (2005).

The situation, I hasten to add, is probably worse in most other African countries as they struggle to address a variety of challenges. The educational system is expected to play a significant role in addressing these imbalances. It is unfortunate that school science has failed to excite and attract many students for further studies in the sciences because science as a discipline is perceived not to be particularly relevant in making decisions about the way they do things individually and collectively that will improve the quality of life. Arguably, there is a crisis: a *crisis of relevance*, which is one of the failures of the educational system to meet the aspirations and interests of both students and society in a rapidly changing environment and competitive landscape.

South Africa, for example, is a country experiencing a period of rapid transformation and democratisation. As I have discussed elsewhere (Onwu 2012a), a positive element in this concomitant social and political change is the intellectual space provided for marginalised indigenous knowledge and voices to be given free rein, to be heard. According to le Grange (2004), South Africa, at a macro level, has been eager to develop its scientific human capital by providing policy frameworks including norms and standards for transforming and reconstructing South African society. South Africa is perhaps one of the very few African countries that have implemented policies that recognise and stress the importance of protecting and promoting indigenous knowledge and other knowledge systems and technologies to address specific problems.

Following the calls for an *African Renaissance* (cf. Odora Hoppers 2002) at the start of the millennium, a key response by government was the inclusion of indigenous knowledge systems (IKS) in the National Curriculum Statement for General Education and Training (GET) and for Further Education and Training (FET). The incorporation of IKS into mainstream curriculum is best clarified by quoting from the National Curriculum Statement document (Department of Education, Pretoria 2003) as follows:

> Now people recognize the wide diversity of knowledge systems through which people make sense of and attach meaning to the world in which they live. Indigenous knowledge systems in the South African context refer to a body of knowledge embedded in African philosophical thinking and social practices that have evolved over thousands of years. The National Curriculum Statement Grades 10–12 (FET) has infused indigenous knowledge systems into the Subject Statements.

Further, an example of a curriculum policy document that explicitly acknowledges the significance of bodies of knowledge or discourses other than that of orthodox science is South Africa's national curriculum policy for natural sciences. It states that "The Natural Sciences Learning Area deals with the promotion of scientific literacy. It does this by the:

- Development and use of science process skills in a variety of settings
- Development and application of scientific knowledge and understanding
- Appreciation of the relationships and responsibilities between science, society and the environment" (Department of Education, Pretoria 2002, p. 4)

The document elaborates on each of these three aspects of scientific literacy, synthesises their intended meaning in three broad learning outcomes and provides an extended discussion of how the three components can be assessed. The third outcome, described as "challenging, with potential to broaden the curriculum and make it distinctively South African" (Department of Education, Pretoria 2002, p. 10), is of special interest because it includes attention to relationships between science and traditional and local practices/technologies as these relate to traditional wisdom and knowledge systems. The document addresses this relationship by suggesting that:

> One can assume that learners in the Natural Sciences Learning Area think in terms of more than one world-view. Several times a week they cross from the culture of home, over the border into the culture of science, and then back again. How does this fact influence their understanding of science and their progress in the Learning Area? Is it a hindrance to teaching or is it an opportunity for more meaningful learning and a curriculum which tries to understand both the culture of science and the cultures of home? (Department of Education, Pretoria 2002, p. 12)

The valuing of indigenous knowledge (IK) and the recommendation of its integration into the national curriculum statements pose for science teachers and indeed practitioners in the field several questions.

Is IK in synch or congruent with science especially when it is more nuanced and more specific to a given locality – i.e. its localness? For example, is "science" in the learning outcome understood as referring to Western modern science only? Such an interpretation, as le Grange (2004) has pointed out, would necessarily view on the one hand indigenous ways of knowing as having inferior explanatory powers of the natural world to that of Western modern science. On the other hand, if science is viewed more broadly as systematic knowledge, then in that case, le Grange (2004) continues, there is the possibility of accommodating and validating aspects of indigenous knowledge as science. Whichever view is taken up in classroom practice, Le Grange (ibid) posits that there are obvious implications for how teachers interpret the integration of IK into the natural sciences teaching, and this (interpretation) could result in radically different science learning experiences in South African classrooms.

From the perspective of the school-community interface, does IK serve as a useful link between home/community and school-based experiences? Is IK meaningful and relevant to students' life experiences? Will valuing of IK boost the morale and enhance the sense of self-worth and scientific literacy of the local indigenous community? Can IK be extrapolated or applied to other contexts or similar situations – such as the school?

These questions are frequently raised, and science educators seek to resolve them in an attempt to justify incorporating IK into school science. The question of whether or not authentic IK exists and whether or not it is complementary with modern science is no longer the issue or a matter of serious debate (Ogunniyi 2011). The answer is at worst a qualified "Yes". What is of particular interest takes on more of a philosophical hue than a pragmatic one, and it centres on how the two frameworks – modern science and IK – are able to complement each other in the context of science classroom teaching and learning. For the integration agenda, the pressing concern is how to develop a method of philosophical analysis, of an epistemology for interfacing IK with school science in a valid and legitimate way. One of the major challenges faced by science educators is to identify or come up with epistemologies of IK practices and technologies in relation to their infusion into school science curriculum. Once a valid method(s) of identifying epistemologies for IK practices has been developed, it would facilitate the interfacing of IK with mainstream science curriculum.

Despite the impressive contributions indigenous knowledge and local technologies have made, particularly in dealing with specific developmental and ecological problems, there appears to be what one might refer to as "intellectual inertia" – a restraining mind-set – on the part of science teachers to actively engage with IKS. The jury is still out as to South African science teachers' perception of the nature of IK in relation to orthodox science and its incorporation into science classroom teaching. The national curriculum does not specify how to integrate IK into mainstream science curriculum and for that reason, teachers, as I noted earlier, would have different interpretations on how best this could be achieved.

Multiculturalists would argue that orthodox science and indigenous knowledge are essentially complementary and that some of the modern inventions have come about as a result of this symbiotic relationship. The discovery of aspirin, it is said, would not have been possible without indigenous knowledge (le Grange 2004). Indigenous knowledge system is a domain or area of study which is in its infancy and has not presently filtered into the curriculum of teacher training institutions locally. Because of its relative newness, science teachers in schools are poorly prepared to meet the challenges of its integration into school science teaching and the curriculum (Onwu and Mosimege 2004). They lack the content, the epistemology/ methodology and the training to comprehensively introduce and use it (IK) in schools. Nevertheless, the inclusion of IK in science lessons offers myriads of investigative opportunities such as debates, critical discussions, arguments, inquiry and decision-making that can invariably form part of the cultural dialectics of the science classroom (Ogunniyi 2011), which could profitably be used by the teacher on the students' behalf to impart knowledge.

The science classroom as contexts for effective and productive communication does raise the question of the effect of language as an aspect of culture in science education. Recent research reports (Centre for Development and Enterprise (CDE) 2004, 2010) have conclusively shown that in the South African setting, students' facility with language of instruction and examination among others is a strong determinant of performance in science and mathematics public examinations.

Again, focusing on South Africa with 11 official languages, for many students, the greatest barrier to learning science is language. The majority of Black students are English as second language (ESL) speakers and users, and the majority may not necessarily have the language facility for comprehending abstract theoretical scientific concepts and may therefore resort to rote memorisation. Like many other African countries, South Africa has developed its science curricula content solely on the basis of orthodox Western trends and science is taught mainly in English or Afrikaans. Black South Africans encounter additional challenges in the sense that there are no direct translations of scientific concepts in mother tongue or vernacular; hence, some of the conceptual difficulties that teachers and students find with science could be attributable to instructional language factor (CDE 2004).

12.4 An Agenda for Action: Context-Based Approach for a Socially Responsible Science Education

Public examination results particularly in science and mathematics education generally attract the attention of various key stakeholders including universities, employers and parents. This is because these are gateway subjects for admission into higher-level institutions of learning, and good performance holds prospects of rewarding future careers in the science and technology. In South Africa, for example, given its antecedents from the apartheid legacy, African students for various reasons do not perform particularly well in public examinations in school science and mathematics and are therefore not inclined to choose science and engineering at tertiary level. Black African students are underrepresented in the science, engineering and technology field of study. The situation is just as worrying in Nigeria, which has the largest economy in the African continent, and where in the latest available published 2012 results, an abysmally low percentage of students passed in physics and chemistry in the public national examinations organised by the National Examination Council (NECO). The situation may not be any different in some other countries of Africa, and these poor performances have serious implications for science and engineering graduate output, sustainable economic development as well as scientific research output.

Existing explanations for this state of affairs range from lack of provision of resources, rote memorisation of complex facts and meaningless data, through to quality of teacher education and availability of specialist teachers. Others include quality of school management and leadership, classroom environment, poverty and lack of family support systems. Language-related conceptual issues, the cultural clashes between modern or orthodox science and African traditions, have been advanced as well. Students' dwindling interest, low motivation and poor performance in science can all be attributable to inadequate preparation, under-resourced science classrooms and the lack of recognisable relevance of science education in their lives (Gilbert et al. 2011; Kazeni and Onwu 2013).

There is no doubt that alienation of students from science may be due in part to the way science is taught and in part to the failure of the curriculum – its purposes in meeting the interests and aspirations of the generality of students. Science as presently taught largely ignores the wider and more complex sociocultural and political factors that impact on the ways schooling is structured, which, as unfortunate, tend to benefit some students more than others. But having said that, Onwu and Kyle (2011, p. 16) note that in recent times, "scientists have come a long way from the ivory tower and many scientists nowadays are well aware of societal needs and constraints associated with the nature of their inquiry". To this end, research activities have emerged focusing on socio-scientific issues and involving interdisciplinary research groups working in such related areas as food security and production, healthcare and sustainability sciences, among others. However, the question remains about where in the science curriculum should real-world socio-scientific issues be located and how can the curricular support for such teaching be provided on a more global basis? These are some of the challenges or remaining issues as we explore and create more opportunities for interdisciplinary learning in the sciences. Clearly we are of the opinion that the aims of school science need to be examined with respect to sustainable development and related social knowledge and responsibility.

Other proponents of cross-national science achievement surveys such as Trends in International Mathematics and Science Study (TIMSS), however, hold contrary views arguing that science education will be transformed by implementing norms and standards that assess students' technical knowledge and engage them in competitive cross-national comparisons of science achievement. Such views have come under criticisms and are currently being challenged. Suffice it to say that such international comparisons according to Kyle (2006) tend to invoke a narrow image of science by placing more emphasis on the technical interests of the empirical-analytical sciences at the expense of the practical emancipatory interests of the hermeneutic-interpretive sciences. Additionally, such a position not only neglects the fundamental issues of the place of science in its larger social context but also fails to acknowledge the political considerations, which inform various aspects of the scientific enterprise. Why this debate is important is because, in essence, students and citizens alike have been unwittingly denied access to the social and political process of science. This debate strikes directly at questions of relevance and the importance of engaging our students as "active agents" in meaningful learning. It is precisely for these reasons, among others, that many science educators are advocating that science education moves progressively towards a real-world, issues-based approach at all levels of the school curriculum. A context-based education focuses on developing knowledge and skills from contexts that are familiar and closely related to students' needs and interests and which reflect the realities of the communities in which they lead their lives.

South Africa's former National Curriculum Statement and the current Curriculum Assessment Policy Statement (CAPS) in physical and life sciences promote and place emphasis on context-based education, notably in the use of real-life issues including socio-scientific issues in teaching the subjects. Empirical and anecdotal evidences, however, point to the fact that science classroom teaching in the South African schools is hardly influenced by contextualised teaching. Instead, the emphasis is more or less as it has always been on structured content-driven approaches. Even though the third learning outcome of the Natural Sciences National Curriculum statement specifically stipulates and deals with the interface of science and society, the recommended learner-centred, issues-based teaching approach(es) which ought to facilitate this interfacing is largely absent from South Africa's science classrooms (Onwu and Stoffel 2005). Science lessons are manifestly dominated by teacher-centred knowledge transmission practices. It can be surmised that although contextualised teaching is recommended in the national curriculum, its application at school level remains just that, a recommendation.

A recent study (Kazeni and Onwu 2013) in the South African setting has shown that failure of science teachers to develop valid teaching approaches that seek to link the day-to-day lived experiences of students to their science classroom experiences has further eroded and undermined the credibility and relevance of science in their lives and, in turn, their performances. Few people, I think, would argue against the general principle of making science education more relevant. And here, relevance would entail the use of local community resources and incorporation of local issues into the school curriculum. To this end, issues-based learning would invariably be seen in the eyes of the student as something worthwhile, meaningful, useful and important.

Kazeni and Onwu (2013), in collaboration with some science teachers located in fairly representative Black township schools in South Africa, sought to investigate the comparative effectiveness of context-based and traditional teaching approaches in determining performance differences in genetics achievement, problem-solving, decision-making and integrated science inquiry skills and attitudes towards life sciences. Context was operationally defined in terms of types of situations or contexts identified by the learners themselves, which they felt would help to make the learning of genetics meaningful, relevant and accessible to them.

The nature of the research questions and the researchers' commitment to issues-based learning and authentic contexts (for relevance) rendered a mixed method research approach an appropriate orientation for the inquiry. The innovative aspects of the study included, first, the use of contexts selected by students themselves in developing the real-life narratives for the genetics materials and, secondly, the application of a five-stage learning cycle model that maximised participatory modes of learning of the students. The context-based teaching approach provided opportunities for the experimental group, which sought to maximise students' issues-based learning through participatory discussions, knowledge acquisition on a need to know basis, hands-on experience, decision-making and problem-solving. The study's findings showed that context-based teaching approach was statistically significantly better than traditional teaching approaches in enhancing grade 11 student's performance in genetics content knowledge, problem-solving, decision-making, inquiry skills of hypothesis formulation and experimental design and in improving students' attitude towards the study of life sciences. Teachers and students who participated in the study voiced their views, which should encourage more research into the use of socio scientific themes (identified by students themselves where applicable) as contexts for designing curriculum learning materials.

Post-intervention focus group interviews held with both the experimental and control group members revealed that, first, students from the experimental group found the study of genetics using issues based approach to be interesting and great fun. The students also expressed confidence about their post-test performance in the genetics themes examined. Secondly, the experimental group mostly attributed their performance to the way the topic was taught using context-based teaching approach, as is evident in the following quotations from students' interview protocols:

- *If all educators taught us the way 'sir (the teacher)' did, we would never fail any subject. I enjoyed looking back at my original ideas.* (Reference to stages 2 and 4 of the learning cycle)

- *The nice thing about the lessons was that we talked about things that happen in our own homes. I now understand why my brother looks so different from all of us.* (Familiar and authentic context)
- *The stories made the study of genetics easy because we managed to understand what was happening and were able to explain the situations* (Use of authentic narrative contexts from the students' perspective).
- *It was fun to learn genetics by using our own experiences. It makes genetics so easy. I am sure I have passed the test.*

Teachers who taught the experimental group expressed similar sentiments as their students, as stated below:

Teacher A	*Students were very excited during lessons, especially during phase 4. Sometimes, it was difficult to control them, because they came up with so many questions and suggestions.*
Teacher B	*For the first time, I did not have to force my students to talk. In fact, I had to control them most of the time. Everyone wanted to say something.*
Teacher C	*Learners who were taught using the new method really understood the lessons, because of relating everything they did to what happens in real life, and the practical activities and discussions. Once you give them what happens in real life, and then teach them the relevant genetics concepts, it becomes easier for them to understand.*

Students from the control group also found the study of genetics to be interesting. However, their comments indicate that they were apprehensive about their post-test performance in the topic. The reasons given for their lack of confidence among others were the difficulty in remembering the many genetics terms or concepts and their inability to relate to some of the materials meaningfully in terms of their daily life experiences.

The study benefits science educators through the provision of a prototype for developing context-based curriculum materials. Specifically, the model developed provides science teachers with practical guidelines for devising storylines as authentic contexts for linking situations or issues to science concepts to be taught, when, for instance, infusing local knowledge, IK or social knowledge into school science curriculum. With this plea for relevance and social responsibility in the context of science education, there is an increasing recognition that science teaching and learning worldwide ought to be fostering engagement with goals of sustainability in various domains such as IK practices and technology, health and environment, poverty alleviation and economic prosperity, which are of interest to students and to their communities.

12.5 Conclusion

In conclusion, this chapter derives from and builds on previous works and discussions of science educators in Africa and elsewhere and is written with the mind of continuing the debate about the kind of relevant science education that is needed to help address the current global challenges facing humanity in general and Africa in particular. It adds to the call for science education to be made more pertinent and empowering to students. Indeed, certain aspects of South Africa's science educational system have been used as exemplars to further highlight some of the cultural barriers to science education as well as to touch on the anticipated challenges to the call for a paradigm shift, if we wish to prepare future citizens able to deal with complex everyday issues in socially responsible ways. Such a vision I believe is what scientifically literate citizenry is really all about.

References

Bennett, J., & Holman, J. (2002). Context-based approaches to the teaching of chemistry: What are they and what are their effects? In J. K. Gilbert, O. DeJong, R. Justi, D. F. Treagust, & J. H. Van Driel (Eds.), *Chemical education: Towards research-based practice* (pp. 165–184). London: Kluwer Academic Publishers.

Carter, L. (2008). Socio-cultural influences on science education: Innovation for contemporary times. *Science Education, 92*, 165–181.

Centre for Development and Enterprise (CDE). (2004). *From laggards to world class: Reforming maths and science education in South Africa's school. Abridged report on research.* Johannesburg: Centre for Development and Enterprise.

Centre for Development and Enterprise (CDE). (2010). *Value in the classroom: The quantity and quality of South Africa's teachers. Abridged report on research.* Johannesburg: Centre for Development and Enterprise.

Department of Education, Pretoria, South Africa. (2002). *Revised national curriculum statement for grades R – 9 schools: Natural sciences*. Pretoria: Department of Education.

Department of Education, Pretoria, South Africa. (2003). *National curriculum statement grades 10–12 (general). Overview*. Pretoria: Department of Education.

Economic and Social Research Council (ESRC). (2008). *Improving take up of science and technology subjects in schools and colleges: A synthesis review* (Report prepared for the Economic and Social Research Council (ESRC)). Newcastle: Newcastle University, Business School.

Gilbert, J. K., Bulte, A. M. W., & Pilot, A. (2011). Concept development and transfer in context-based science education. *International Journal of Science Education, 33*(6), 817–837.

Glenn, J. C., Gordon, T. J., & Florescu, E. (2009). *2009 state of the future*. Washington, DC: The Millennium Project.

Glenn, J. C., Gordon, T. J., & Florescu, E. (2011). *2011 state of the future*. Washington, DC: The Millennium Project.

Gray, D., Colucci-Gray, L., & Camino, E. (Eds.). (2009). *Science, society and sustainability: Education and empowerment for an uncertain world*. New York: Routledge.

Habermas, J. (1972). *Knowledge and human interests* (trans: Shapiro, J. J.). London: Heinemann (Original work published in 1968).

Holbrook, J. (2009). *Meeting challenges to sustainable development in Africa through science and technology education: 2009 conference concept paper*. Abuja: ICASE African Regional Office.

International Council of Associations for Science Education (ICASE). (2009). *Increasing the relevance of science and technology education for all for the 21st century: Framework document.* Retrieved January 10, 2010, from http://www.icaseonline.net/the.html

Kazeni, M., & Onwu, G. O. (2013). Comparative effectiveness of context-based and traditional approaches in teaching genetics: Student views and achievement. *African Journal of Research in Mathematics Science and Technology Education, 17*(1–2), 50–62.

Kyle, W. C., Jr. (2006). The road from Rio to Johannesburg: Where are the footpaths to/from science education? *International Journal of Science and Mathematics Education, 4*, 1–18.

le Grange, L. (2004). Multicultural science in South Africa's national curriculum statement. *Africa Education Review, 1*(2), 204–219.

Lyons, T. (2006). Different countries, same science classes: Learners' experiences in their own words. *International Journal of Science Education, 18*(3), 311–320.

Malcolm, C., Gopal, N., Keane, M., & Kyle, W. C., Jr. (2009). Transformative action research: Issues and dilemmas in working with two rural South African communities. In K. Setati, R. Vithal, C. Malcolm, & R. Dhunpath (Eds.), *Researching possibilities in mathematics, science and technology education* (pp. 193–212). New York: Nova Science Publishers.

Mudaly, R. (2011). Risking it: Entering uneven socio-scientific spaces in life sciences classroom. *African Journal of Research in Mathematics Science and Technology Education, 15*(3), 27–40.

National Research Foundation Report (NRF). (2005). *Key performance indicator report 2004/2005.* Available at: http://www.nrf.ac.za/publications/annrep/annualreport04_05.pdf

O'Donoghue, R. (2010, January). Science, mathematics and technology education research unlocking barriers across lived world and expert mediated frontiers of learning for a sustainable future. In *Proceedings of the 18th annual meeting of the Southern African Association for Research in Mathematics, Science and Technology Education* 2010 conference (pp. 57–69). Durban.

Odora Hoppers, C. (2002). Indigenous knowledge and the integration of knowledge systems. In C. Odora Hoopers (Ed.), *Indigenous knowledge and the integration of knowledge systems: Towards a philosophy of articulation* (pp. 2–22). Johannesburg: NAE.

Ogunniyi, M. B. (2011). The context of training teachers to implement a socially relevant science education in Africa. *African Journal of Research in Mathematics Science and Technology Education, 15*(3), 98–121.

Onwu, G. O. (2000). How should we educate science teachers for a changing society? *South African Journal of Higher Education, 14*(3), 43–50.

Onwu, G. (2012a). Cultural issues in science education in South Africa. In *African Culture and International Understanding* (IACIU) *1*(2). 18–24 Oct–Dec. Nigeria: UNESCO Institute for African Culture and International Understanding.

Onwu, G. (2012b). Towards a culturally relevant and socially responsible science education. *ECTN Association/(European Chemistry and Chemical Engineering Education Network (EC2E2N)* 13(05), November – Special Issue Africa, pp. 1–4.

Onwu, G., & Kyle, B., Jr. (2011). Increasing the socio-cultural relevance of science education for sustainable development. *African Journal of Research in Mathematics Science and Technology Education, 15*(3), 5–26.

Onwu, G. O., & Mosimege, M. (2004). Indigenous knowledge systems and science and technology education: A dialogue. *African Journal of Research in Mathematics, Science and Technology Education, 8*(1), 1–12.

Onwu, G. O., & Stoffel, N. (2005). Instructional functions in large under resourced science classes: Perspectives of South African teachers. *Perspectives in Education, 23*(3), 79–91.

Ratcliffe, M., & Grace, M. (2003). *Science education for citizenship.* Maidenhead: Open University Press.

Roberts, D. A. (1982). Developing the concept of "curriculum emphasis" in science education. *Science Education, 66*, 243–260.

Roberts, D. A. (1988). What counts as science education? In P. Fensham (Ed.), *Development and dilemmas in science education* (pp. 27–54). Philadelphia: Falmer Press.

Roberts, D. A. (2007). Scientific literacy/science literacy. In S. K. Abell & N. G. Lederman (Eds.), *Handbook of research on science education* (pp. 729–780). New York: Routledge.

Roth, W.-M. (2007). Toward a dialectical notion and praxis of scientific literacy. *Journal of Curriculum Studies, 35*, 9–24.

Sadler, T. D. (2009). Situated learning in science education: Socio-scientific issues as contexts for practice. *Studies in Science Education, 45*, 1–42.

Sinnes, A, T., Kyle, W. C. Jr., & Alant, B. P. (2010, January). What is socially responsible science education? Perspectives from students in Project SUSTAIN. *Proceedings of the 18th Annual Meeting of the Southern African Association for Research in Mathematics Science and Technology Education* (pp. 101–108). Durban.

Sinnes, A, T., Kyle, W. C. Jr., Alant, B., Kazima, M., Nampota, D., & Onwu, G. O. M. (Eds.). (2011). Editorial. *African Journal of Research in Mathematics, Science and Technology Education, 15*(3), 1–4.

Sterling, S. (2009). Sustainable education. In D. Gray, L. Colucci-Gray, & E. Camino (Eds.), *Science, society, and sustainability: Education and empowerment for an uncertain world* (pp. 105–118). New York: Routledge.

UNESCO (2005). *United Nations Decade of Education for Sustainable Development (2005–2014): International implementation scheme*. Retrieved April 18, 2012, from http://unesdoc.unesco.org/images/0014/001486/148654E.pdf

van Eijck, M. (2009). Scientific literacy: Past research, present conceptions, and future developments. In W.-M. Roth & K. Tobin (Eds.), *Handbook of research in North America* (pp. 245–258). Rotterdam: Sense Publishers.

Chapter 13
Curriculum Conception, Implementation and Evaluation: An Experience

Cecília Galvão, Cláudia Faria, Sofia Freire, and Mónica Baptista

13.1 Introduction

We live today in difficult, uncertain and complex times. Secured jobs no longer exist and citizens have to be prepared to adapt to different contexts. So, in this rapidly changing world, citizens' knowledge must also be comprehensive and based on multiple capabilities. According to AMA (2010), the accelerated technological change and globalisation of the market require individuals with skills of creativity, communication skills and the ability to learn throughout life. In this scenario of change, of complexity and of unpredictability and in order to promote students' personal fulfilment and active citizenship, schools have to provide learning experiences that facilitate the development of communication and digital skills; of basic competences in mathematics, science and technology; of social and civic competences; of a sense of initiative and entrepreneurship; and of metacognition, cultural awareness and expression (European Parliament and Council 2006).

Science education can play a crucial role in enacting this agenda. In a society marked by science and technology, in which citizens are increasingly called to make decisions concerning scientific issues that affect their lives and the society they live in, it is essential to equip citizens with fundamental scientific knowledge and knowledge about the processes of scientific activity. In seeking to achieve these purposes, science education can also be a means to foster the development of communication skills, critical thinking, problem-solving and decision-making, among others.

This way of understanding the goals of science education requires a new understanding of how to organise curricula. And indeed, there has been a recent trend towards competence-based teaching and learning, which requires significant changes in school science curricula. International recommendations point to the

C. Galvão (✉) • C. Faria • S. Freire • M. Baptista
Instituto de Educação, Universidade de Lisboa, Lisboa, Portugal
e-mail: cgalvao@ie.ulisboa.pt

B. Akpan (ed.), *Science Education: A Global Perspective*,
DOI 10.1007/978-3-319-32351-0_13

importance of teachers organising challenging learning experiences, which are meaningful and related to students' social reality (EU 2004; UNESCO-ICSU 1999). In addition, there are recommendations that teachers should provide students with careful guidance in learning situations related to problem-solving and decision-making in order to promote self-regulated learning (Autio et al. 2007; Osborne and Dillon 2008). Inquiry-based strategies have been highlighted as having the potential to increase students' engagement with science at all levels and to provide opportunities for the development of competences (European Commission and High Level Group on Science Education 2007; Osborne and Dillon 2008). More recently, some authors have been advocating the need to develop science teaching as practised by making students engage with common practices of science (Osborne 2014).

Aligned with the emerging goals of science education and with the new trends of curricula studies, the Portuguese science curriculum was reorganised as part of a broader curriculum reorganisation, in an extended process that started in 1997 and ended in 2001. This was a particularly important process that introduced significant changes regarding the conceptions of teachers' roles and teaching-learning processes. The curriculum reorganisation proposed a more student-centred orientation, focused on the development of essential competences. Teachers were conceived as active elements that were expected to interpret the curriculum guidelines and to adapt them to different contexts for improving students' successful learning. These were major changes in a very traditionally centralised system that previously expected that the curriculum would be uniformly implemented across the country, based on the idea of teachers as passive players of a curriculum organised according to fixed guidelines and goals, with long lists of content that they were supposed to transmit to students.

In addition to introducing new ideas and principles of curriculum organisation, the process of curriculum reorganisation also sought to trigger a thorough discussion of those ideas and principles, involving universities, schools and teachers, in order to facilitate change. From this process emerged a new curriculum organisation and the curriculum documents that guide Portuguese basic education. In particular, with respect to basic science education, the main documents are the national curriculum of basic science education (DEB 2001a) and the national curriculum guidelines for physical and natural sciences for the third cycle (DEB 2001b).

In this chapter we describe the process of conception and implementation of the Portuguese basic science education curriculum as it introduced several innovations, such as the ideas of flexible management of the curriculum and of educational experiences for developing competences, adapted to the unique Portuguese context. Furthermore, the analysis of the process of implementation of the new curriculum provides relevant data about the process of appropriation of new principles, ideas and practices; and therefore it adds to the discussion about educational change.

13.2 The Portuguese Science Curriculum: Conception and Implementation

13.2.1 Science Curriculum Characterisation

The science curriculum for the third cycle of basic education (12–14 years old) was officially implemented in 2002/2003 (Galvão and Lopes 2002), with the main goal of motivating the development of learning environments that would encourage students to raise questions about the natural world around them, as well as about human action over the world. It also expects that students develop ideas about the environmental and cultural impact of science and technology and that students gain a general understanding of broad explanatory frameworks of science as well as of its procedures (Galvão et al. 2004). Furthermore, this curriculum assumes a constructivist perspective, and it emphasises science-technology-society-environment (STSE) approaches and an inquiry-based learning approach (Galvão and Freire 2004). Finally, it encourages the interdisciplinary exploration of the topics in order to promote an integrated vision of the natural world (DEB 2001b) and the creation of complex learning experiences facilitating the development of essential competences (Galvão et al. 2004).

In accordance with these goals and principles, the curriculum underwent reorganisation, bringing together two disciplines that traditionally were treated as separate subjects: (i) natural sciences (NS) (which incorporates biology, geology, environmental education and health education) and (ii) physics and chemistry sciences (PCS) (which incorporate physics, chemistry and environmental education) (Galvão et al. 2007). Furthermore, in order to overcome a disciplinary logic, it was organised around four topics: "Earth in space", "Earth in transformation", "sustainability on Earth" and "better living on Earth". These topics are meant to be explored flexibly (and when possible, also interdisciplinary) during the 3 years of the third basic education period (DEB 2001b). While exploring these themes through complex learning experiences, students are expected to construct substantive knowledge concerning science and also to develop other important competences, such as reasoning and critical thinking and communication (DEB 2001a, b). So, other important changes introduced were learning experiences for enacting students' competences and to rethink assessment accordingly and the idea of flexible management of the curriculum, requiring a culture of collaboration and articulation between the teachers.

The overall national curriculum was based on Perrenoud's (1997) definition of competences as "the integration of knowledge and skills developed in complex learning experiences" (Galvão et al. 2007). Within a competence-based curriculum framework, the goal is to promote the integrated development of both capacities and attitudes that enable the use of knowledge in different situations, especially complex ones (DEB 2001a), through the creation of contexts that enact students' comprehension, personal meaning construction and enjoyment with learning (Abrantes 2001).

In what concerns the science curriculum, students are expected to develop a set of competences, such as substantive, methodological and epistemological knowledge, reasoning and communication, social and scientific attitudes and life-long learning skills (DEB 2001a, b; Galvão et al. 2004). The development of these competences enables students to act autonomously in the outside world, to participate critically in society and to make informed and reasoned decisions, as is the explicit goal of the science curriculum (Galvão et al. 2004). In order to accomplish this, the science curriculum proposed a set of learning experiences, such as observing the surrounding environment, gathering and organising materials, planning and developing different types of research, designing projects, carrying out experimental activities and using different instruments for observation and measurement, applying scientific knowledge to everyday situations and carrying out discussion on controversial and contemporary issues (DEB 2001a, b).

Within this new context centred on the development of competences, teachers were encouraged to rethink students' assessment. The idea is that assessment should not only be a means of certifying learning, but it should also work as a regulatory tool of pedagogical practices (Abrantes 2001). In science, assessment was conceived as a way to help students regulate their own learning and, at the same time, as a means to facilitate teachers' decisions concerning their own practices, strategies and learning experiences (DEB 2001a, b). So, the emphasis was no longer only on the product or content knowledge but also on the processes of learning, following the idea of formative learning (Black and Williams 1998a, b).

The logic of this competence-based curriculum places a heavy burden on teachers, as interpreting and developing curriculum guidelines are seen as within the teacher's responsibility. Indeed, in order to facilitate competence's development, teachers have to strategically and intentionally design complex and interdisciplinary learning experiences, which involve students in problem-solving, decision-making discussions and negotiation processes (DEB 2001a, b; Galvão et al. 2004), and to adequately assess the process of developing competences. So, within this framework, the teacher is expected to act as a reflective practitioner, who interprets the guidelines, who articulates with other educational agents, who diversifies strategies and who manages the situations in a flexible and local way in order to reach out to all students and to improve their learning (Abrantes 2001; DEB 2001a, b).

Thus, the reorganisation of the national curriculum introduced several new ideas and principles, such as the notion of competence, the development of adequate learning experiences, an emphasis on formative assessment of students and the idea of flexible management of the curriculum. In order to make teachers familiar with the concepts and principles and also to promote changes in teachers' practices and conceptions, the reorganisation of the Portuguese curriculum of basic education involved a complex and extended process of implementation, as described below.

13.2.2 Science Curriculum Implementation

The process of implementation of the science curriculum was characterised by a model that stood between the mutual adaptation perspective and the curriculum enactment perspective (Galvão et al. 2004). According to the first approach, the

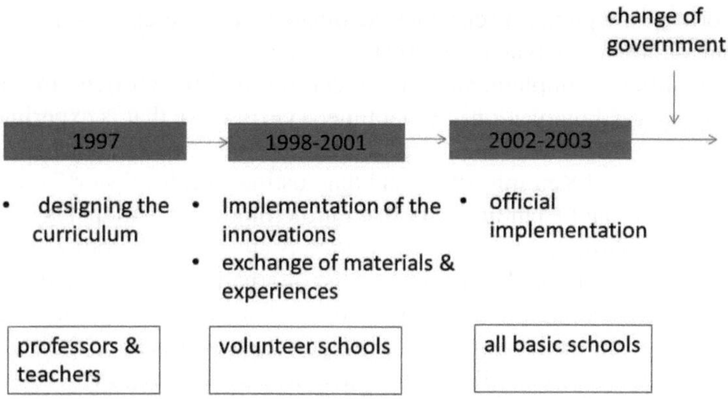

Fig. 13.1 Process of the curriculum reorganisation

implemented curriculum results from mutual adaptations emerging from the users, given their interests, needs and competences and also emerging from the central agencies. Thus, the adjustments that occur in the curriculum are made not only by central agencies but also by the schools and in the context of the classroom and thus imply "a certain amount of negotiation and flexibility on the part of both designers and practitioners" (Snyder et al. 1992, p. 410). In comparison, in the curriculum enactment approach, the curriculum is understood "as the educational experiences jointly created by student and teacher" (Snyder et al. 1992, p. 418). Thus, teachers play the role of curriculum makers who, together with their students, are increasingly responsible for developing educational experiences (Snyder et al. 1992).

In the Portuguese case, the movement of curriculum reorganisation took place across the country under the direction of the Ministry of Education, but simultaneously it sought to involve schools and teachers in the process (Fig. 13.1). So, initially, a team of university professors and school teachers was formed, with the goal of designing a curriculum. While the documents were being produced, several consultants belonging to other universities and other primary and secondary schools and representatives of scientific societies and associations gave their opinion (Galvão et al. 2004). At this stage, concepts such as competence, flexible management of curriculum and formative assessment were exhaustively discussed.

At a second stage, the proposed innovations were carried out in some Portuguese schools before being extended to the whole country. This process began with 93 volunteer schools from various locations in Portugal, and this number increased after 2 years to 184 schools (Galvão and Lopes 2002). Over these years, meetings were held to exchange materials and experiences of the schools involved, with the objective of discussing the difficulties, observations, criticism and the different arguments for and against the curriculum proposals (Galvão et al. 2004). However, soon after the entry of the national curriculum guidelines for physical and natural sciences, this process of curriculum change was stopped abruptly due to a change of government. As a result there was no opportunity to make a systematic monitoring of schools and teachers nor time to develop an assessment process after 3 years of implementation as originally planned. Moreover, the learning process which developed from the public discussions, seminars and conferences was not released to the general public (Galvão et al. 2004).

A decade after its implementation, we can say that the curriculum was in line with the educational innovations of that time: a curriculum that is experimental and negotiated (Elbaz 1983) and a curriculum that attaches value to teachers' practical knowledge (Carr and Kemmis 1986) and that justifies teachers' work as curriculum makers (Connelly and Clandinine 1986). Furthermore, current analysis of the science curriculum shows that this reorganisation is aligned, even today, with the curricula of the best positioned countries' on the international assessment of students' science competences (OECD 2007). Indeed, an analysis of the science curricula of five of the best positioned countries (Hong Kong, Taiwan, Finland, The Netherlands, New Zealand) revealed that all of them value the active involvement of students in their own learning processes and recommend science teachers use diverse teaching strategies, such as inquiry activities, problem-solving and field work, to develop a set of transversal competences that allow students to make informed decisions related with science and society.

Despite the characteristics of the Portuguese science curriculum and despite significant improvements in international assessment (OECD 2010), Portuguese students still do not perform as well as expected considering the characteristics of the intentional curriculum. This apparent contradiction motivated the development of a national study to evaluate the science curriculum of the third cycle of basic education, with the goal to understand how teachers interpret and implement the curriculum and to identify students' achieved and experienced curriculum. Particularly, this chapter's goal is to understand the process of appropriation of new curriculum concepts. Thus, we will focus on teachers' perspectives and describe how teachers interpret and implement the science curriculum of the third cycle of basic education and identify what factors affect the way they interpret and implement it. Furthermore, as we intend to capture the process of appropriation of the new concepts and of the enactment of the new principles, we next present data concerning a set of studies centred on different aspects of the curriculum reorganisation.

13.3 Interpretation and Implementation of the Science Curriculum: Teachers' Perspective

13.3.1 Methodology

In order to know how teachers interpret and implement the science curriculum and to understand the process of appropriation of new curriculum concepts, we will present and discuss data from three types of studies:

- A nationwide study focused on teachers' perspectives regarding the science curriculum (2010/2011)

- A meta-analysis of several comprehensive studies, spread over the country and performed since the official adoption of the documents until the present (Table 13.1)
- A multiple case study involving five schools spread over the country (2011/2012)

Table 13.1 Description of the studies included in the meta-analysis study

References	Date of data collection	Method of data collection	Participants	District	Focus
Viana (2003)	2001/2002[a]	Case study	PCS teachers ($n=6$)	Lisbon	Classroom practices
Galvão et al. (2004)	2003/2004	Survey study	NS and PCS teachers ($n=53$)	Lisbon	Concepts and practices about curriculum
Sítima (2005)	2003/2004[a]	Case study	Pedagogical pair (NS and PCS teacher) ($n=2$)	Lisbon	Collaboration between teachers
Abelha (2005)	2004/2004	Case study	NS department ($n=1$)	Aveiro	Collaboration between teachers
Correia (2006)	2004/2005[a]	Case study	PCS teachers ($n=3$)	Lisbon	Concepts and practices about assessment
Ferreira (2006)	2004/2005	Case study	Pedagogical pair ($n=1$)	Aveiro	Practice of co-teaching
		Survey study	Principals ($n=126$)		
Júlio (2006)	2004/2005[a]	Interview	NS and PCS teachers ($n=8$)	Lisbon	Inquiry activities
Raposo (2006)	2004/2005[a]	Case study	PCS teachers ($n=6$)	Faro	Concepts and practices about assessment
Negrais (2007)	2005/2006[a]	Survey study	NS teachers ($n=75$)	Aveiro	The STS approach
Abelha (2011)	2006/2007	Survey study	NS and PCS teachers ($n=1122$)	Aveiro	Collaboration between teachers
	2007/2008	Case study	NS and PCS teachers ($n=12$)		
Martins (2012)	2006/2007	Survey study	NS and PCS teachers ($n=1122$)	Aveiro	Professional knowledge, curriculum concepts and practices
		Interview	NS and PCS teachers ($n=8$)		
			Teachers of intermediate leadership levels ($n=3$)		

[a]Date of data collection not confirmed

The nationwide study aimed at knowing the perspectives of teachers who were implementing the science curriculum; in particular, it aimed at describing how teachers interpret the curriculum documents and how they implement the proposals made. Data was collected through the application of questionnaires that were administered to a representative national sample of 789 teachers (395 PCS teachers, 394 NS teachers). Most of the teachers who responded to the questionnaire were females (80 %) and had a background in education (75 %). Their ages ranged from 25 to 50 years. Ninety-five percent of teachers acknowledged that they knew the national curriculum document, and 97 % said they knew about the national curriculum guidelines for physical and natural sciences for the third cycle. About 69 % of teachers stated their satisfaction with the science curriculum guidelines document.

The meta-analysis of the comprehensive studies aimed at capturing the evolution of the process of appropriation of the curriculum ideas since its official implementation in order to understand the complexity of the relationships and factors that interfere with the interpretation and implementation of the curriculum and its impact on students' learning. So, we analysed data collected by several studies conducted in Portugal (see Table 13.1). These studies are mainly at master's and doctorate levels, existing in institutional repositories (ten studies analysed). We also analysed a study conducted in the early stages of the process of curriculum reorganisation (Galvão et al. 2004), which had as a main objective to disseminate new ideas underlying the curriculum reorganisation and to collect information about the perspectives of teachers on new proposals, as well as on the difficulties that they anticipated or that they experienced while they implemented the new proposals. It should be noted that the studies analysed followed different research methodologies. While some studies were based on significant teachers' samples, other studies were qualitative in nature, focusing on a deep exploration of only a few cases (teachers). Thus, in the interpretation of the results, we were especially careful and took into consideration that these studies are like pieces of a puzzle, which together allow us to create an image of the process of appropriation of the new curriculum by the teachers.

Finally, the multiple case study (which took place in 2011/2012) encompassed five separate schools in the country through which we explored in more depth all mentioned aspects. The case studies involved interviews with all the science teachers that were teaching the ninth grade, the principals of the schools as well as teachers in charge for the science department. We collected several documents, such as lesson plans, instruments of assessment and documents concerning the school politics. We also interviewed groups of ninth grade students and collected data on students learning.

13.3.2 Results

The results are organised according to two major innovations advocated by the new curriculum reorganisation: (a) the idea of flexible management of the curriculum, which requires a culture of collaboration and articulation between the teachers, and (b) the development of learning experiences for enacting students' competences and of a new perspective of assessment.

13.3.2.1 Flexible Management of Curriculum

One of the intentions of the curriculum reorganisation was that teachers would break away from their routine and individualistic way of working and would develop collaborative work with other teachers. The idea was that by assuming themselves as curriculum makers, teachers would make strategic and intentional decisions about curriculum organisation and learning experiences, using their professional knowledge (DEB 1999). For this, teachers were expected to break away with the traditional model of school, based on a transmissive pedagogy and with an individualistic culture of teaching (Formosinho and Machado 1998). In order to facilitate this change, several curriculum management structures were created, namely, the class council responsible for designing and managing the class curriculum project (DEB 1999). It was expected that the class council would facilitate, among other aspects, the coordinated management of the different academic subjects and, at the same time, that it would facilitate teachers' collaboration in common projects (DEB 1999).

The results that emerged from our study reveal that the curriculum reorganisation did enhance some changes in what concerns teachers' collaboration with each other in making curriculum decisions. However, the observed changes are frequently not fully aligned with the original intentions of the curriculum documents.

Data obtained by the nationwide study shows that a large part of the teachers usually work in collaboration only with colleagues of the same academic subject (52 %) and preferably with colleagues who teach the same subject at the same grade (63 %). Furthermore, data shows that they mainly collaborate in planning lessons (78 %), in defining criteria for students' assessment (89 %), in planning field study (79 %), in preparing materials (71 %) and in planning laboratory activities (70 %) and interdisciplinary activities (71 %).

Similarly, the comprehensive studies analysed and the multiple case study reveal that, in general, the collaboration of teachers is based primarily on the exchange of learning materials and experiences, field study organisation and assessment instruments. Furthermore, those teachers who teach the same subject, either at the same grade or different grades, tend to collaborate more often with each other. As one teacher of the multiple case study mentions, "We share information. Nobody hides anything. Nobody is afraid to share". In addition, these studies reveal that teachers' collaboration is occasional and informal, as can be observed in the following excerpt:

> *When anything happens, we meet in the staff room, frequently informally. As we are only three, it is very easy for us to come together in a corner for discussing our ideas. Many times, we do it in an informal way.* (Multiple case study)

Nevertheless, formal and institutional collaboration occurs. The study of Abelha (2011) shows that the class curriculum project is an important management instrument that creates formal moments for teachers meeting with each other. However, as teachers understand the class curriculum project as being under the exclusive responsibility of the class director, they do not get fully involved with the class curriculum project. The same lack of involvement with the meetings, where teachers would supposedly make joint decisions concerning the curriculum, was observed in

Martins' (2012) study. Based on these observations, both authors concluded that the formal curriculum management instruments are devoid of meaning and are experienced by the teachers as one more bureaucratic instrument.

Finally, most times, flexible management of the curriculum is understood as a means for the teachers to jointly decide when and at what level of depth each subject should be addressed and explored, in order to circumvent what teachers consider to be the problem of the extension of the curriculum and its repetition of topics. This operating philosophy is well captured by one of the interviewed teachers of the multiple case study. He stated:

> *Articulation is usually discussed during the meetings. Each one of us shares with the others the current topic that he is exploring on that moment. Then, we try to cross information, in order to, as far as it is possible, save time. Indeed, some topics are addressed in geography, science, physics and chemistry. Our intention is to save time. And also we want to show the universality of knowledge.* (Multiple case study)

Consistently, results from comprehensive studies and multiple case study point to an articulation between NS and PCS subjects focused primarily on adjustments to the content level and the informal exchange of materials, experiences and ideas, as can be observed in the following excerpt:

> *We combine more or less what each one of us teaches, for not repeating themes. We have to work like that due to some repetitions of the curriculum. For instance, in natural sciences (NS) we first explore the genesis of the heliocentric theory, but in physics-chemistry sciences (PCS) it is the other way round. We, for example, explain pollution in the eighth grade, and then my geography colleague explains it again later!* (Multiple case study)

One important innovation of the reorganisation of the science curriculum of basic education was the establishment of a bridge between NS and PCS. Its intention was that teachers would explore the NS and PCS curricula in an articulated way, making coordinated decisions on how to organise the content and planning of educational situations and jointly develop activities with students (DEB 2001b). These were positive aspects indicated by the teachers in the early stages of the process of reorganising the curriculum (Galvão et al. 2004). However, simultaneously, teachers anticipated it as one of the greatest difficulties in implementing the new curriculum. In their view, the difficulties were related to the organisation of the school (e.g. schedules, lack of resources) and related to the lack of time required to accomplish curriculum (Galvão et al. 2004).

These same difficulties have been identified in several of the comprehensive studies developed afterwards. In addition, teachers repeatedly pointed out that the absence of more concerted practices among them was related to the characteristics that they attached to the national curricula guidelines for PCS and NS, namely, the difficulty of articulating the content and curriculum's extension (Abelha 2011; Ferreira 2006; Martins 2012; Sítima 2005). Results from the multiple case study are coherent with these findings in that the teachers pointed to the vagueness of the curriculum documents and mentioned in contrast that the textbooks provide guidelines that are far more concrete and objective.

13.3.2.2 Learning Experiences for Developing Competences

This way of understanding and enacting the curriculum, in the specific case of science, seems to be rooted in a specific view of science education and its purposes, which is not in accordance with the ideas advocated by the new curriculum proposal.

One of the innovations of the national curriculum was to challenge the conception of knowledge as something static and external to people, knowledge as a product and knowledge as unquestionable truth, organised in a compartmentalised and a linear way and non-contaminated by values (Alonso 2005). However, studies suggest that frequently teachers' understanding of some of the ideas of the curriculum is superficial, or it is not aligned with the original intentions, thus affecting their practices in a very precise way.

The study by Martins (2012) reveals a very peculiar situation. Most teachers are familiar with the term competence. However, this familiarity with the term is not followed by deep changes in these teachers' practices. In comparison, the multiple case study shows that even though the term competence is frequently used in the various school documents, it is largely absent from the discourse of the interviewed teachers. Of the 20 interviewed teachers, only 5 used the term competence in their discourse. Furthermore, competence is understood as *something* vague that students must acquire and that teachers have a duty to provide the students with. For example, one of the teachers stated "if we demand too much from the students, many of them will not *achieve* the competences in the end. They will leave school only with some general competences related with the school subjects" (multiple case study; *emphasis ours*). Another teacher mentions: "There is a whole set of competences that they [students] could *acquire* and that would be useful for day-to-day applications" (multiple case study; *emphasis ours*). The emphasis is on acquiring competences and on equipping students with them, suggesting an underlying perspective of "learning as acquisition", rather than learning as a developmental process constructed by the student (as is the underlying assumption of the curriculum).

So, despite the wide dissemination of the terms in the school documents, and even despite some familiarity with the term, in general, the notion of competence is not fully understood by the teachers, who have not changed their practices accordingly.

In terms of assessment practices, the study by Viana (2003) shows that despite teachers having introduced some new practices derived from the new curriculum guidelines, their views and beliefs concerning the purposes of teaching science had not changed. Indeed, these teachers considered that their lessons were less expositive, that they used new materials and that they developed more experiments, stimulated more debate and provided more inquiry activities. However, their assessment instruments and methods remained unchanged. Similarly, Raposo (2006) and Correia (2006) observed a mismatch between the assessment practices and the development of competences that teachers claimed to promote. So, the studied teachers used written tests, worksheets, observation and mental records (to assess attitudes). In addition, they still overvalued the summative assessment of content.

Indeed despite emphasising some procedural and communicative competences, teachers mainly focussed on the assessment of knowledge.

The nationwide study shows that only a small proportion of teachers teach as recommended in the curriculum documents. Indeed, only 32 % of teachers frequently use classroom discussions, concept maps, students' observations, students' oral responses and portfolios (this group differs statistically from the rest of the teachers ($p < 0.001$)). This small proportion of teachers also use more students' self-assessment practices, written reports on experimental activities, research work ($p < 0.01$) and homework ($p < 0.05$) for assessing students. Regarding written tests, these teachers more often than the rest of the overall sample included items requiring an explanation; elaboration of text; construction of graphs ($p < 0.01$); interpretation of experimental results; argumentation; interpretation of graphs, tables and texts; planning investigations; and enunciation of hypotheses ($p < 0.001$). Finally, these teachers more frequently inform students about their learning objectives, involve them in the identification of objectives, more frequently assign different work depending on students' results on assessment, use performance level descriptors for assessment and provide more frequent opportunities for students to reflect on their work ($p < 0.001$).

Regarding the strategies recommended in the science curriculum, most teachers of the nationwide study indicate that they implement inquiry activities (86 %), project work (68 %) and experimental work (85 %) only in some classes (Fig. 13.2). Results are less consensual concerning the activities related to decision-making and problem-solving. In the first case, 50 % of teachers say that they used decision-making in few classes, and about 25 % of the teachers say that they use this type of activities in almost every class. In the case of problem-solving activities, around 49 % of teachers state that they used these types of activities in almost every class.

So, the nationwide study reveals that most teachers (68 %) seem to follow a more traditional methodology, i.e. they rarely use (a) strategies related to inquiry activi-

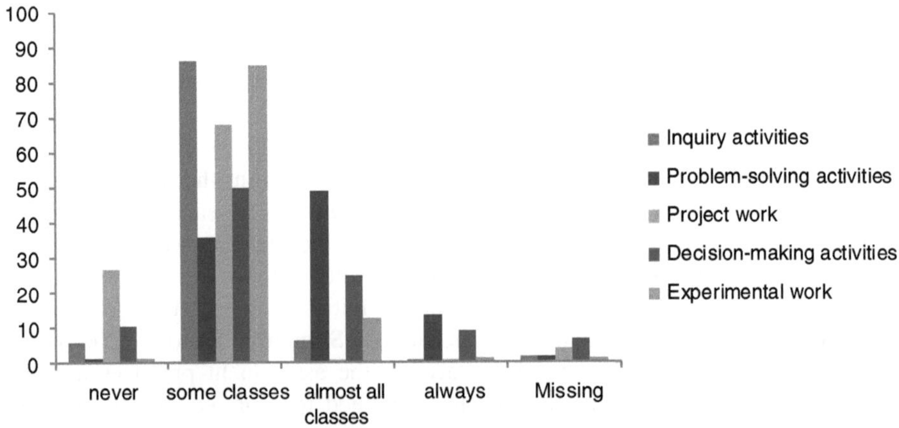

Fig. 13.2 Frequency of implementation of the different strategies

ties (such as identifying a problem, formulating hypotheses, interpreting data, producing texts, using models, constructing graphics based on data, arguing and debating ideas, observing natural phenomena, planning investigations), (b) strategies related to STSE issues (such as discussing controversial issues, exploring issues that affect society welfare, reading texts such as biographies and/or scientific discoveries) and (c) strategies related to experimental activities (such as conducting experiments, reporting on experimental activities and oral presentation of the results of the experiments). Opposing these teachers, mention should be made that 32 % of the surveyed teachers indicated that they frequently use (in almost all the classes) such strategies, mainly those related to inquiry activities and to STSE issues. These teachers more frequently use inquiry ($p<0.001$), problem-solving ($p<0.001$), project work ($p<0.01$), decision-making ($p<0.001$) and experimental work ($p<0.001$) strategies than the other teachers.

These results are consistent with those obtained in other studies. For instance, in the study of Martins (2012), about half of the respondents indicated they had changed their practices, particularly in terms of: (i) the use of information and communication technologies according to the different situations, (ii) organisation of debates related to STS situations, (iii) integration of multidisciplinary knowledge in the context of the classroom, (iv) the use of strategies that relate science, technology and society and that promote a higher degree of autonomy in students, (v) implementation of teaching methods and strategies that promote the development of competences and (vi) promotion of critical analyses of news published by the media.

However, these changes revealed, according to Martins (2012), a scant appropriation of the concepts, in the context of developing articulated and strategically defined curriculum practices, as well as in the context of creating learning experiences that promote the development of competences in students. In fact, results obtained in her study reveal that teachers' decision-making is strongly influenced by aspects related to the fulfilment of the curriculum.

In line with the study by Martins (2012), the teachers surveyed in the other comprehensive studies showed a positive attitude regarding some of the proposals of the new curriculum (namely, the STSE dimension, the experimental, inquiry and problem-solving activities, debates and bibliographic research). However, as in the study by Martins (2012), these studies show that the appropriation done by teachers is very superficial and even different from the original intentions. For example, in her study focused on the STSE dimension, Negrais (2007) noted that the surveyed teachers valued the dimension STSE as a way to develop competences and critical citizenship and that they referred to doing more debates and more problem-solving activities and exploring more about the connections between scientific concepts and the daily world. However, these teachers did not value activities developed outside the context of the classroom, did not explore topics in an interdisciplinary way, nor have they changed the form of assessment, and as such still placed great emphasis on content.

Similarly, Júlio's (2006) study shows that teachers expressed positive attitudes towards experimental activities. However, the type of experimental activities developed was restricted to those proposed by the textbook, relied on closed protocols, did not involve the control of variables and did not include reflection and debate.

The same practices and underlying conceptions were identified in the multiple case study. Indeed, many of the surveyed teachers mentioned that they carried on debates and bibliographic research activities and problem-solving activities and that they implemented some practical classes where they developed demonstration activities, experimental work, inquiry activities and some field work. Other teachers reported the importance of linking science to students' daily lives as a way to motivate learners. For instance, one of the teachers explained:

> *Sometimes [students] do not realize why they are going to study those things. And then, I usually bring ... for example, we are now studying earthquakes and volcanoes. I'll get news and reports. "So, now you know what is meant to say that: -'on the Richter scale,...'. So, it is for them to understand the news...* (Multiple case study)

But again, the choice of this type of activity does not seem to involve real strategic and pedagogical options, but essentially the motivating aspect. This aspect is very prominent in the way the following teacher described a chemistry experiment: "Then, in chemistry when we study the alkaline metals, I do those explosions, and they consider it very amusing, isn't it?" (multiple case study). Another teacher exposed his ideas concerning field study: "I like to take in students and go to the garden. We pick the ants and it is very funny. It turns out to be a practical lesson" (multiple case study).

Where there is a concern about the learning, the focus is on facilitating the understanding of the concepts and theories. Again, teachers emphasise the products of science to the detriment of students gaining an understanding about processes and the nature of scientific knowledge and scientific activity, which is an explicit goal in the Portuguese science curriculum.

This overemphasis on content seems associated to a traditional and limited conception about science and therefore science education, which devalues the development of attitudes (of wonder, questioning, respect for evidence), the understanding of the nature of scientific knowledge and scientific activity (with its practical component, but also with its rational component of questioning, inductive and deductive reasoning and argumentation) and the understanding of the complex relationships that are established between science, society and technology. These are the explicit aims of the current curriculum documents.

The study conducted with teachers early in the process of curriculum reorganisation (Galvão et al. 2004) presents similar results. In Galvão et al.'s study, the teachers criticised how scientific knowledge seemed to be undervalued and that the curriculum was too focused on providing student learning experiences. They had misinterpreted the original intentions of the curriculum. The authors of the study concluded:

> [There is a] lack of understanding of what is considered nowadays to be the actual concept of science teaching. In their criticism, the teachers reveal a static perspective of science, overvaluing the products of science and undervaluing other components. The objection of some teachers shows their criticism of the proposed educational experiences, because they feel these undervalue the transmission of scientific knowledge and give more importance to the learning experiences that are more pupil-centred, based on a constructivist perspective of learning. Sometimes, in the criticism made by certain teachers regarding the Curricular

Guidelines, they value previous practices and do not seem open to the potential of teaching oriented towards the development of competences. They adopt a different jargon, but do not change their practices. (Galvão et al. 2004, p. 64)

The tension between content (facts and theories) and competences works as a justification for not implementing a set of learning experiences more often. Indeed, Martins (2012) concluded that:

The extension of the programmes, the curriculum organisation by disciplines, the limited cooperation among educational stakeholders, including students' families, and an alleged lack of working conditions, were the main constraints appointed for the development of a curriculum organised by competences. (p. 206)

Indeed, there is a constraint related to the extension of the curriculum programme and the need to comply with it, as evidenced in the following statement by a teacher in the study by Martins (2012):

Indeed competences acquisition is not performed with the desired frequency. This is due to the fact that the curricular programmes are very extensive. Having the obligation to apply them, then it is not always possible to analyse day-to-day problematic situations involving curricular content knowledge. Furthermore, the acquisition of competences implies that students acquire the knowledge necessary for the resolution of the problematic situation, and students do not always master the content taught. (p. 258)

This argument, once again, reflects the excessive preoccupation of teachers with the content, as reflected in the following excerpt from an interview with a teacher:

I do problem-solving and experimental activities on some occasions. Expository lessons too, more frequently. The situation is as such: or we teach the entire programme, and in that case we have to rely on expositive lessons; otherwise we will not accomplish it. Or we do according to the proposed innovations and we will not be able to teach the entire program. Dialogues and debates are OK. In what concerns exercises, students usually take them home as homework and I have only to correct it in the class, and so, I do not waste so much time. It is to be like this, for not failing to fulfil the programme. (Multiple case study)

Both statements reveal a very usual situation. Teachers frequently confuse curricular programme (which enumerates all contents to be explored in a certain cycle of education or grade) with curriculum. Curriculum, as understood within the national curriculum reorganisation, is a set of apprenticeships that students are expected to undertake, a set of guidelines concerning how those apprenticeships should be organised and how they contribute to students' educational development (Abrantes 2001; Roldão 1999). So, in this sense, the curricular programme is a tool of the curriculum; it is not the end in itself but provides examples of how to enact a set of apprenticeships, the context from which teachers can design further learning experiences (Abrantes 2001; Roldão 1999). However, teachers' excessive preoccupation with teaching all the topics presented in the programme distracts them from the real intentions of the curriculum: to create learning experiences and to facilitate the development of competences. Other constraints pointed out in several studies are students' difficulties and lack of motivation. Indeed, in general, teachers feel that students have very significant gaps in different areas (both in terms of their specific basic science knowledge and in interdisciplinary transfer such as mathematics, reading and interpreting texts). However, students' difficulties are never seen as the result of certain actions or teaching practices, but as a reason for them not to do according to what is recommended by the curriculum. The results obtained in the multiple case study suggest that students' difficulties with writing, reading and interpretation of texts, rather than motivating the creation of learning experiences that facilitate the development of these competences, motivate teaching strategies focused on the transmission of simplified content.

13.4 Final Considerations

The reorganisation of the basic education curriculum has introduced a number of innovations, starting with the assumption that teachers reinterpret political guidelines in the light of their knowledge, experiences and conceptions. Having in mind this assumption, it sought to involve from the beginning the teachers and schools in the process of designing and implementing the curriculum. And indeed, there was an initial phase aimed at helping teachers to overcome difficulties, to clarify concepts, to rethink their conceptions, to develop new knowledge and competences and also to take account of their comments and appreciations and, based on these, to rethink adjustments to the curriculum (Galvão et al. 2004). After this initial phase, the curriculum development process was broken abruptly as a result of a shift in priorities and political agenda. This meant that the formal monitoring of the process of change ended, despite some occasional initiatives of collaboration among teachers/schools and teacher training institutions aimed at supporting teachers with the implementation of the new guidelines. As a result, most of the Portuguese teachers ended up being imposed with a set of top-down guidelines, and they were left alone in this process of change.

Being left alone in the process and in order to deal with the new requests, teachers took up resources that they were familiar with. And indeed, without other support to help them reorganise ideas and the school system, teachers assimilated some of the concepts, i.e. they interpreted and enacted changes according to their underlying assumptions (Corbun 2004). As a result, teachers modified certain practices not according to the original intentions of the curriculum, but according to how they perceived education and the aims of science education, according to their knowledge and competences and how they perceived and experienced some of the constraints and even their context, where they often had to fight against a rigid and bureaucratic school organisation that was suspicious of the new proposals.

Huberman, in 1973, mentioned that "history shows that education is a domain in which the gap between the new and old is almost never clearly filled", giving meaning to the idea that in order for the changes to succeed, they cannot radically put into question previous acquisitions (Crozier 1982). These findings contradict some psychosociological theories, according to which individuals (or groups or institutions) have an innate need to alter the individual balance moved by a sense of curiosity and invention. In comparison, systems theory states that social systems are stable and homeostatic, meaning that after minor disturbances, they return to an equilibrium state similar to the previous condition. Systems are characterised by the principle of self-regulation through which they can meet the demands of the environment without undergoing profound changes. Thus, schools as institutions are, by nature, confronted with this contradictory aspect: the will to remain unchanged and the willingness to change. They are oscillating systems, according to the concept of Enriquez (1972). Therefore, to break away from tradition, schools require strong impulses that may emerge simultaneously from external pressure (at the level of educational policy) and internal pressure (the movement of teachers and school management directions).

From the perspective of the individual, Spillane's work (1999) gives a similar account. His results show that changing the core aspects of teachers' practices depends on a set of internal personal resources of the teacher, such as willingness, capacity and their previous practices, as well as on a complex set of opportunities and incentives for learning (connected to teachers' external context). In what concerns teachers' internal personal resources, recent studies confirm that beliefs that they are able to deal with challenging new situations are strongly associated with the involvement in learning activities, such as experimenting and reflecting and keeping up to date (Geijsel et al. 2009; Runhaar et al. 2010). Accordingly, several authors (Altrichter 2005; Snyder et al. 1992) point out that successful implementation of the curriculum tends to increase if teachers feel that an innovation is relevant. So, teachers have to be given clear and meaningful objectives and be clear about the purpose of the new vision. Another essential dimension is to monitor the process of change, creating structures capable of responding to the training needs of teachers, related to the proposed innovations of the curriculum, by providing materials and resources to accomplish these innovations in the classroom and by providing them with feedback of the work they are developing (Fullan 2008).

From a different angle, the study of Luttenberg et al. (2013) shows that teachers are constantly engaged in a process of sense making of reforms, that their interpretations of reforms change over time and that it is affected by their motivation to maintain consistency between their frame of reference and the perceived frames of reference of reforms. So, requiring teachers to think radically differently from a traditional view entails promoting reflection anchored on teachers' experience, by confronting teachers' previous ideas with students' learning (Capps et al. 2012; Galvão et al. 2011a).

For that, teachers need to have time and space for looking into their own limitations and for taking ownership of what is proposed (Galvão et al. 2011b). It is also necessary that training institutions are open minded, so that they too can learn from school practice, validating the practical knowledge of teachers, fostering collaborative networks of researchers and teachers and creating an organisation based on systemic and interdisciplinary collaboration. The connection between schools and training institutions, in a mutual learning relationship, can be one of the ways to create communities of practice in which researchers and teachers are able to build a curriculum that should be open, flexible and aligned with current perspectives in science education.

Acknowledgements Part of this study was supported by Fundação para a Ciência e a Tecnologia as part of the Project "Evaluation of the Physics and Natural Sciences Curriculum, third cycle of Basic Education" (PTDC/CPE-CED/102789/2008).

References

Abelha, M. C. (2005). *Cultura Docente ao nível do Departamento Curricular das Ciências: um estudo de caso*. Tese de dissertação de mestrado Departamento de Ciências da Educação. Departamento de Didáctica e Tecnologia Educativa. Universidade de Aveiro, Aveiro.

Abelha, M. (2011). *Trabalho colaborativo docente na gestão do currículo do Ensino Básico: do discurso às práticas*. Tese de Doutoramento em Didática, Universidade de Aveiro, Aveiro.

Abrantes, P. (2001). *Reorganização curricular do Ensino Básico: Princípios, medidas e implicações*. Lisboa: Departamento do Ensino Básico. Ministério da Educação.

Alonso, L. (2005). *Reorganização curricular do ensino básico; potencialidades e implicações de uma abordagem por competência. Actas do 1° encontro de educadores de infância e professores do 1° ciclo do ensino básico* (pp. 15–29). Porto: Porto Areal editores.

Altrichter, H. (2005). Curriculum implementation: Limiting and facilitating factors. In P. Nentwig & D. Waddington (Eds.), *Context based learning of science* (pp. 35–62). Münster: Waxmann Verlag.

AMA. (2010). *Critical skills survey*. Retrieved in 29 March, 2013, from, http://www.amanet.org/news/AMA-2010-critical-skills-survey.aspx

Autio, O., Kaivola, T., & Lavonen, J. (2007). Context-based approach in teaching science and technology. In E. Pehkonen, M. Ahtee, & J. Lavonen (Eds.), *How Finns learn mathematics and science* (pp. 202–214). Rotterdam: Sense Publishers.

Black, P., & William, D. (1998a). Assessment and classroom learning. *Assessment in Education: Principles, Policy & Practice, 5*(1), 7–75.

Black, P., & William, D. (1998b). Inside the black box: Raising standards through classroom assessment. *Phi Delta Kappan, 80*(2), 39–148.

Capps, D. K., Crawford, B. A., & Constas, M. A. (2012). A review of empirical literature on inquiry professional development: Alignment with best practices and a critique of the findings. *Journal of Science Teacher Education, 23*, 291–318.

Carr, W., & Kemmis, S. (1986). *Becoming critical: Education, knowledge and action research*. London: Falmer.

Connelly, M., & Clandinin, J. (1986). On narrative method, personal philosophy, and the story of teaching. *Journal of Research in Science Teaching, 23*(4), 293–310.

Corbun, C. (2004). Beyond decoupling: Rethinking the relationship between the institutional environment and the classroom. *Sociology of Education, 77*, 211–244.

Correia, M. (2006). *Concepções e práticas de avaliação de professores de Ciências F/Q do EB*. Tese de dissertação de mestrado. Universidade de Lisboa, Lisboa.

Crozier, M. (1982). Mudança individual e mudança colectiva. In R. KnoKe & E. Figueiredo (Eds.), *Mudança Social e Psicologia Social* (pp. 69–81). Lisboa: Horizonte.

Departamento da Educação Básica [DEB]. (2001a). *Currículo Nacional do Ensino Básico – Competências Essenciais*. Lisboa: Ministério da Educação.

Departamento da Educação Básica [DEB]. (2001b). *Ciências Físicas e Naturais – Orientações curriculares para o 3° ciclo do Ensino Básico*. Lisboa: Ministério da Educação.

Departamento de Educação Básica [DEB]. (1999). *Gestão flexível do currículo*. Lisboa: Ministério da Educação.

Elbaz, F. (1983). *Teacher thinking: A study of practical knowledge*. London: Croom Helm.

Enriquez, E. (1972). Problématique du changement. *Connexions, 4*, 3–45.

European Commission, & High Level Group on Science Education. (2007). *Science education NOW: A renewed pedagogy for the future of Europe (EUR 22845)*. Brussels: DG Research.
European Parliament and Council (EP&C). (2006). Recommendation of the European Parliament and of the Council on key competences for lifelong learning. *Official Journal of the European Union, L 394*, 10–18.
European Union [EU]. (2004). *Europe needs more scientists. Report by the high level group on increasing human resources for science and technology in Europe*. Brussels: European Union [EU].
Ferreira, A. M. (2006). *A co-docência na área das ciências Físicas e Naturais: um estudo de caso*. Tese de dissertação de mestrado. Universidade de Aveiro, Aveiro.
Formosinho, J., & Machado, J. (1998). *Autonomia e Gestão das Escolas. Virtualidades e Contradições de um Compromisso Político*. Braga: Universidade do Minho.
Fullan, M. (2008). Curriculum implementation and sustainability. In M. Connelly, M. He, & J. Phillion (Eds.), *Handbook of curriculum and instruction* (pp. 113–123). Thousand Oaks: Sage Publications.
Galvão, C., & Freire, A. (2004). A perspectiva CTS no currículo das ciências físicas e naturais em Portugal. In I. Martins, F. Paixão, & R. Vieira (Eds.), *Perspectivas Ciência-Tecnologia-Sociedade na inovação da educação em ciência. Actas III Seminário Ibérico CTS no Ensino das Ciências* (pp. 31–38). Aveiro: Universidade de Aveiro.
Galvão, C., & Lopes, A. (2002). Os projectos curriculares de turma no contexto da Gestão Flexível do Currículo. In *Gestão flexível do currículo – reflexões de formadores e de investigadores* (pp. 97–115). Lisboa: Ministério da Educação, Departamento da Educação Básica.
Galvão, C., Freire, A., Lopes, A., Neves, M., Oliveira, T., & Santos, M. C. (2004). Innovation in Portuguese science curriculum: Some evaluation issues. In ME (Ed.), *Flexibility in curriculum, citizenship and communication/Flexibilidade curricular, cidadania e comunicação* (pp. 341–357). Lisboa: Ministério da Educação, Departamento da Educação Básica.
Galvão, C., Reis, P., Freire, A., & Oliveira, T. (2007). Science curriculum in Portugal: From the development to the evaluation of students' competences. In D. Waddington, P. Nentwig, & S. Schanze (Eds.), *Making it comparable. Standards in science education* (pp. 237–253). Münster: Waxmann.
Galvão, C., Reis, P., Freire, S., & Almeida, P. (2011a). Enhancing the popularity and the relevance of science teaching in Portuguese science classes. *Research in Science Education, 41*(5), 651–666.
Galvão, C., Reis, P., Freire, S., & Faria, C. (2011b). *Ensinar ciências, aprender ciências*. Porto: Porto Editora, IE.
Geijsel, F., Sleegers, P., Stoel, R., & Krueger, M. (2009). The effect of teacher psychological and school organizational and leadership factors on teachers' professional learning in Dutch schools. *The Elementary School Journal, 109*(4), 406–427.
Júlio, V. N. (2006). *Actividades investigativas no ensino das ciências. Tese de dissertação de mestrado*. Lisboa: Universidade Nova de Lisboa.
Luttenberg, J., van Veen, K., & Imants, J. (2013). Looking for cohesion: The role of search for meaning in the interaction between teacher and reform. *Research Papers in Education, 28*(3), 289–308.
Martins, I. (2012). *O currículo das Ciências Físicas e Naturais na perspetiva docente, Saberes profissionais e possibilidades de acção*. Tese de dissertação de mestrado. Universidade de Aveiro, Aveiro.
Negrais, M. J. (2007). *Percepções dos professores de ciências naturais cobre o ensino no âmbito CTSA*. Tese de dissertação de mestrado. Universidade de Aveiro, Aveiro.
OECD. (2007). *PISA 2006: Science competences for tomorrow's world, volume 1: Analysis*. Paris: OECD Publishing.
OECD. (2010). *PISA 2009, results: Executive summary*. Paris: OCDE Publishing.
Osborne, J. (2014). Teaching scientific practices: Meeting the challenge of change. *Journal of Science Teacher Education, 25*(2), 177–196.

Osborne, J., & Dillon, J. (2008). *Science education in Europe: Critical reflections*. London: The Nuffield Foundation.

Perrenoud, P. (1997). *Construire des compétences dès L'école*. Paris: ESF éditeur.

Raposo, P. (2006). *Concepções sobre avaliação das aprendizagens – um estudo com professores de FQ*. Tese de dissertação de mestrado. Universidade de Lisboa, Lisboa.

Roldão, M. C. (1999). *Os professores e a gestão do currículo. Perspectivas e práticas em análise*. Porto: Porto Editora.

Runhaar, P., Sanders, K., & Yang, H. (2010). Stimulating teachers' reflection and feedback asking: An interplay of self-efficacy, learning goal orientation, and transformational leadership. *Teachingand Teacher Education, 26*, 1154–1161.

Sítima, M. A. (2005). *Implementar colaborativamente o currículo de ciências físicas e naturais*. Tese de dissertação de mestrado. Universidade de Lisboa, Lisboa.

Snyder, J., Bolin, F., & Zumwalt, K. (1992). Curriculum innovation. In P. W. Jackson (Ed.), *Handbook of research on curriculum* (pp. 402–435). New York: Macmillan.

Spillane, J. (1999). External reform initiatives and teachers' efforts to reconstruct their practice: The mediating role of teachers' zones of enactment. *Journal of Curriculum Studies, 31*(2), 143–175.

UNESCO-ICSU. (1999). *Declaração sobre a Ciência e o uso do saber científico*. Paris: UNESCO.

Viana, P. (2003). *Perspectivas dos professores relativamente ao ensino de Física e Química preconizado pelas orientações curriculares para as ciências físicas e naturais*. Tese de dissertação de mestrado. Universidade de Lisboa, Lisboa.

Chapter 14
Indigenous Knowledge and Teachers' Professional Development in a West Brazil Context

Christiane Gioppo, Michelle Bocchi Gonçalves, and Cristiane Vertelino Marques

14.1 Introduction

This chapter discusses in-service teachers' professional development program established after an invitation by the Technological Institute of Transportation and Infrastructure (ITTI) at the Federal University of Paraná (UFPR) with a science/environmental focus on three cities affected by a highway construction in West Brazil. In this sense, ITTI signed a memorandum of understanding (MOU) with the National Department of Infrastructure and Transportation (DNIT) of the Brazilian Ministry of Transportation, to fulfill the demands of a mitigation implementation plan regarding the construction of the road between the cities of Corumbá and Anastácio (292 km in total). To fulfill some environmental requirements as well as assist the vulnerable communities affected by the road construction on these municipalities, ITTI constituted the Environmental Education Section (PEA). This section has many actions, among them a teacher professional development program which, in this case, was developed with a group of indigenous people of Terena ethnicity.

In order to understand the program developed, we introduce the context, first explaining ITTI and its function to clarify why a Transportation and Infrastructure Institute is connected with teacher education actions. Then we move to the Brazilian context of Pantanal to describe one of the most intriguing biomes in the world and why we worked in this area. Then we discuss the group of Terena teachers,

C. Gioppo (✉)
The Federal University of Paraná, Education College, Paraná state, Brazil
e-mail: cgioppo@yahoo.com

M.B. Gonçalves
The Federal University of Paraná, Coastal Section, Paraná state, Brazil

C.V. Marques
Marcolino Lili Indigenous Elementary School, South Mato Grosso State, Brazil

B. Akpan (ed.), *Science Education: A Global Perspective*,
DOI 10.1007/978-3-319-32351-0_14

an indigenous ethnic group who lives in the area and who are the collaborators on the educational program.

The theoretical background uses two concepts: culturally responsive teaching (Gay 2000) and resistance (Adorno 1980; Benjamin 1994). After describing the grounds of the teacher professional development program, we proceed to a more detailed description that includes three steps:

1. *Understanding teacher needs and demands*: This shows how we sought their interests and made connections with the road issues.
2. *Tackling immediate needs*: This discusses two moments with teachers – firstly, for teachers to reveal their own knowledge on the topic and build up their knowledge using scientific environmental concepts going beyond the idea of modeling and, secondly, when teachers tried to translate (Saramago n.d.) traditional knowledge and scientific environmental knowledge into classroom plans and activities with the support of a mentoring process. This second step also included sharing and evaluating the translated activities and materials.
3. *Empowering teachers to become more autonomous*: This was done collectively during meetings that they had requested to tackle another important issue.

In addition, we present to the reader two texts/stories written by Terena teachers about a mythical entity of Saci. The theoretical background that guided this part was those of discourse analysis, discussed by Eni Orlandi and Michel Pêcheux.

The chapter finishes with some considerations and discussion about the professional learning processes.

14.2 The Context of the Study

14.2.1 What Does ITTI Stand for?

ITTI is the Technological Institute of Infrastructure and Transportation. It is part of the Transportation Department on the College of Technology at the Federal University of Paraná (UFPR), a Research One (According to Carnegie Foundation for the Advancement of Teaching (n.d.) is a Highest Research Activity with more than 20 doctorates. It is also Known as Doctoral Univeristy) in the south of Brazil. The institute operates in the preparation, implementation, and monitoring of programs and studies for the environmental management of engineering constructions, especially in the area of transport, such as highways, railways, harbors, and other modes. In the case of this chapter, the construction site is an area on BR 262 highway and covers 292 km (181,5 mi) in length, from the Municipality of Corumbá (border of Colombia, passing through the municipalities of Miranda finishing in the city of Anastácio, all located in the state of South Mato Grosso, West Brazil. In this sense, ITTI signed a memorandum of understanding (MOU) with the National Department of Infrastructure and Transportation (DNIT) of the Brazilian Ministry of Transportation, to fulfill the requirements of implementing a mitigation plan regarding this highway construction site.

Services provided by ITTI included: priority for research and development; consultancy services, engineering special services, inspections, and supervision of construction sites; education programs including the organization and implementation of workshops and distance education courses; a professional development program for teachers; seminars, lectures, and preparation of written education materials; and several scientific and technological activities for the in-site workers and communities affected by the construction. Moreover, projects of this institute also focused on environmental issues, the use of natural resources, and land territorial management. ITTI crew included professionals from different fields, such as faculty members from the university, environmental and civil engineers, cartographers, biologists, educators, undergraduate and graduate students, journalists, graphic designers, and others.

To fulfill the environmental education requirements as well as assist the vulnerable communities affected by the road construction, ITTI has an Environmental Education Section (PEA) which analyzes the mitigation requirements for the region, and it developed one out of 13 programs designed for the environmental management of the BR 262 highway construction site. The Environmental Education Section aims to assist communities at large, construction site workers, and educators. It also helps teachers organize their knowledge and traditional knowledge along with scientific and environmental education requirements as indicated either by the National Science and Environmental Education Standards or by the Brazilian Institute of the Environment and Renewable Natural Resources (IBAMA). Thus, the teacher professional development program described in this chapter is just one action of the Environmental Education Section (PEA) of ITTI/UFPR.

BR 262 highway cuts Center Brazil from the east coast to west crossing four states (Espírito Santo, Minas Gerais, São Paulo, and South Mato Grosso) finishing on the border of Colombia, as visible in Fig. 14.1a. The west most state on this area is South Mato Grosso (Fig. 14.1b) where one of the most important Brazilian biomes in terms of diversity of life and natural resources is located (Fig. 14.1c).

In the next section, we provide a description of Pantanal and where it is located. Then we describe the group selected for this study, the indigenous group of Terenas.

14.2.2 *The Pantanal Wetland Biome in South Mato Grosso: Area Where the Teacher Professional Development Program Was Implemented*

Pantanal is described (Embrapa Pantanal 2000) as one of the largest continuous extensions of wetlands on the planet. It is located in the heart of South America, in the Upper Paraguay River Basin. The wetland has about 250,000 km^2 (25,000,000 ha^2), most of it is in West Brazil, in the Brazilian states of Mato Grosso

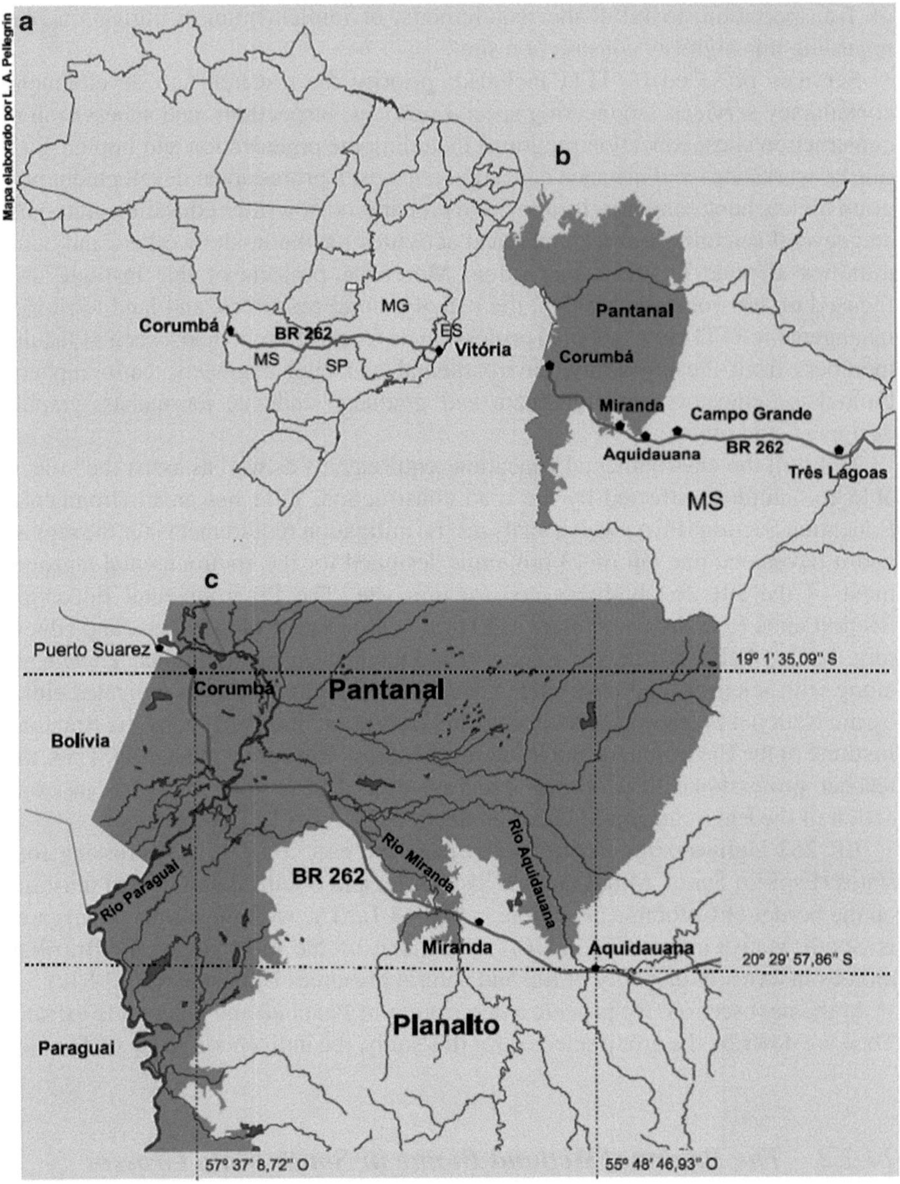

Fig. 14.1 BR 262: (**a**) starts in the city of Vitoria (ES) and finishes on the city of Corumbá (MS), (**b**) part of the highway on the state of South Mato Grosso starting on the city of Três Lagoas and finishing on Corumbá, (**c**) area of the construction site 292 km from Aquidauana/Anastácio to Corumbá with areas of plateau and floodplain of South Mato Grosso Pantanal (Source: Catella et al. 2010, p. 8)

and South Mato Grosso. The remaining area is mainly in Bolivia and a small portion in Paraguay. Paraguay River and its tributaries run through Pantanal, forming extensive wetlands that serve as shelter for many fish such as pintados, dourados, and pacus and also of other animals, such as alligators, capybaras, and giant otters, among others.

The Brazilian portion of Pantanal covers an area of 138,183 km^2; 65 % of it is located in the state of South Mato Grosso and 35 % in the state of Mato Grosso. The area is a floodplain influenced by rivers that drain the basin of the Upper Paraguay. Pantanal is also influenced by four major biomes: Amazon, Cerrado, Chaco, and Atlantic Forest. The fauna and flora developed there is of rare beauty and abundance.

The United Nations Educational, Scientific and Cultural Organization (UNESCO) recognized Pantanal as a biosphere reserve in year 2000, due to its characteristics and importance, as this biome is considered one of the most diverse natural and lush reserves on Earth. According to UNESCO (n.d.), Pantanal has an amazing diversity of flora and fauna: there are 1132 species of butterflies, 656 birds, 122 mammals, and 263 fish and 93 species of reptiles. Many species still have big populations in the region: Pantanal deer, capybara, caiman, and the jabiru. The climate is hot and humid in summer and mild and dry in winter. During the rainy season, between October and February, it is almost impossible to cross the flooded areas by land. As for the soils, most of them are sandy and support native pastures used for grazing by cattle, introduced by settlers in the region.

According to Embrapa Pantanal (2000), there is not only one Pantanal; eleven wetland types have been identified; all of them have their own characteristics of soil, vegetation, and climate. The different Pantanals are Cáceres, Poconé, Baron Melgaço, Paraguay, Paiaguás, Nhecolândia, Abobral, Aquidauana, Miranda, Nabileque, and Puerto Murtinho (see Fig. 14.2). The same study showed the whole Pantanal biome with its eleven environs is being gradually changed by socioeconomic developments with a lack of planning to ensure the sustainability of the natural resources. Thus, a teacher professional development program, with a special focus on environmental education, was designed to consider the fragility of this environ and find ways to develop a broader awareness about conservation.

Figure 14.3 shows part of South Mato Grosso state with its Pantanal in green and the indigenous villages close to the road (in red). The municipalities where this research is being conducted are also visible (Corumbá, Miranda, Anastácio, and Aquidauana). The teacher professional development program worked with 15 indigenous villages in South Mato Grosso. The second part of this chapter describes the program developed with the Terena indigenous people.

It is important to notice in Fig. 14.3 that in the same area, there are also other indigenous ethnicities such as Guanás, Kadiwéu, Chamacocos, and Kinikinaos, since South Mato Grosso state has one of the biggest indigenous populations in the country.

The next section provides some background about Terenas' ethnicity and culture.

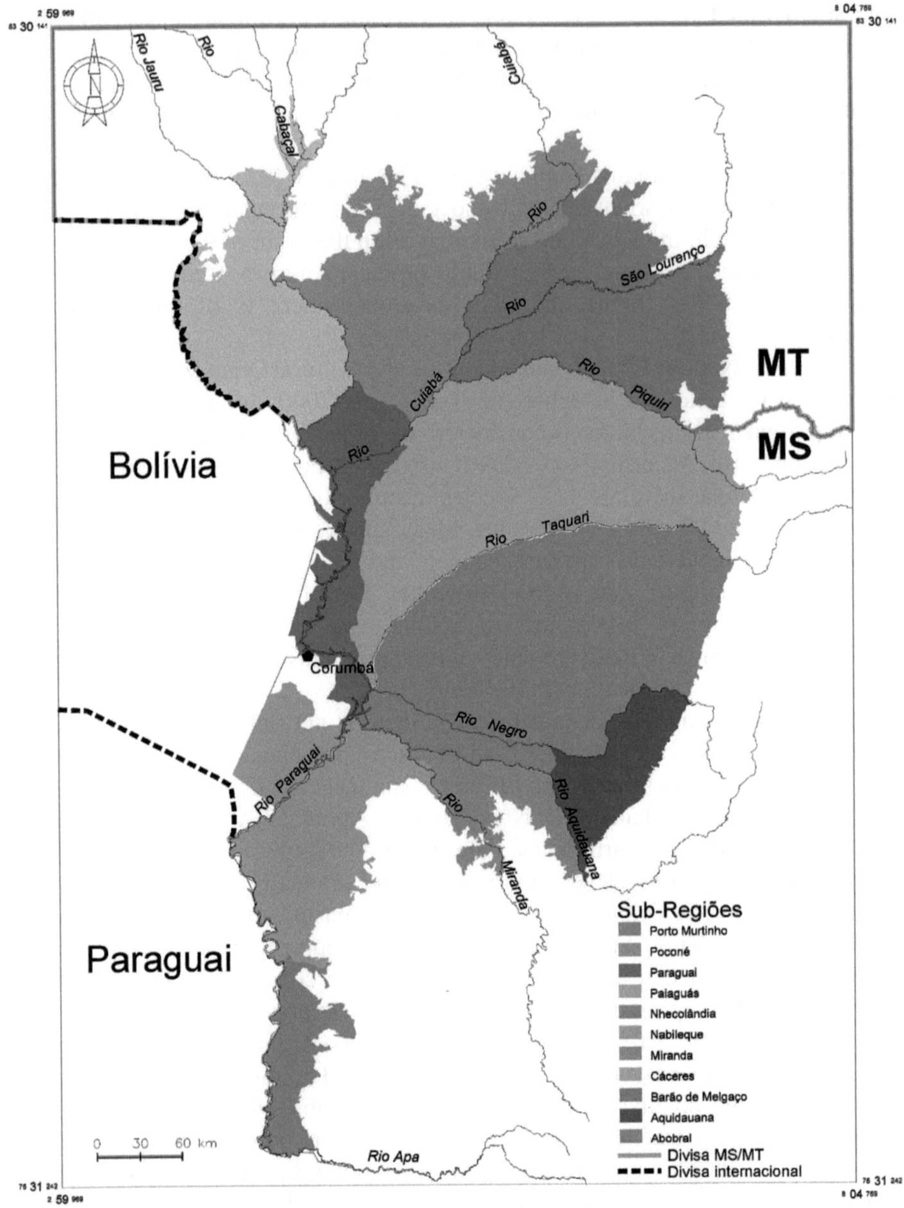

Fig. 14.2 Pantanal environs or subregions. On the *left map* shows Brazilian border with Bolivia and Paraguay. On the *right* of the map, all 11 Pantanal subregions distributed in Brazilian territory between the states of Mato Grosso (MT) and South Mato Grosso (MS) (Source: Pellegrin in Gioppo 2014, p. 13)

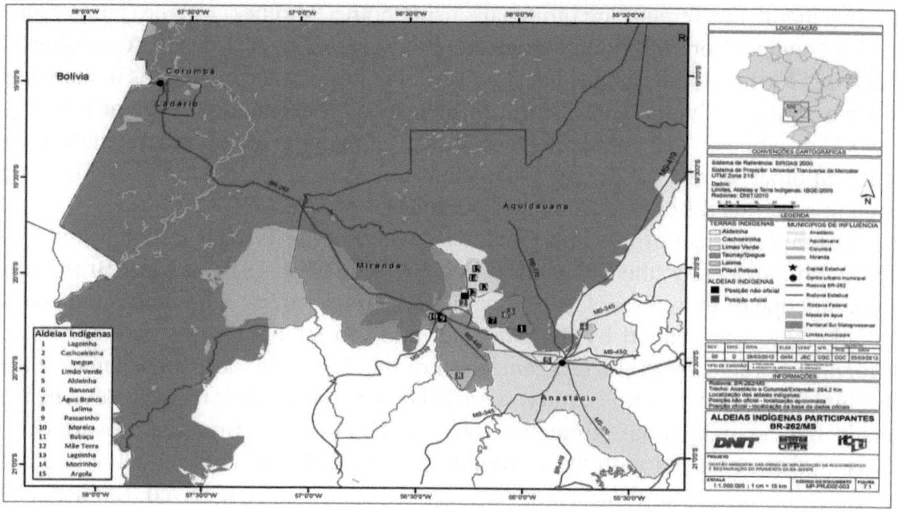

Fig. 14.3 Part of South Mato Grosso state in West Brazil with the municipalities of Corumbá, Miranda, Aquidauana, and Anastácio that are close to the highway site in *red*. It indicates the 15 Terena villages participating in the Teacher Education Program (Source: ITTI 2013)

14.2.3 Indigenous Ethnicity of Terena: Collaborators on This Study

As mentioned earlier, South Mato Grosso state has one of the largest indigenous populations in the country. Terenas are an indigenous ethnicity with a very large population who maintain intense contact with the regional population, either through women vendors or the cane sugar cutters, in addition to temporary workers on farms or in sugar and alcohol mills. Through their intense involvement with the surrounding society, Terenas are also known as "urban Indians." However, this misleading view may mask their struggle through generations to keep their culture alive.

Terenas began interacting with surrounding society in the eighteenth century. According to Oliveira (1976, p. 26), at this time along with other ethnicities – such as Guanás, Layanas, and the Kinikinau Exoaladi – Terenas crossed Paraguay River, in successive waves, and settled between Miranda and Aquidauana rivers. Terenas are a subgroup of the Guaná ethnicity and belong to the Arawak linguistic family. With a population of about 25,000 inhabitants distributed throughout the state, about 25 indigenous lands occupy discrete areas located in the municipalities of Anastácio, Aquidauana, Miranda, Dois Irmãos do Buriti, Nioaque, Sidrolândia, Campo Grande, Terenos, and Dourados. Some Terena indigenous lands are also installed in municipalities of Avaí and Braúna, in São Paulo state (Leitão 2005, p. 43). Teachers who participated in this study are limited to Terena from villages located in the heart of South Mato Grosso state.

In general, the Terenas are bilingual – using both a "mother" tongue (indigenous) and a "contact" or "adoption" language (Portuguese, in this case). According to Ladeira (2001), mother tongue has no socializing importance to Terenas in the sense of integrating individuals in a world of their own, conceptually different from the "world of whites." The author suggests its use is linked to an affective socialization only. In other words, the Terena language is not used in these societies as diacritic sign to assert its difference from non-indigenes.

Several historical events interfered, decisively, in the Terenas' relationship with the Brazilian society; the Paraguayan War (from 1864 to 1870) against Brazil was certainly one of these events. Since that time, Terenas, along with other indigenous populations, were called by the Brazilian army to strengthen the defense of the Brazilian borders. With the reorganization of the Terena in indigenous lands at the beginning of the twentieth century, schools became one of their main achievements, made by the indigenous protection service (SPI) of the Brazilian government. Arising from contacts with the surrounding society, the demand for schooling has opened the doors to Protestantism: since 1912, Terenas received English and German Protestant missions and later from the USA. In order to evangelize them and provide the education they wanted, the missions developed different schooling projects (Carvalho 1995).

In 1936, SPI also started its activities in the Bananal Terena village where a public school was installed, "Rondon General School," whose language of instruction was Portuguese. Today this school operates in three shifts, and the final grades of primary and secondary schools are taught mainly by non-indigenous teachers. Alongside the development of this school began the process of public schooling with a peculiarity: the teaching was conducted in Portuguese by non-indigenous teachers. Therefore, schooling among Terenas, and also among other ethnic groups in Brazil, has a very different context compared with the current one whose national guidelines for indigenous education indicate respect for cultural differences. It is important to note that although it has been used as an agency of acculturation (because the first goal was to integrate the indigenous to the Brazilian society), for Terenas, this school was also, paradoxically, their instrument of resistance, since from it the non-indigenous world and codes could be understood. It was vital for the people to learn Portuguese, and the school was the privileged locus for that. Nowadays, indigenous schools still have non-indigenous teachers, and the indigenous teachers are still not the majority; however, a teacher education program from a local university is trying to change this profile.

When the highway construction started, ITTI became aware of the proximity of the indigenous lands and villages with the road, and we decided to cooperate with them to create a teacher professional development program, to meet their requests. The next section poses a rationale for the program based on Terenas' needs and interests.

14.3 Teachers' Professional Development Program

14.3.1 Theoretical Framework

The teacher professional development program on socio-environmental issues was created mainly to develop the awareness on the big diversity of animals and plants connected to the problems of vehicles running over animals; to better understand the uniqueness of Pantanal biome and its changes with increasing cattle and soy farming on the region, the fragile stability of water pulse between flood and dry seasons and the big problems caused by changes on river flow for navigation and farming, as well as the introduction of new fish species; and also to tackle the cultural and social changes on the communities that live fairly close to the road, especially the Terena traditional villages. The program was framed into two major interconnected research fields: culturally responsive teaching and resistance.

Gay (2000) described culturally responsive teaching as using students' cultural knowledge, to make learning more relevant and appealing to them. Culturally responsive teaching empowers students. Thus, teachers foster and validate culture-based education on their planning, pedagogy and everyday teaching.

UNESCO (2003) publicized guidelines for language education with the emphasis on the use of culturally appropriate educational materials. "Education should raise 'awareness of the positive value of cultural diversity' in order to do that curriculum promotes a realistic and positive inclusion of the minority history, culture, language and identity" (p.33). In this study, we consider Terena indigenous people as the minority, since they are almost 30 % of the local population in the cities of Aquidauana, Anastácio, and Miranda, besides their own villages.

Castagno and Brayboy (2008) suggested that culturally responsive teachers have a role in developing full educational potential through developing an awareness of students' knowledge, values, norms, beliefs, and world views. Stephens (2003) also suggested that teachers should consider and acknowledge students' beliefs instead of denying and trying to fit them into the dominant culture. So, the challenge is to share practices and understandings within the cultural context in the classroom. McConaghy (2000) emphasizes that to teach in a culturally responsive manner, one should overcome the idea of naturalizing the prevailing cultures and anything that divides cultures into good/bad or right-wrong dualities or that ranks them from more cultured to a non-cultured.

Resistance was discussed by Adorno (1980) and Benjamin (1994) but is not a concept clearly or systematically stated by them. On the contrary, this concept is subtly related to aspects of contemporary life as the disappearance of experience, the diminishing of narratives, the loss of history followed by memory suppression, and the raising of cultural industry. Therefore, incorporating indigenous knowledge into a teacher professional development program can only be an approximation. We understand resistance as a way for individuals not to be transformed in a "thing" or a piece of a gear or be swallowed by a mass culture. It is a way to emphasize ones' experience and uniqueness and even a way to reframe and review cultural injustices that have historically and arguably continued to this day to subjugate indigenous knowledge in teacher education, as well as classroom pedagogies, and students' learning specificities for indigenous people. By using the resistance concept, the intent is to reexamine teacher professional development in the deconstruction/reconstruction of their practices in order to work beyond the established social order and reframe it as a reconceptualization of what could and should be done for schools, students, and the local community where the indigenous teachers are from.

Grounded in these two concepts, we created a professional development program in three steps:

- Understanding teachers' needs and requirements
- Addressing teachers' immediate needs
- Empowering and developing teacher autonomy

These steps are described in the next section.

14.3.2 The Program

14.3.2.1 Understanding Teachers' Needs and Requirements

To define the needs and priorities, we visited the area. We started at the local education board, knowing the local standards, and then we visited schools where we met teachers and principals and talked about their expectations in relation to the highway, its problems, and how ITTI could be a source of help to teachers' professional development. Then talked to community leaders and stakeholders, our conversations with them provided information on their expectations and their views on schools' weaknesses. From these visits we learned the following:

Communication and Timing Communities are an average of 20 km from each other. The secondary roads were all unsealed and some are partially flooded during the rainy season. Therefore, communication is an issue because there are not enough cell phone masts and the village has only a one public phone on the main street. Many teachers work in more than one school and sometimes three shifts on the same day. Terenas have their own timing and priorities. These must be considered when asking them to create activities, deadlines, and so on. Besides this, the national government is offering a lot of in-service professional development opportunities; thus, a course calendar must remain flexible, and some changes were expected along the process.

Leadership and Decision-Making The main leader is the head of the village (cacique) who should be heard and should be the first one to be contacted about all activities and intents. Caciques are elected and each village has a different cacique and a tribe council. They discuss most of the issues among themselves and may also call the whole tribe for an important decision. Caciques can change principals and teachers in the city schools and even on the state schools. They have the final decision about village priorities and gives permits to hold any activity in the village, including school activities.

Considering constraints of communication and timing and distances from each tribe to downtown and costs for bringing people to the course, we decided it was better to have a 1-day workshop instead of two half-days. However, some teachers who worked for municipality and for the state schools needed to negotiate with their principals to participate for a full-day workshop. To communicate, we had to develop a menu of alternative ways and find a connection inside the village authorized by the head. The greatest concern about the impact of building the highway was that the fauna would get run over, thereby killing small animals such as armadillos and all sorts of birds as well as big animals like tapirs, capybaras, caimans, Pantanal deers, guará wolfs, ant eaters, and even jaguars, on the highway.

From the talks with heads of the villages, principals, teachers, and community, we realized that the Terenas were willing to have more teaching and learning materials in their own language. They have some texts translated to literacy level but very few books that discussed scientific concepts and that would connect their culture and world view with science and environmental issues. However, we decided not to start with the textbooks because the program was financed by the national government with very strict timelines and the Terenas' conception of time might clash with that. Besides, this a collective textbook production which demands huge time and effort; and in order to do that, many teachers needed to first develop confidence in us. Thus, we decided that a textbook written in Terena language would be developed later in the course of the program. However, the collective textbook production is beyond the scope of this chapter.

14.3.2.2 Tackle Immediate Needs

To tackle immediate needs, first we realized that a person from the local community should be hired to bridge the gap between ITTI and the indigenous teachers. Hence, we created the figure of the "local organizer" and conceived the role as: community people selected by the institute for temporary work, whose function is to provide the link with the institute, giving organizational support in terms of making contacts, locating places for the activities, and arranging all the needed infrastructure. Beyond organizational support, we also expected this person to be able to check teachers' needs and difficulties; give feedback to teachers; provide support for mentoring, discussions, and follow-ups; receive evaluations; and be able to pay attention to teachers' needs and demands in order for proposed activities to be effective. In other words, this person plays a key role as a bridge between ITTI and teachers. Thus, we needed to be very careful while choosing this person.

After selecting the "local organizer," we developed a small three-part plan. The first was a 1-day workshop (8 h) to find out the teachers' ideas and connections to indigenous knowledge (not only in a modeling format). The second included many distance education ideas in which teachers would adapt, plan, and implement a series of activities connected to their lessons (assisted by a mentoring process). The third moment was a 1-day exhibit at school campus for teachers and their students to present developed activities to other children, parents, and other colleagues in the program, who would be sharing and evaluating results.

The One-Day Workshop At first, teachers and faculty were introduced to each other. Then, we started by discussing about mythical entities of Brazilian folklore focusing mainly on Saci, a mythical entity well known in Brazil and usually described as an African-Brazilian boy with only one leg, who smokes a pipe and wears a magic red cap. Wearing this cap Saci is able to appear and disappear from places. Saci is also known as a forest protector because he can scare the ones who come to cut wood or pollute rivers. However, in each part of Brazil and surrounding countries, this boy is described differently; thus, the idea was to read a children's book entitled *Decomposing Saci* (Gioppo and Gioppo 2010). The book mentions several cultural and physiognomic characteristics of Sacis throughout Brazil and abroad, showing that there is not only one possible description for Saci or a single source for this folklore.

The goal was to deconstruct (Deleuze and Guattari 1992) conceptions that mythical entities of folklore have a single origin or version and that different perspectives can be aggregated in each region bringing new folds and reconstructions, shifting predetermined senses (Derrida 1972) for these stories. The idea behind the stories was to emphasize environmental education as a way to bring contributions to prepare the reader to critical thinking.

Since the book presented many different characteristics of Saci, we called the group's attention to the lack of a book page on the characteristics of a Saci for the Pantanal region. With this absence we urged everyone to remember bedtime stories describing as much as possible the physical characteristics of Saci that had not been mentioned in the book. Thus, teachers were able to tell their stories and talk about their fears and knowledge about Saci and its influence on teachers' own education. After hearing many different Saci stories and features, we emphasized the idea that Saci has no Manichean dichotomy character divided between good and bad, as it was usually presented.

As a second activity, we read a game book entitled *A new beginning* (Gioppo 2011) (Fig. 14.4), in which Saci is in a forest when suddenly a fire starts by a lit cigarette thrown into the bush. With the forest in ashes, Saci realizes he needs to find another place to live. The book is a sort of role-playing game (RPG), so-called solo adventure, with each page or half a page having a numbered scene. The reader starts reading scene zero, and at the end of each page, there are usually two options to choose from. Each decision takes the reader to different paths or stories. In this way, the game has several different stories, some are interconnected, while some are not; and usually, there is more than one possible ending.

The reading/game was held in pairs so that both teachers could discuss every decision and come to a consensus on what to do at any given time. In this story, readers play the role of Saci and should choose the new place to live. However,

Fig. 14.4 (**a**) Game book cover; (**b**) game book inside: an example of numbered scenes (Source: Gioppo 2014)

choosing the Pantanal wetland was not an option offered in this story even if it is a path that one could go through. The goal of this activity was to realize that there are several possible paths and that different decisions lead to different results. In this way we emphasized the idea that environmental education could also be conducted in several ways and that individual's decisions are extremely important for learning. On the other hand, the portion related to Pantanal in the book is rather small and shallow. Thus, teachers were left puzzled: why didn't Saci choose Pantanal to live?

Based on the two readings, discussions, and findings, teachers realized Pantanal is weakly presented in the material, so maybe that was the reason Saci did not choose Pantanal, because he did not know enough about it. From this inference, we have proposed to the group to complement the information about this biome (at least, one little piece of the 11 types of Pantanal), connecting scientific and traditional descriptions of a small area. To help create this description, we suggested a field study. Its goal was to select a small study area, to know it better, geo-referring it. This activity was adapted from the Mapping Our School Site (MOSS) project developed by Stubbs et al. (2003). After analyzing the area, teachers would choose some items on the site to photograph. We understand photography as a personal selection of the frame (Barthes 1984; Cartier-bresson 1968); thus, teachers' background knowledge and world view would play a strong role in this case.

We wanted teachers to have a rationale for taking each picture. To help them develop it, we asked teachers to write a letter to Saci, telling features of the studied area; they could also give more information based on their own knowledge about Pantanal and include their photographs to the letter. Teachers demarcated the study area by dividing it into four quadrants, using aspects of mapping, and then they used grids and colored pencils to create captions. In locus, the teachers identified the existing plants and animals and mapped the area. They photographed what they considered important and discussed why it was important based on their traditional knowledge. Furthermore, we discussed Pantanal biome animals and plants and also discussed the problems of running over fauna close to the highway. Photography was used to discuss their selection and connection with their own knowledge and traditional knowledge; thus, letters written to Saci included a description of Pantanal and an invitation for Saci to come and live in the area, plus maps and photos. All these materials were presented to the class. At the end of the workshop, all materials were collected and analyzed.

Hence, we wanted to use teachers' traditional knowledge as a core part of the information that provided their focus and rationale. The subsequent discussions considered the many possibilities for adapting the use of photography to link environmental education, traditional knowledge, and scientific concepts on classroom activities. Trying to connect these ideas, teachers should create an activity with as many sessions (or days) as needed to be validated in their own schools. At the end of the workshop, teachers were assigned days when they would receive the visit of our "local organizer" who would mentor the activity with teachers in the schools.

The second moment was to validate the activity created in each teacher's own school where they collected traditional stories about Saci or the Whistler (Saci's name among Terenas). Teachers then had time to check aspects from the activities discussed during the workshop such as time and feasibility. They then rewrote, adapted, and applied the ideas to their planning. We allocated about 2 months for this part; thus, activities could be adapted to different grades, class levels, and school schedules. During this time our "local organizers" visited schools, heard teacher doubts, and supervised them during their teaching, observing connections and "translations" (Saramago n.d.) of the acquired experience on their teaching specially focusing on the connections between traditional and scientific knowledge.

The idea of translation in its various forms (Saramago n.d.) was core to developing teachers' self-confidence, autonomy, and empowerment. During their teaching it became obvious that interdisciplinary work created a new woof on the environmental fabric. This interdisciplinary work included aspects of children's literature and Brazilian folklore, writing, geography, geology, mathematics, biology, ecology, and photography. Furthermore, we discussed teachers' traditional knowledge as fulcrum of all of these disciplines.

Some teachers performed the exact activity they had as an example on the workshop with their classes; some limited their work to folklore, whereas others related to the fauna getting run over. Some teachers used the mapping activity to discuss area in math and started connecting with geography. Traditional knowledge was discussed sometimes in relation to the uses of certain plants. For instance, in one school, students found a manioc plant on the study, and then they talked about it. Later on they asked elders for a traditional manioc recipe, wrote it down, cooked, and shared portions. They noted cooking differences from what they have seen on TV, cooking shows, or other places.

In the third moment, teachers and students organized an exhibition of work done through classroom activities. The exhibition included pictures taken throughout the process of developing an environmental look beyond the mapping process, the texts produced, and complementary activities resulting from the adaptation that teachers made for their planning. The goal was, once again, to give voice to teachers and appreciate the work done in schools. Students also had time to present their findings and have their work valued.

Fig. 14.5 Shows the book used to inspire teachers (first one on the *left*) and three books published at the end of the program, with teacher stories: Whistler Saci – Bilingual Terena and Portuguese (on the *left*) – Bridge Saci, Pantanal Saci (Source: Gioppo 2014)

As for tackling immediate needs as a result of the workshops, we have published three children's books including stories written by these teachers (Fig. 14.5). One of them is bilingual, translated into Terena language by themselves. Printed books came back to the teachers and their schools. Thus, the need for materials written in Terena language started to be fulfilled by their own stories and world views related to Saci folklore, who they call Whistler.

In the next part, we analyzed two stories published on Whistler Saci book (Fig. 14.5) produced by Terena native teachers about the legendary figure of Saci. The theoretical background that framed these analyses was from discourse analyses, discussed by Eni Orlandi and Michel Pêcheux.

Discourse Analyses

Below we present statements taken from the stories and their analyses.

Text 1: "Saci on village":

So far away begins the story of Saci on Bird village: on my Street, there were few houses, kills dense, barton was closed by. The nights were dark and quiet, silence dominated the night, only the nightlife reigned; the Potoo with its melancholic sing, the owls, the vampires of the night. In the meantime the old Indian (my grandfather), was putting fire, while telling a story of a Saci who lived behind their old House of straw […] At that time every night we could hear the whistle of Saci, because there was a lot of greenery, today there is no Woods in the village thus is rare to hear the whistle of the famous Saci.

In the text above this teacher does a reference to his village, a Terena village. It was important for him to present information about his roots before telling the story on how his village was organized in the past. There is a need in this teacher to report on his origin, his roots, which are indigenous. He described details of straw houses, dense vegetation, reigned as the nightlife, Besides this the teacher alsomentioned an "old Indian," his grandfather, who told the story of Saci, showing that in Terena tradition, eldersare respected for their wisdom. This speech shows how the teacher is subject to his history, ideology, and culture – in this case Terena culture. Thus, the discursive subject is influenced by ideology and is flooded by the words of others – this teacher has his discourse amended, and it can be modified by the appropriation of pre-existing sayings (Pêcheux 2009).

Text 2: "The Whistler Saci," in full:

The elders of the Bird village say they do not know for sure whether Saci is an adult or a child. Sometimes he appears as a child, usually he appears during the evening when everyone is asleep. Furthermore if a mother forgets children's toys on the yard or children's clothing on the clothesline, then her child will not be able to sleep because Saci can appear, until the mother goes to the backyard to collect all the kid's toys. The child will continue to have horrible nightmares. Sometimes, Saci appears in an adult form. He smokes a pipe, wears a cap and likes drinking cachaça,[1] at the village people usually say that in order for Saci not to bother anyone, elders offer him tobacco and cachaça and then he will take it and leave.

Text 2 presents directions related to the history of Saci. Words are showing how the author is subject to his historicity and is still strongly marked by the creed of his ancestors, which discourse analysis calls discursive memory (Orlandi 2011).

This teacher falls into a set of sayings that circulate socially, the one of another figure called Saci Pererê, which is from African heritage and became very famous and common in some parts of Brazil. Saci Pererê is usually described as an: "adult man who smokes a pipe, wears a cap and like cachaça." Thus, this teacher's speech brings us a picture of a contradictory Saci, sometimes harmless and playful child Saci, other times dangerous adult Saci. In the first part of the text, Saci is a child, who is not so dangerous and scary, yet playful, who "appears during the evening when everyone is asleep." At the end of the text, he describes a threatening Saci that always needs to be indulged by the elders: "In order for Saci not to bother anyone, elders offer him tobacco and cachaça and he will take it and leave." This discourse has shown other similarities with well-known African heritage legend of Saci Pererê which states the need to please Saci with "smoke and cachaça," for example.

On this teacher's words, there is an absence of amalgam in which Whistler legend is related to the African heritage legend of the Saci Pererê, which is well-known and pre-built circulating story in children's literature. Children's literature books read at school, mostly, want to unravel Saci legend as an African-descendent, one-legged boy who smokes a pipe and wanders on the woods scaring people and playing tricks. A quick Internet search for images using the word "Saci" is enough to understand that Saci's image is mostly associated with this legendary figure found in the words of the teacher.

[1] Caçhaça is an alcoholic beverage made out of sugar cane.

The dispersion of meanings attributed to Whistler Saci, materialized in the words of this Terena teacher, converses with existing references in his memory. Thus, even if in his speech there is no amalgam, it is possible to assume a relationship between the school speech and the speech of his ethnicity and his indigenous culture, with all myths and legends, infiltrated by enunciations of other Terena teachers, and, in addition, the interaction loci of these speeches are teachers' memoirs; in this case, the teacher is also an indigenous person.

Teacher refers to words that have been spoken somewhere in his social interaction (whether in his training as a teacher, namely, in his life experience): they have a sense, even before being pronounced. Often, these words are not transmitted nor learned or taught "formally," but they are internalized by the subject without any explanation. Thus, our discussion looks forward to what discourse analysis calls *discursive memory*. The discursive memory can be understood as one that has already been spoken before, in other situations, in other places, and in other independent historical, cultural moments.

Discourse analysis defends the idea that relationships among thoughts, world, and language are never direct, but mediated by speech. In this perspective, this teacher produces senses in a relationship between knowledge and language. Furthermore, it is from his subject position, as a public school teacher and as an indigenous teacher living at a traditional Indian village of South Mato Grosso, that his sayings are always traversed by other speeches, by other words.

The school is a place for learning, full of people who use their own languages to communicate, since language and learning are social processes; discourse analysis does not deal with language and does not deal with grammar. It treats discourse, bringing the idea of in course, in movement. Speech is word itself in motion. And in the context of this research, the word in movement is about the legendary figure of Saci, enunciated by Terena teachers who live in the state of South Mato Grosso.

The two text analyses show us some loss on Terena's culture. Besides this, the published stories arouse teachers' consciousness to the need to improve their indigenous identity lost over the years. Hence, after finishing the second step, publishing, and giving them back in the form of books, teachers asked for a meeting with the Institute's (ITTI) personnel. During the meeting, they emphasized the need for more teaching materials written in Terena. They wanted to write lessons for children based on traditional stories told by the elder. This new demand enabled us to start the third step: empowering and developing teacher autonomy.

14.3.2.3 Empowering and Developing Teacher Autonomy

The third step started when teachers asked for the meeting. At the beginning, they wanted to develop bilingual teaching materials for first-grade children to teach Terena language. Teachers knew exactly what they wanted and asked for help to formalize the proposal and find funds to publish it. ITTI personnel then encouraged them to discuss book themes and topics and create a plan on how lessons would be developed. One constraint was the status of Terena's language because each village has its own dialect and way of writing. Thus, the proposal of a new book was way beyond a planning and writing process and included discussions of language agreement and negotiation on how to write certain words. At the beginning we organized meetings with teachers from different villages to decide how to write certain words. They foresee this new book could bond elders because Terena culture would be embraced, and if traditional knowledge would be written, then it was a way of enduring, preserving, and respecting their knowledge.

After this we held a series of meetings; the first one was to draft design of book plan for the first grade. Then teachers went back to the elders to ask them traditional stories. Analyzing stories, they drew conceptual maps on the themes and made connections with subject areas. After this, the meeting focused on planning lessons.

One example of a lesson is related to a type of brick they prepare with mud and straws that are used to build houses. Their brick is much larger than a regular brick, not cooked or burned, and is called "adobe." To develop the lesson, they discussed geology to specify the type of mud they needed, biology to determine which type of straws to collect in order to prepare *adobe's* recipe and what time of the year they should be collected, and math to figure out how to calculate the number of bricks necessary to construct a room. They were also involved in calculating how long it takes to prepare adobe bricks to construct a room and so on. Furthermore, they tried alternative recipes to create better bricks, for the wet, hot, and humid climate common in Pantanal. Hence, one entire lesson on *adobe* involving mathematics, biology/geology, and Portuguese and Terena language was prepared. Teachers were very satisfied realizing they could include their way of life, culture, and language to teach regular content at school, not only science but other subjects, too. Thus, the ideas of resistance (Adorno 1980; Benjamin 1994) and culturally responsive teaching (Gay 2000; Castagno and Brayboy 2008; Stephens 2003) were empowering these teachers to let them discuss their culture without losing any school content.

The professional development model we developed with them was spiral, since the third step led us to another and different discussion that would go to another cycle of teacher development as shown in Fig. 14.6.

14.4 Some Considerations

This is work in progress and probably will be for some years since the third step is leading us toward another cycle of teachers' professional development that requires different knowledge and approach.

It is important to know that the program we created for teachers' development for this situation was not fully developed in advance. Some things were discussed during the process and along the way. At the end, we realized the full cycle and hence could draw the model shown in Fig. 14.6. However, we do not see it as finished or frozen to be used step by step as it will probably be modified in different situations.

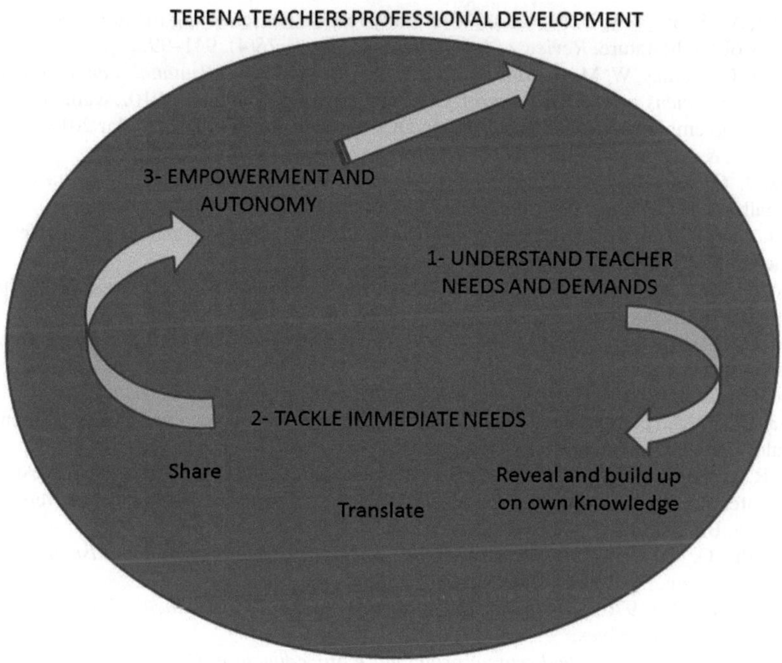

Fig. 14.6 Teachers' professional development developed with Terena indigenous teachers (Source: Authors 2012)

The Terena teachers' autonomy is growing fast, and there is still a long way for them to go. We keep giving them support and feedback, but the most important issue is that they feel their culture and knowledge are being respected, kept, and useful in their everyday lives. This is quite different from what they thought years ago when, as children, they were instructed not to speak in Terena or not to keep old beliefs and traditions; otherwise, they would have no chance in a non-indigenous world. Now time is against us to recover a little piece of the humongous damage we as non-indigenes did to Terena's culture.

References

Adorno, T. (1980). Posições do narrador no romance contemporâneo. In *Os Pensadores*. São Paulo: Abril.

Barthes, R. (1984). *Camera Lucida: Reflections on photography*. London: Flamingo.

Benjamin, W. (1994). O Narrador. In W. Benamin (Ed.), *Magia e técnica, arte e política: ensaios sobre literatura e história da cultura*. Translation: Sérgio Paulo Rouanet; prefácio de Jeanne Marie Gagnebin. (Selected Pieces; vol 1) 7ªed. São Paulo: Brasiliense.

Cartier-Bresson, H. (1968). *The world of Henry Cartier-Bresson*. London: Thames and Hudson.

Carvalho, R. F. (1995). *Subsídios para a Compreensão da Educação Escolar Indígena Terena no Mato Grosso do Sul*. Master thesis, UFSM, Santa Maria.

Castagno, A., & Brayboy, B. M. (2008). Culturally responsive schooling for Indigenous youth: A review of the literature. *Review of Educational Research, 78*(4), 941–993.

Catella, A. C., Tomás, W. M., & Mourão, G. M. (2010). *BR 262 no Pantanal: cenários de encontros entre homens e animais silvestres*. Corumbá: Embrapa Pantanal, 2010. Available in: http://www.cpap.embrapa.br/publicacoes/online/DOC111.pdf. Retrieved in 25 Mar 2013.

Deleuze, G., & Guatarri, F. (1992). *O que é filosofia?* Rio de Janeiro: Editora 34.

Derrida, J. (1972). Marges de la Philosophie. Paris: Minuit. Trad. Joaquim Costa e Antonio Magalhães. *Margens da Filosofia*. [S.D.]. Porto: Ed. Rés.

Gay, G. (2000). *Culturally responsive teaching: Theory, research, and practice*. New York: Teachers College Press.

Gioppo, C. (2011). *Um novo começo. Coleção Aventuras pelo Mundo da Biologia. V. 1.*. Curitiba: Setor de Educação.

Gioppo, C. (2014). *Pesca amadora:uma perspectiva conservacionista*. Curitiba: Departamento de Transportes.

Gioppo, C., & Gioppo, M. (2010). *Decompondo o Saci*. Curitiba: Setor de Educação.

Ladeira, M. E. M. (2001). Língua e história: Análise sociolingüística em um grupo Terena. São Paulo: USP. 179 P, Doctoral dissertation.

Leitão, R. M. (2005). Escola, Identidade Étnica e Cidadania: comparando experiências e discursos de professores Terena (Brasil) e Purhépecha (México). Doctoral dissertation, Universidade de Brasília, Brasília.

McConaghy, C. (2000). *Rethinking Indigenous education: Culturalism, colonialism and the politics of knowing*. Flaxton: Post Pressed.

Oliveira, R. C. O. (1976). *Do Índio ao Bugre: o processo de assimilação dos Terêna*. Rio de Janeiro: Francisco Alves.

Orlandi, E. (2011). *Análise de discurso: princípios e procedimentos*. Campinas: Pontes.

EMBRAPA Pantanal. (2000). III Simpósio sobre Recursos Naturais e Sócio-Econômicos do Pantanal. *Os Desafios do Novo Milênio – resumos*. Corumbá: Embrapa Pantanal. 496p.

Pêcheux, M. (2009). *O discurso: estrutura ou acontecimento. Tradução de Eni Pulcinelli Orlandi*. Campinas: Pontes.

Saramago, J. (n.d.) *Traduzir.* Available in: http://caderno.josesaramago.org. Retrieved in 28 Mar 2011.

Stephens, S. (2003). *Handbook for culturally responsive science curriculum*. Fairbanks, AK: Alaska Science Consortium and the Alaska Rural Systemic Initiative. Retrieved April 1, 2013, from http://www.ankn.uaf.edu/publications/handbook/

Stubbs, H., Hagevick, R., & Hessler, E. (2003). Investigating ants: Projects for curious minds. *Green Teacher, 71*, 34–42.

The Carnegie Classification of Institutions of Higher Education. (n.d.). Carnegie Foundation for the Advancement of Teaching.http://carnegieclassifications.iu.edu/definitions.php. Retrieved in 27 Apr 2016.

UNESCO. (2003). *Education in a multilingual world (UNESCO education position paper)*. Paris, France: United Nations Educational, Scientific and Cultural Organization. Available in http://unesdoc.unesco.org/images/0012/001297/129728e.pdf. Retrieved in 31 Mar 2013.

UNESCO. (n.d.). *Pantanal conservation area. World Heritage Committee (WHC)*. Available in http://whc.unesco.org/en/list/999. Available in: http://whc.unesco.org/uploads/nominations/999.pdf. Retireved 30 Mar 2013.

Part IV
Development and Future Studies in Science Education

Chapter 15
International Perspectives and Recommendations on Equity and Gender: Development Studies in Science Education

Teresa J. Kennedy and Cheryl Sundberg

15.1 Background

The 1990 World Conference on Education for All (WCEFA) set in motion significant international efforts to provide access to quality educational experiences for every child throughout the world (Mundy 2006; UNESCO 1990). The initial declarations produced after the WCEFA event concentrated on reducing the numbers of out-of-school children, primarily consisting disproportionately in South Asia and sub-Saharan Africa, and focused on matriculating girls in these regions into school in equal numbers to boys.

> The *UNESCO World Atlas of Gender Equality in Education* is a valuable resource that provides visual information about the educational pathways of girls and boys in terms of access, participation, and progression from preprimary to tertiary education.

Although the problem still exists in some countries today, the Education for All (EFA) initiatives that sought to ensure gender parity in enrollment did successfully inspire international efforts to promote matriculation of all children into primary schools, yielding a significant decrease in the percentage of out-of-school children and reducing the global gender gap (Chisamya et al. 2011). According to the EFA Global Monitoring Report, while the gender gap of out-of-school children has

T.J. Kennedy (✉)
College of Education & Psychology and College of Engineering,
The University of Texas at Tyler, Tyler, TX, USA
e-mail: tkennedy@uttyler.edu

C. Sundberg
Science Education Research, Ronin Institute, Montclair, NJ, USA

B. Akpan (ed.), *Science Education: A Global Perspective*,
DOI 10.1007/978-3-319-32351-0_15

decreased, at least 54 % of these children are girls (UNESCO 2014, Progress). For a comprehensive view of the extent to which gender disparities in education have changed since 1970 and are shaped by factors such as national wealth, geographic location, investment in education, and fields of study, refer to the United Nations Educational, Scientific and Cultural Organization (UNESCO) World Atlas of Gender Equality in Education (2012) produced by the UNESCO Institute for Statistics. The atlas features over 120 maps, charts, and tables on gender-disaggregated educational indicators through the 2012 school year.

While oftentimes gender equity is seen primarily as an issue associated with access (enrollment and/or matriculation) to education, the EFA goals clearly acknowledge that gender equity also encompasses the overall "quality" of education provided to all (UNESCO 2003). Miske, DeJaeghere, and Adolwa noted that "inequalities frequently are measured by raw numbers or percentages to show the extent to which parity has been achieved in women's and men's participation in economic, political, and social life" (2012, p. 3). However, equal access to education does not necessarily mean the education provided to girls and boys has the same rigor and academic expectations. Delineation of a *gender-equitable, quality education program* was described by UNESCO Bangkok, UNESCO EAPRO, and UNICEF ROSA to include not only equal access (*parity*) but equal opportunity to quality educational programs. A gender-equitable, quality education includes:

> … the capacity to promote the value of gender equality in and through education in the wider society. Quality education demands that teaching processes, curricula and learning materials allow for boys and girls to engage in and benefit from learning equally. (UNESCO Bangkok et al. 2012, p. 11)

According to Halai (2011, p. 44), "an issue with numerical indices and measures of gender parity in education is that they do not reflect the inherent biases and imbalances in classroom practice which often manifest wider cultural and societal patterns of behaviors and traditional stereotypical gender roles of women and men." In a similar fashion, Ito (2012) also indicated EFA parity must include quality of education:

> While I argue that the focus on universal primary education, especially for disadvantaged girls in rural areas, was a necessary step for promoting EFA at that time, the EFA coordination mechanisms have struggled to switch from this idea that EFA is primarily about access to primary education and gender equality to the original concept of EFA as a good quality basic education for all. (2012, p. 5.)

Providing equitable access is a step toward reducing barriers to high achievement. However, reform programs need to consider other typical obstacles to EFA: teachers' expectations and instructional models. Oftentimes, teachers' expectations from girls in areas of science, technology, and mathematics are lower as compared to expectations from boys. Basic assumptions about what is appropriate for boys and girls to learn can undermine aspirations for equality in pedagogy, oftentimes favoring the dominant gender perspective (Aikman et al. 2005; UNESCO Institutes for Statistics 2011). Many high-level reports include similar findings in regard to educational outcomes based on instructional models. A few of these reports, citing gender-related pedagogical approaches aimed at achieving equity in educational environments, are summarized below.

15.2 Promoting Gender Equity in the Classroom

The Organisation for Economic Co-operation and Development (OECD), an international organization established in 1961, promotes policies aimed at improving the economic and social well-being of people around the world, conducting research on the economic, social, and governance challenges of a globalized economy. The 2011 meeting of the OECD Council at the ministerial level was "the first milestone in the Gender Initiative, which was launched by the OECD to help governments promote gender equality in Education, Employment and Entrepreneurship (the "three Es"). Reducing persistent gender inequalities is necessary not only for reasons of fairness and equity but also out of economic necessity" (OECD (2011, p. 2). The executive summary resulting from the meeting stated the following on the promotion of gender equality in education:

> Investing in formal education is essential to promote equality of employment opportunities and strengthen economic growth. It increases cognitive and non-cognitive skills, it improves productivity and it provides individuals with a greater ability to further develop their knowledge and skills throughout their lives. Increased education participation is also associated with better health, and more investments in the education and health of children – especially among women and particularly in developing countries. Gender equality in terms of participation in, and attainment of, education has been achieved in most OECD countries: girls have on average better grades and often outnumber boys among new college graduates. However, in many developing countries, girls still have poorer educational attainments, especially at the secondary and tertiary levels. Achieving gender equality in education in these countries will not only promote greater equality in employment outcomes but also help postpone early-marriages, reduce infant mortality rates and improve health and education of future generations. Gaps in cognitive skills of boys and girls around age 15 are similar across countries: boys perform better than girls in mathematics in most countries, and girls outperform boys in reading in all countries. In terms of science literacy, there are no significant gender differences. But young women are much less likely than young men to choose Science, Technology, Engineering, or Mathematics (STEM) as field of study at graduate level; the share of women in these fields further declines at the post-graduate level. Gender differences in educational choices appear to be related to student attitudes (motivation, interest) in studying a particular subject rather than their ability and school performance. Gender gaps in performance are smaller than gender gaps in fields of tertiary study, indicating that young women often do not translate their good school performance into field of studies for higher education that offer better employment prospects, such as STEM studies. Furthermore, even when women complete STEM studies they are less likely than men to work in these sectors. While it is difficult to separate innate and learned behaviors and to assess the influence of stereotypes, the effect of this gender imbalance is very clear. It hinders women's careers, lowers their future earnings levels, and deprives OECD economies of a source of talent and innovation. It is also an inefficient use of investment in education. (OECD 2011, p. 2)

According to the report, gender gaps occur in participation, attainment, performance in education, and field of study. However, it also concluded that "while educational outcomes vary across and within countries, there is no one country that consistently has large gender gaps (with an advantage to either men or women) or a near gender parity across all indicators" (p. 15). A subsequent OECD report further stated:

> Greater gender equality in economic opportunity contributes to stronger, more sustainable economic growth. Similarly, investing in formal education and training increases individuals' lifelong skills sets, so improving the employment and entrepreneurial opportunities of both men and women. Gender equality also increases labour productivity and the available talent pool, affording businesses greater opportunities to expand, innovate and compete, which in turn provides governments with additional, much needed tax revenue and social security contributions as the population ages. (OECD 2012, p. 24)

The OECD Program for International Student Assessment (PISA) was launched in 1997 and aims to evaluate education systems worldwide every 3 years by assessing 15-year-olds' competencies in three key subjects: reading, mathematics, and science. To date over 70 countries and economies have participated in PISA. To access the list of participating countries, visit http://www.oecd.org/pisa/participatingcountrieseconomies/.
For more information about PISA performance outcomes, visit www.pisa.oecd.org

The UNESCO Institute for Statistics (2011) reported that from 1999 to 2009 a 9 % worldwide increase in access to a primary education occurred. However, conclusions indicated that gender gaps in educational outcomes differ greatly between advanced economies and developing countries, citing that in advanced economies, girls outperformed boys, whereas they lag behind in developing countries. And interestingly, in advanced economies, coming from a disadvantaged socioeconomic background results in a larger negative effect for male students, while in developing countries the negative effect is larger for girls. For example, in a study of 24 countries involved in PISA, Sikora and Pokropek (2012) reported paternal employment in a STEM field enhanced their sons' interest in any field of mathematics, science, or engineering, while girls whose mothers were employed in STEM careers were influenced more in a field of biology, agriculture, and health and less in STEM fields of computing, mathematics, and engineering. PISA studies have made considerable findings related to gender, enrollment, retention, and achievement in countries around the world. PISA considers students' knowledge in relation to their ability to reflect on past experience and understandings and to apply them to real-world issues. A common metaphor used in the United States to describe the underrepresentation of women in science is the "pipeline model." According to Stephen et al. (2013), a key to understanding the persistence of occupational segregation is gaining an understanding of the processes occurring much earlier in the life course, the processes that lead both females and males to choose their personal educational pathways into STEM majors. Science educators have a responsibility to change the factors that remove women from the STEM pipeline and promote equal participation (Blickenstaff 2005). One barrier to ethnic and gender EFA involves local culture and customs. Claussen and Osborne (2013) indicated local culture and customs can have a significant impact on students' choice to pursue a STEM career. They wrote:

> The theoretical lens offered by Bourdieu and his collaborators helps to establish what the real value of a science education is for students, teachers, and policy makers. More importantly, it helps to identify why the failure to ask what is the cultural capital that science education offers its students has led to a set of emphases which have no apparent value for many of today's young people (p. 79).

Even with equitable access to higher education in a STEM field, women often do not continue the formal education requirements for a career in research. Only 30 % of the world's STEM researchers are women, even in countries with equal access to higher education in STEM fields (UNESCO Institute for Statistics, 2014, Women in Science). According to O'Brien et al. (2014), STEM fields tend to be associated with independence and self-reliance, a characteristic frequently linked with masculinity, resulting that oftentimes, women's lack of interest in STEM fields can be explained by stereotypes and misconceptions linked to gender disparity.

Language can be a barrier to educational parity since learning is mediated through language (Vygotsky, 1987, in Lee and Smagorinsky 2000, p. 6). In addition, language is a sociocultural construction. Thus, science educators need to be mindful of the cultural linguistic norms. Lemke (1990) described the social nature of classroom communication. He wrote:

> The classroom is not isolated from the attitudes, values, and social interests of the larger community. Teachers and students bring these with them into the classroom. Science education itself tries to teach certain values, and those values may not always agree with students' values or with students' views about their own interests (p. xi)...Students also need to find the science in the dialogue. If they don't they may learn how to play the classroom game, but they won't learn how to talk physics or biology (p. 11).

In a study of 78 school-age Bangladesh children (grades 3–5) who had dropped out of school the previous year, Dajaeghere and Lee (2011) cited girls were twice as likely to report they felt unsafe in school and traveling to school. Despite acknowledging that females had a right to an education, the girls and women in the Bangladesh community were aware that the gendered cultural norms helped to maintain discrimination.

While cultural-sensitive EFA is crucial, educators need to consider how to affect equitable access to a quality education within the local culture. In addition to cultural/ethnic disparities for EFA, gender inequities are often noted. For example, in a study of 100,000 students in the United Kingdom, Rothon (2006) reported *significant inequalities in educational achievement by ethnicity* (p. 2). However, she also noted that girls outperformed boys in each ethic subgroup on O-level-style tests and General Certificate of Secondary Education (GCSE) exams used in England, Wales, and Northern Ireland.

These life attitudes and values must be instilled in all students beginning in the primary classroom and continue through their tertiary educational experiences. Cobern (1993) wrote:

> It is important for science educators to understand the fundamental, culturally-based beliefs about the world that students bring to class, and how these beliefs are supported by students' cultures, because science education is successful only to the extent that science can find a niche in the cognitive and socio-cultural milieu of students. (p. 57)

15.3 Structuring Successful and Equitable Educational Environments

EdQual is a partnership between six academic institutions, two from the United Kingdom (the University of Bristol and the University of Bath) and four from Africa (the University of Dar es Salaam, Tanzania; Kigali Institute of Education, Rwanda; the University of Cape Coast, Ghana; and the University of the Witwatersrand, South Africa). Their extensive research has produced outstanding working papers that detail interim findings and advance debate on education quality. For more information, see http://www.edqual.org/.

There are many challenges related to equity in education.

> Equity in education has two dimensions. The first is fairness, which basically means making sure that personal and social circumstances – for example gender, socio-economic status or ethnic origin – should not be an obstacle to achieving educational potential. The second is inclusion, in other words ensuring a basic minimum standard of education for all, for example, that everyone should be able to read, write and do simple arithmetic. The two dimensions are closely intertwined: tackling school failure helps to overcome the effects of social deprivation which often causes school failure" (OECD 2008, p. 2).

The OECD cites three key policy areas that can affect equity in education: design of the educational systems, the practices encouraged in and out of school, and the resources that are allocated for both. UNESCO indicated the progress to provide access to an education had stalled (UNESCO 2012, *Reaching Out-of-School Children.*)

EdQual RPC, a Research Consortium led by the University of Bristol in the United Kingdom (UK) and sponsored by the UK Department for International Development (DFID), conducted five research projects in the areas of school effectiveness, language and literacy, ICTs in basic education, implementing science and mathematics curriculum change, and leadership and management for quality improvement. In addition, three additional, smaller-scale projects in the areas of inclusion, school buildings, and the use of ICTs in education to support community empowerment were also conducted.

Researchers at EdQual have devised the following definition of a good quality education:

> A good quality education is one that enables all learners to realize the capabilities they require to become economically productive, develop sustainable livelihoods, contribute to peaceful and democratic societies and enhance wellbeing. The learning outcomes that are required vary according to context but at the end of the basic education cycle must include threshold levels of literacy and numeracy and life skills including awareness and prevention of disease.
>
> Creating a good quality education involves paying attention to the interface between each environment and ensuring that enabling inputs and processes have the effect of closing the gaps that often exist between them creating greater synergy and coherence. Closing the 'expectations gap' between the outcomes of education and what parents and communities expect education systems to deliver requires paying attention to the relevance of the curriculum, listening to the voices of parents and of communities in national debates and developing greater accountability within the system.
>
> (Tikly 2010, p. 13–14)

Barwell et al. (2007) wrote, "The significance of mathematics and science is recognized as areas of curriculum that would lead to improving the prospects of the learners and the community" (p. 8). They cite instructional issues related to the language of instruction leading to "excluding learning from accessing mathematics and science ideas and concepts" (p. 33), as well as teacher levels of preparation and conceptual understanding of the materials that they teach causing gender bias "prevalent not only in the prescribed curriculum materials but in teachers own perception of these disciplines as male" (p. 34). Since gender issues are rooted in culture and traditions, they further recommend that concepts like "gender-sensitive curriculum" highlighted in Western classrooms should be considered as follows:

> The way forward for a research program in science and mathematics curriculum would certainly need to take into account these issues and complexities:
>
> - Recognize the centrality of the teacher in the implementation of a new curriculum and empower the teacher through critical reflection on issues which inhibit students' learning of mathematics and science.
> - Enable teachers to interpret the national curriculum for a conceptual learning and not banking and memorization of facts. Promote problem-solving, critical thinking, and reasoning in mathematics and science as means to reduce poverty.
> - Develop and promote gender-sensitive curriculum materials, enable teachers to practice a gender-inclusive pedagogy, and become aware of implicit gender bias in the curriculum, curriculum materials, and generally in the society (Barwell et al. 2007, p. 34).

15.4 Increasing Gender Enrollment to Gender Retention in Science Education

"The increasingly complex changes in the nature and amount of knowledge and demands in the field of science and technology necessitate an understanding of how students perceive science and technology classes in terms of their gender" (Gömleksiz 2012, p. 116). As described by Magno et al. (2003, p. 14), "educational systems shape, and are shaped by, the societies that they serve." In a summary of the efforts made by the government of Bhutan to increase the numbers of all children and youth attending and completing school, Schuelka (2012) indicated culture plays a major role in the development of educational curricula and programs supporting the culture of Bhutan. While gender equity in education is generally a mainstream goal for modern educational systems, the realities of gender inequity in society are uniquely replicated in educational systems around the world. Research conducted in Central and South Eastern Europe and the Former Soviet Union found that the traditional indicators of education equity – access, participation, and academic achievement – emerged across both gender groups and are most often related to many different factors, including geographical locations, income levels, ethnicity, and school practices:

> Schools have a significant impact on the gender equity in society through school structure, curricula, textbooks and teaching materials, teacher attitudes, the overall atmosphere of the school and leadership structures. Children develop their ideas about gender roles from a variety of sources, but the authority that accompanies formal education makes the information, ideas, and values transmitted particularly powerful…Textbooks and curricula replicate narrow and often outdated gender stereotypes through which women are represented primarily in the private sphere or in low-paid or low-prestigious roles. Men are presented as having more options and shown in more prestigious roles. …The curriculum often fails to teach the knowledge and skills that could help to reduce women's disadvantage. In addition, teachers' ignorance of gender issues leads to biased expectations about how boys and girls will perform academically. (Magno et al. 2003, p. 33)

Toglia (2014) reported dismal statistics of equality in career and technical education, as well as STEM fields for females; "more than 30 years after Congress outlawed sex discrimination in education, the gender divide in career and technical education (CTE), has narrowed barely at all" (p. 14). While the average wage is $9.52 per hour in CTE fields traditionally selected by females like cosmetology, only 6 % of electricians are women, where the typical hourly wage is $20.33 (Toglia 2014, p. 15.). Summarizing research from the American Association of University Women and others, Toglia recommended school counselors receive training to provide females with a full range of career options.

In order to provide an equitable education, Bybee (2007) emphasized the need for instructional support materials. He stated:

> Support for new instructional materials should include several programs for all grade-levels (e.g., elementary school, middle school, and high school), incorporation of educational technologies (e.g., use of simulations), and addressing the needs of all students (e.g., college bound and workforce). (p. 456)

In a similar manner, reform efforts can shape society. Bybee (2009; 2011) indicated the adoption of the standards naturally supports educational equity for all students. He reported national standards present policies for all students. By their very nature, national standards are policies that embrace equity. Answering the question – what should all students know and be able to do – identifies the standards as a clear statement of equity (Bybee 2011, p. 11).

Learning is mediated through language (Vygotsky, 1987, in Lee and Smagorinsky 2000, p. 6); thus, classroom dialogue is critical to the learning process. However, demographics typically impact classroom dialogue, possibly resulting in a barrier for an equitable education. In addition to language, social class status frequently impacts achieving educational parity. Perceived skill level of students oftentimes results in biases that teachers have difficulty overcoming in regard to the level of instruction they provide. Sorhagen (2013) conducted a longitudinal study of first-grade teachers' expectations of US children from ten sites. The results indicated that:

> ...students' academic achievements in high school are affected by early teacher expectations, such that high school students whose first-grade teachers underestimated their abilities performed significantly worse on standardized tests of mathematics, reading comprehension, vocabulary knowledge and verbal reasoning than would have been predicted on the basis of their early test scores. Conversely, when early abilities were overestimated, high school students performed better than expected. The findings of the present study demonstrate that misperceptions of abilities early in students' schooling continue to exert an effect on academic achievement 10 years later (p. 472).
>
> ... On the other hand, teachers' overestimation of abilities seemed to disproportionally help low-income students, suggesting that knowledge of self-fulfilling prophecies in the classroom could be relevant to policies aimed at ameliorating the achievement gap between low- and high-income students, especially considering the persistence of the achievement gap in America. (pp. 475–476)

Social class status regularly influences classroom dialogue. According to Fisher and Frey (2007), the amount of teacher versus student talk in the classroom varies according to the demographics of their students. Likewise, Lingard et al. (2003) reported that in classrooms where there are increased numbers of students living in poverty, teachers tend to dominate discussions and students talk less. Furthermore, studies have shown that teachers of struggling students usually offer their students less challenging tasks and spend less time on meaning and conceptualization and, instead, tend to employ more rote drill and practice activities than do teachers of high-performing or heterogeneous groups and classes (Cotton 1989; Rothon 2006). As well as class status, differing conversation patterns are noted with gender. Research has shown that gender plays a role in how much communicative interaction occurs in the classroom. While there are debates regarding which gender is at a greater risk due to lack of engagement and overall academic failure (van Langen et al. 2006; Wilhelm and Smith 2005), Orenstein (1994) provided clear evidence that the amount of time that girls spend participating orally in class decreases as they get older. In addition, studies have shown that teachers tend to call on boys more often than girls, ask boys higher-order questions, use longer wait time with boys than with girls, and provide boys with more extensive feedback (Sadker and Sadker 1995).

Equity pedagogy, as defined by Key (1995), utilizes techniques and methods designed to facilitate the academic achievement of students from diverse racial, ethnic, and social class groups. At its most basic level, educators must take into consideration that individual student achievement is influenced by many factors, including attitudes, interests, motivation, type of curricula, relevancy of materials, as well as the culture of the students. The five dimensions of multicultural education, as described by Banks and Banks (1995), provide basic foundational components from which to build an inclusive scientific culture in the classroom. These include:

- *Content integration* – encompasses the extent to which teachers use culturally relevant examples, data, and information from a variety of cultures and groups to illustrate key concepts, principles, generalizations, and theories in their subject areas or disciplines
- *Knowledge construction* – involves the procedure by which social, behavioral, and natural scientists create knowledge and the manner in which the implicit cultural assumption, frames of reference, perspective, and biases influence the ways that knowledge is constructed within each discipline
- *Prejudice reduction* – describes the characteristics of children's racial attitudes and suggests strategies that can help students develop more democratic attitudes and values
- *Equity pedagogy* – consists of using techniques and methods that facilitate the academic achievement of students from diverse racial, ethnic, and social class groups
- *Empowering school culture and social structure* – is used to describe the process of restructuring the school's culture and organization so that students from diverse racial, ethnic, and social class groups will experience educational equality (Banks and Banks 1995)

The importance of building learner-centered classrooms has been increasingly cited, drawing attention to the fact that successful initiatives share characteristics of "structured pedagogy," promoting careful lesson planning, providing clear introductions linking to the previous lesson structured around specific learning objectives, and consistently using formative assessment (Tikly 2010; Barrett et al. 2007). The issues related to gender retention in schools must not only focus on keeping students in school to receive a basic education but rather need to promote pedagogical approaches that are gender inclusive.

15.5 Gender and Science Enrollment Rates and Achievement

According to Ejifugha and Ogueri (2011), gender equity focuses on the socially constructed and reconstructed roles of males and females, through social, cultural, and psychological lenses. Their research described enrollment statistics in science teacher education dominated by females, while male-dominated enrollment remained in mechanical, electrical, chemical, and petroleum engineering. They concluded that "enrollment statistics into tertiary education in Nigeria portray gender biases," stating that "courses at the tertiary level are erroneously and inadvertently categorized by gender" (p. 2). They further stated that females tend to enroll in courses that are gender compliant and where perceived opportunities for success may exist, such as in the teaching profession.

Enrollment rates in Taiwan follow similar trends. According to Chen (2012), even though the percentage of female students is almost the same as males, their majors differ greatly, with males opting for science and technology disciplines while women favor the humanities and social sciences. Furthermore, Liu and Chen (2007) studied trends in gender segregation in Taiwan from 1972 to 2003 and found that the expansion of higher education did not reduce gender segregation, specifically reporting that gender segregation increased due to technology colleges upgrading to science and technology universities in the 1990s. Regarding employment statistics, Lui reported that in Taiwan, graduates in science and technology have better chances to secure employment after graduation, and with better pay, adding that female graduates tend to have more difficulties in finding jobs and typically earn less than their male counterparts.

An examination of UK performance patterns in mathematics and science, by gender, social class, and ethnicity, found girls in each ethnic subgroup (Black, Indian, Pakistani, and Bangladeshi) outperformed the boys, and ethnic minority students scored significantly lower than their white counterparts on the General Certificate of Secondary Education (GCSE) (Rothon 2006).

After 40 years of emphasis on gender equity in science, women and girls in England are still underrepresented in STEM disciplines, particularly the physical sciences (Archer et al. 2012). Equally disheartening results of surveys of over 9000 elementary students, ages 10–11, indicated science was "not for me" (Archer et al. 2012). Gender inequity in science is an issue in other countries as well. Research conducted in 20 different elementary schools in Turkey revealed that student perceptions of science and technology differed significantly by gender (Gömleksiz 2012). Interestingly, significant differences were found between gender groups in response to the teaching strategies and classroom management techniques used in the science and technology classes. While male students were successful in response to the teaching strategies used, the female students found them insufficient.

15.6 Gender Differences in Selecting and Completing a STEM Baccalaureate

Ma (2011) indicated three factors that impacted women's successful completion of a STEM baccalaureate education: high school achievement, positive attitudes, and taking prerequisite STEM course in high school such as physics and calculus. In particular, these three factors impacted the women's successful completion of

STEM careers in physical sciences and engineering (Ma 2011). Perez-Felkner et al. (2012) also reported the impact of high school course selection on subsequent enrollment in a physical science, engineering, or computing science baccalaureate program; women who completed more mathematics and physical science courses in high school were more likely to select and complete a career in physical science STEM-related fields.

15.7 Recommendations

In highlighting some of the research that may contribute to understandings of gender inequities in the science classroom, the general focus of this chapter has been to discuss sample findings around the world leading to action items that can be undertaken by the classroom teacher to reduce gender bias in the science classroom and provide equal learning opportunities to all students.

Although gender biases differ from community to community due to cultural constructs, general strategies that promote a more inclusive learning environment can be implemented regardless of the location. The focus of gender equity in enrollment statistics must also concentrate on retention rates and quality educational preparation for all students. If highly qualified educators are to be charged with ensuring that no child is left behind regardless of socioeconomic status, race, or gender, then they must be given the training and tools to assist them in this task (Thomas and Stockton 2003). However, it is extremely important to ensure that the curriculum is not overloaded in order to allow teachers adequate time to engage with students and to focus more on developing scientific thinking than on content (Barwell et al. 2007). Recommended strategies, gleaned from the literature previously cited, to ensure that all students, regardless of gender, are given equal opportunity to achieve sustained academic progress in the science classroom include the following:

- Check for appropriateness of the content and activities to ensure alignment with students' ages, interests, expectations, and environment.
- Identify the key concepts from the curriculum standard to guide instruction, evaluation, and intervention.
- Focus on the key concepts by utilizing the 5E model: engage, explore, explain, elaborate, and evaluate.
- Utilize multiple teaching strategies and design cooperative learning environments (consisting of mixed gender groups) that encourage the participation of all students.
- Employ a variety of learning activities to create student-centered learning environments based on cognitive and constructivist learning approaches rather than on behaviorist approaches.

- Use inquiry-based approaches focused around students' interests structured in a manner that leads students to gain a deeper understanding of the way science is authentically carried out.
- Employ problem-based learning (PBL) experiences focused around current events and authentic activities to promote thinking strategies and domain knowledge.
- Create examples and assignments that emphasize the ways that science can improve the quality of life for all living things.
- Ensure students are provided with equal access to the teacher(s) and all classroom resources.
- Implement assessment and evaluation of quality while allowing students the opportunity to monitor their own progress and track their own results in order to gain more ownership of their personal educational outcomes.
- Maintain school, home, and community links while inviting and including parent and community voice in curriculum design.
- Include occupational options and educational pathway planning within the curricular model implemented.

15.8 Conclusion

The purpose of this chapter was to provide a sampling of perspectives surrounding one of the most important issues of our time – ensuring quality educational opportunities for all – in the mindset of preparing science teachers, from primary to tertiary education, who in turn, are able to prepare future generations of scientific leaders, representing both male and female interpretations and insights. Despite implementation of programs and interventions designed to increase gender parity in STEM fields over the last 40 years, young women still tend to be less likely to select a STEM career than young men, especially in less developed communities/countries. Even though there is a lack of a statistically significant evidence of gender differences in science literacy, young women often select a vocation in humanities and social science instead of a more lucrative STEM career. One factor identified as a barrier for STEM career choice for young women, especially in the physical science, engineering, mathematics, and computing science fields, appears to be attitudinal (motivation or interest). The challenge is, therefore, that science educators vigilantly monitor their classroom activities and adjust pedagogical practices to promote gender parity. There is a strong need to continue to seek innovative approaches and methodologies that can actively include the voices and perspectives of the marginalized to ensure our understandings are "better able to reflect the experiences of these groups and begin to shift discourse and policy in a way that more accurately reflects their interests" (Tikly 2010, p. 20). A desire for EFA and provision of suitable instructional materials offers a potential for educational equity. However, there needs to be a belief that parity can be achieved.

As Kofi A. Annan, Nobel Peace Prize awardee and seventh secretary-general of the United Nations from 1997 to 2006, acclaimed:

> There can be no long-term security without development; and there can be no long-term development without security; no society can remain prosperous for long without respect for human rights and the rule of law. This is what we have to teach our students and future leaders…Education is the catalyst for progress, in every society, in every family. Most of us can point to a teacher who changed our lives. In my case there can be no doubt that that teacher was Francis Bartels. I can remember his tireless efforts to broaden our horizons, to open our eyes, speak our minds, and engage with the issues of the day and the world at large while never forgetting the traditions and values of our own society. (Loveland 2013, p. 26)

References

Aikman, S., Unterhalter, E., & Challender, C. (2005). Achieving gender equality through curriculum change and pedagogy change. *Gender and Development, 13*(1), 44–55.

Approach. *Cultural diversity and ethnic minority psychology*. Advance online publication. Retrieved from: http://dx.doi.org/10.1037/a0037944

Archer, L., Dewitt, J., Osborne, J., Dillon, J., Willis, B., & Wong, B. (2012). Balancing acts: Elementary school girls' negotiations of femininity, achievement and science. *Science Education, 96*(6), 967–989.

Banks, J., & Banks, C. (Eds.). (1995). *Handbook of research on multicultural education, research into practice: Science*. New York: Macmillan.

Barrett, A., Ali, S., Clegg, J., Enrique Hinostroza, J., Lowe, J., Nikel, J., Novellis, M., Oduro, G., Pillay, M., Tikly, L., & Yu, G. (2007). *Initiatives to improve the quality of teaching and learning: A review of recent literature*. EdQual Working paper 11. Retrieved May 26, 2013, from www.edqual.org, and used with permission.

Barwell, R., Bishop, K. Erduran, S., Halia, A., Iyamuremye, D., Nyabanyaba, T., Rizvi, N.F., Rodrigues, S., Rubagiza, J., & Uworwabayeo, A. (2007). *Implementing curriculum change – Literature reviews: South Africa, Rwanda, and Pakistan*. EdQual Working Paper 6. Retrieved May 30, 2013, from www.edqual.org and used with permission.

Blickenstaff, J. C. (2005). Women and science careers: Leaky pipeline or gender filter. *Gender and Education, 17*(4), 369–386.

Bybee, R. W. (2007). Do we need another Sputnik? *American Biology Teacher, 69*(8), 454–457.

Bybee, R. W. (2009). *The BSCE 5E instructional model and the 21st century skills*. Washington, DC. Retrieved May 30, 2013, from http://itsisu.concord.org/share/Bybee_21st_Century_Paper.pdf

Bybee, R. W. (2011). Scientific and engineering practices in K-12 classrooms: Understanding a framework for K-12 science education. *Science Scope, 35*(4), 6–13.

Chen, D. I. (2012). Higher education reform in Taiwan and its implications on equality. *Chinese Education and Society, 45*(5–6), 134–151.

Chisamya, G., Dejaeghere, J., Kendall, N., & Aziz Khan, M. (2011). *Gender and education for all: Progress and problems in achieving gender equity*. Retrieved November 3, 2015, from http://www.cehd.umn.edu/olpd/people/pubs/dejaeghereGendereducation2012.pdf

Claussen, S., & Osborne, J. (2013). Bourdieu's notion of cultural capital and its implications for the science curriculum. *Science Education, 97*, 58–79. doi:10.1002/sce.21040.

Cobern, W. (1993). Construction of knowledge and group learning. In K. Tobin (Ed.), *The practice of constructivism in science education* (pp. 51–70). Hillsdale: Lawrence Erlbaum Associates.

Cotton, K. (1989). *Expectations and student outcomes*. Portland: Northwest Regional Education Laboratory. Retrieved May 30, 2013, from http://educationnorthwest.org/webfm_send/562

Dejaeghere, J. & Lee, S. K. (2011). What matters for marginalized girls and boys in Bangladesh: A capabilities approach for understanding educational well-being and empowerment. *Research in Comparative and International Education*, *6*(1), 27–42. Retreived from: doi: http://dx.doi.org/10.2304/rcie.2011.6.1.27

Ejifugha, A., & Ogueri, E. (2011). Gender equity in science teacher education for sustainable development: An emerging perspective. In *Proceedings of the 1st international technology, education and environment conference (c) African Society for Scientific Research (ASSR), Federal College of Education (Technical), Omoku, Rivers State, Nigeria.*

Ethnic Variation in Gender-STEM Stereotypes and STEM Participation: An Intersectional

Fisher, D., & Frey, N. (2007). *Checking for understanding: Formative assessment techniques for your classroom.* Association for Supervision and Curriculum Development (ASCD).

Gömleksiz, M. N. (2012). Elementary school students' perceptions of the new science and technology curriculum. *Educational Technology & Society, 15*(1), 116–126.

Halai, A. (2011). Equality or equity: Gender awareness issues in secondary schools in Pakistan. *International Journal of Educational Development, 31*(1), 44–49. Retrieved May 30, 2013, from http://www.msu.ac.zw/elearning/material/1347798015Gender%20equality%20and%20equity.pdf

Ito, H. (2012). Jomtien to Jomtien: The evolving coordination process of education for all 1990–2011. *International Education Studies*, *5*(5). Retrieved from: http://dx.doi.org/10.5539/ies.v5n5p1

Key, S. G. (1995). *Research into practice: Diversity in science education*. Pearson Scott Foresman. Retrieved May 30, 2013, from http://assets.pearsonglobalschools.com/asset_mgr/current/20109/Diversity_in_Science_Education.pdf.

Lee, C. D., & Smagorinsky, P. (2000). Introduction: Constructing meaning through collaborative inquiry. In *Vygotskian perspectives on literacy research: Constructing meaning through collaborative inquiry*, 1–15. (C.D. Lee and P. Smagorinsky, Eds.).

Lemke, J. L. (1990). *Talking science: Language, learning, and values* (p. 11). New Jersey: Alex.

Lingard, B., Hays, D., & Mills, M. (2003). Teachers and productive pedagogies: Contextualizing, conceptualizing, utilizing. *Pedagogy, Culture and Society, 11*, 399–424.

Liu, J., & Chen, J. J. (2007). The pattern and trends of sex segregation on fields of study for higher education, 1972–2003. *Journal of Education and Psychology, 30*(4), 1–15.

Loveland, E. (2013, May–June). A champion for development, security and human rights: An interview with Kofi Annan. *International Educator*, *2013*, 24–28. Retrieved May 30, 2013, from http://www.nafsa.org/_/File/_/ie_mayjun13_voices.pdf and http://www.nafsa.org/Explore_International_Education/In_the_News/A_Champion_for_Development,_Security,_and_Human_Rights/

Ma, Y. (2011). Gender differences in paths leading to a STEM baccalaureate. *Social Science Quarterly, 92*(5), 1169–1190.

Magno, C., Silova, I., Wright, S., & Demeny, E. (2003). *Open minds*. Opportunities for gender equity in education: A report on Central and South Eastern Europe and the Former Soviet Union. Open Society Institute, NY. Retrieved May 30, 2013, from http://www.academia.edu/754796/Open_Minds_Opportunities_for_gender_equity_in_education

Miske, S., DeJaeghere, J., & Adolwa, J. (2012). *Equality at the intersection of gender and poverty in education: Conceptual and methodological considerations for measurement.* Addressing inequalities: The heart of the post-2015 development agenda and the future we want for all. Retrieved May 30, 2013, from www.worldwewant2015.org/file/283247/download/307078

Mundy, K. (2006). Education for all and the new development compact. *Education and Social Justice*, 13–38.

O'Brien, L. T., Blodorn, A., Adams, G., Garcia, D. M., & Hammer, E. (2014, September 22). *Ethnic variation in gender-STEM stereotypes and STEM participation: An intersectional approach.* Cultural Diversity and Ethnic Minority Psychology. Advance online publication. Retrieved May 30, 2013, from http://dx.doi.org/10.1037/a0037944

OECD. (2008). *Ten steps to equity in education*. Organization for economic co-operation and development policy brief, OECD Observer. Retrieved November 3, 2015, from http://www.oecd.org/education/school/39989494.pdf

OECD. (2011). *Report on the gender initiative: Gender equality in education, employment and entrepreneurship*. Meeting of the OECD Council at Ministerial Level, Paris, 25–26 May 2011, OECD Publishing, Paris. Retrieved November 3, 2015, from http://www.oecd.org/education/48111145.pdf and used with permission.

OECD. (2012). Closing the gender gap: Act now, Paris. Retrieved November 13, 2015, from http://www.oecd.org/gender/closingthegap.htm and used with permission.

Orenstein, P. (1994). *Schoolgirls: Young women, self-esteem, and the confidence gap*. New York: Doubleday.

Perez-Felkner, L. M., Schneider, S.-K., & Barbara-Grogan, E. (2012). Female and male adolescents' subjective orientations to mathematics and the influence of those orientations on post-secondary majors. *Developmental Psychology, 48*(6), 1658–1673.

Rothon, C. (2006). The importance of social class in explaining the educational attainments of minority ethnic pupils in Britain: Evidence from the Youth Cohort Study. Sociology Working Papers Paper Number 2006-02. Department of Sociology. University of Oxford. Retrieved November 13, 2015, from http://www.sociology.ox.ac.uk/working-papers/the-importance-of-social-class-in-explaining-the-educational-attainments-of-minority-ethnic-pupils-in-britain-evidence-from-the-youth-cohort-study.html

Sadker, M., & Sadker, D. (1995). *Failing at fairness: How America's schools cheat girls*. New York: C. Scribner's Sons.

Schuelka, M. (2012). Inclusion education in Bhutan: A small state with alternative priorities. *Current Issues in Comparative Education, 15*(1), 145–156.

Sikora, J., & Pokropek, A. (2012). Intergenerational transfers of preferences for science careers in comparative perspective. *International Journal of Science Education, 34*(16), 2501–2527. Retrieved May 30, 2013, from http://dx.doi.org/10.1080/09500693.2012.698028

Sorhagen, N. S. (2013). Early teacher expectations disproportionately affect poor children's high school performance. *Journal of Educational Psychology, 105*(2), 465–477.

Stephen, L., Morgan, S. L., Gelbgiser, D., & Weeden, K. A. (2013). Gender, occupational plans, and college major selection. In Press: *Social Science Research*. Retrieved May 30, 2013, from http://www.kimweeden.com/wp-content/uploads/2013/02/Morgan%20Gelbgiser%20Weeden_SSR_2013.pdf

Thomas, J., & Stockton, C. (2003). Socioeconomic status, race, gender, and retention: Impact on student achievement. *Essays in Education, 7*. Department of Education at the University Of South Carolina Aiken. ISSN: 1527–9359. Retrieved May 30, 2013, from http://www.usca.edu/essays/vol72003/stockton.pdf

Tikly, L. (2010). Towards a framework for understanding the quality of education. EdQual Working Paper No.27. Retrieved May 26, 2013, from www.edqual.org and used with permission.

Toglia, T. (2013, February). Gender equity issues in CTE and STEM education: Economic and social implications. *Tech Directions*, 14–17. Retrieved May 26, 2013, from http://illinoiscte.org/PDF/Gender_Equity_Issues_in_CTE_and_STEM_Education.pdf?lbisphpreq=1

UNESCO. (1990). World Conference on EFA, Jomtien, 1990. Retrieved May 26, 2013, from http://www.unesco.org/new/en/education/themes/leading-the-international-agenda/education-for-all/the-efa-movement/jomtien-1990/

UNESCO. (2003). EFA Global Monitoring Report 2003–4 – Gender and education for all: The leap to equality, Regional Overview South and West Asia. Retrieved May 26, 2013, from www.unesco.org/education/efa_report/zoom_regions_pdf/soweasia.pdf

UNESCO. (2012). *World atlas of gender equality in education*. Paris: UNESCO. Retrieved from: http://unesdoc.unesco.org/images/0021/002155/215522E.pdf

UNESCO Institute for Statistics. (2011). *Global education digest 2011 comparing education statistics across the world: Focus on secondary education*. Montreal: UNESCO Institute for Statistics. Retrieved from: http://www.uis.unesco.org/Education/GED%20Documents%20C/GED-2011-Book-EN-web2.pdf

UNESCO Institute for Statistics. (2012). *Reaching out-of-school children is crucial for development, June 2012, Fact Sheet, No. 18*. Montreal: UNESCO Institute for Statistics. Retrieved from: http://www.uis.unesco.org/Library/Documents/fs18-reaching-out-of-school-children-crucial-developmenteducation-en.pdf

UNESCO Institute for Statistics. (2014). *Women in science*. Retrieved from: http://www.uis.unesco.org/ScienceTechnology/Pages/women-in-science-leaky-pipeline-data-viz.aspx

UNESCO Institute for Statistics. (2014, June). *Progress in getting all children to school stalls but some countries show the way forward*. New York: UNESCO. Retrieved May 26, 2013, from http://unesdoc.unesco.org/images/0022/002281/228184E.pdf

van Langen, A., Bosker, R., & Dekkers, H. (2006). Exploring cross-national differences in gender gaps in education. *Educational Research and Evaluation, 12*(2), 155–177.

Wilhelm, J. D., & Smith, M. (2005). Asking the right questions: Literate lives of boys. *The Reading Teacher, 58*, 788–789.

Chapter 16
The Promise of Science Education Research

Lindsey N. Conner

16.1 Introduction

This chapter proposes and considers some directions for future science education research, how we engage students with science, scientists, and communities in ways that enthuse and excite them to develop a restless curiosity and fascination with the world and how it works, as well as to see science as a way of generating new ideas and knowledge that can contribute to the quality of life. While much research on science education assumes that teachers are able to make changes to their teaching to enhance student outcomes, the research on learning environments has recently highlighted the social inequities related to access of information in many urban/rural schools around the world and how other conditions often have to be redressed before the reform agenda in science education can be advanced. While I acknowledge the importance of informal learning situations, I have restricted the discussion here to formal "schooling" as we currently recognize it, rather than presuppose alternative forms of learning environments such as those proposed by the contributions in Rodrigues (2010). No doubt the emphases presented here will be contested. So, the purpose of this chapter is to provide points for further discussion about the directions of research in science education that can be pursued.

Up front is a caveat that this chapter cannot provide a comprehensive collection nor discussion of the extant research such as that covered in previous handbooks (e.g., Abell and Lederman 2009; Tobin 2006). Instead, it contains selected works to highlight trends and issues and the implications for research. Some of these issues are continuances of previous ones (e.g. conceptions about science and learning-specific science content, learners and learning, specific characteristics of students and how we meet their diverse needs, problem-based and project-based learning),

L.N. Conner (✉)
University of Canterbury, Christchurch, New Zealand
e-mail: lindsey.conner@canterbury.ac.nz

B. Akpan (ed.), *Science Education: A Global Perspective*,
DOI 10.1007/978-3-319-32351-0_16

while others relate to trends in changes to the learning environment, particularly access to digital sources and the role of teachers and students as collaborative and codrivers as well as participants of research in action.

The chapter firstly discusses the general shifts and trends in science education, from *knowing, doing, and being* to the integration of these constructs. Then, I discuss systems evaluation, as these contribute to our understanding and identification of potential issues or interesting connections between attributes at a macro level and provide information for policy and practice. Thereafter, I discuss research on aspects of learning science that sit alongside trends in learning more generally. The integration of the *nature of science* (NoS) more explicitly is covered in a separate subsection of research on learning to give it emphasis, even though teaching and learning approaches may rely on each other. The next section discusses research for teaching and teacher development with a separate subsection on the importance of research on integrating the use of digital technologies. All of the above have implications for policy, curriculum development, and teacher professional learning and how we propel science education to the fore for enhancing future lives.

16.1.1 From Knowing, Doing, and Being to Integration Approaches in Science Education

In a very general sense, historically teaching science has tended to focus on what was already known – the scientific knowledge, theories, and applications and how they were generated – *knowing science*. To a large extent through assessment practices, many education systems still highly value the knowledge base that can be classified as science knowledge. Then after *Sputnik*, in response to a need to engage students in *doing* science, new curriculums in many countries around the world also provided practical work or demonstrations to illustrate science concepts (Roth 2006) with the promise that *doing* science would motivate students and provide them with hands-on examples and skills to become scientists. Unfortunately *doing* science, by itself, does not equate to understanding science. Lemke (1992) called this approach to include practical work "simulacrum," which means a sham or a hollow shell posing as something that it is not. Gough also criticized the misalignment of school science with what scientists actually do as being like a "theater of the absurd" (Gough 1998). Other education psychologists have indicated that most practical work in science assumes "an epistemology that is entirely at odds with the epistemology of real science" (Chinn and Malhorra 2002, p. 204). Given that scientists think and work in a myriad of ways and if the purpose of science education was to develop scientists, then the "recipe" approaches to practical work for confirming science facts were not necessarily reflecting the critical and creative thinking processes that scientists needed to evaluate their work, nor did they help students to develop next steps to investigate phenomena, let alone create new science knowledge. A focus on content and verifying existing declarative knowledge through practical work in school science did not capture the discovery element of science, the "eureka!," nor the importance of asking driving questions (Krajcik and Mamlock-Naaman 2006). There was a need which continues to incorporate the innovative way people, who have relentless curiosity about the world, inherently make connections between disparate ideas. While doing practical work can enable students to acquire process skills, by itself it does not necessarily do this. Often students experience practical work in science as a form of recipe. As Tasker and Freyberg (1985) indicated, at that time, school science did not adequately reflect what it was like to *be a scientist*. Therefore, it became important to incorporate the basic premises on which science is based more explicitly, within the constructs of the nature of science (Lederman 2009). Ford and Forman (2006) proposed a deceptively simple model for conducting even very simple investigations in ways that explicitly afford insights into the disciplined nature of knowledge construction in science. In this model, investigative activity is structured to overtly simulate, so far as this is realistically possible, the types of activities in which scientists themselves engage. Based on a synthesis of insights from the field of science studies, the model posits two roles that every scientist plays as they work to build new knowledge. Scientists are simultaneously *constructors* of claims about the natural world and *critiquers* of claims made by other scientists in their field. When preparing a knowledge claim, the constructor scientist keeps the critiquer in mind. Then, when in the critiquer's role, they use their own deep knowledge to look for flaws or possible alternative explanations for the work of others. Ford and Forman (2006) proposed providing students with experiences similar to those of *being* a scientist. These include the simultaneous and dynamic roles of being "constructors" of knowledge and "critiquers" of knowledge. As individuals or small groups of investigators, they carry out and create convincing evidence to support their claims. As part of the community of peers, they learn how to take part in the more social and collective activities of critique. Students must be able to put these two types of experiences together so that they can see how both roles are codependent for constructing valid and reliable knowledge claims. If this approach is used, it is more likely that students would gain a sense of what it means to *be* a scientist. Furthermore, scientists have to be independent problem-solvers and collaborators, as well as being able to critically analyze and evaluate their findings, as part of *being* a scientist.

The *Science for All* movement in the late 1980s and 1990s also indicated the need to make science relevant to students' lives so they could apply science to a multitude of personal life and public decisions as well as a multitude of careers (Aikenhead 2006). The change in emphasis in some science classes in the 1980s and 1990s was to include a range of inquiry approaches alongside learning science content knowledge (Abd-El-Khalick et al. 2004) where students had some choice about how they designed and conducted investigations. This required students to integrate content and investigative knowledge to find out answers to their own questions. Linn et al. (2004) define science inquiry as "engaging students in the intentional process of diagnosing problems, critiquing experiments, distinguishing alternatives, planning investigations, revising views, researching conjectures, searching for information, constructing models, debating with peers, communicating to diverse audiences, and forming coherent arguments" (p. 518).

In parallel was the recognition that students who developed *learning to learn* skills and aligned their use of appropriate learning strategies for particular learning intentions, were more likely to develop as independent learners and thinkers who could self-regulate their learning and achieve better (Conner2014a, b). However, there was recognition that teachers and learners had to undergo a huge shift in thinking to move from activity-based approaches to student-centered practical inquiry approaches, where students determine what is studied and how they will learn (Roth 1995; Roth et al. 2008).

More recently, there has been recognition of the cognitive, affective, and social benefits of students becoming more involved in determining what and how they learn. Much of the work on cognitive science research is related to the need for students to make connections between conceptual understanding and problem-solving in physics. An instructional theory – cognitive load theory – is helping inform the design of instruction that seems to make a difference to student learning outcomes (Sweller et al. 2011).

What has also been underleveraged in science learning contexts is the huge capacity for students to bring intellectual capacity to developing new science knowledge, as part of learning about science. In other words, they can experience the intellectual processes of *being* scientists through participating in meaningful inquiry (Bielaczyc 2011; Linn et al. 2004). Engaging students in thinking about the future and their role in designing it provides some promise and is worthy of further investigation (Jones et al. 2011).

A large part of the literature and research on science education has also been devoted to various types of teachers' knowledge and its application to teaching. The simultaneous development of science content, practical skills, understanding NoS, and developing thinking and cooperative skills requires integrating a range of approaches to teaching and learning. Good teachers know how to integrate aspects of the discipline knowledge with aspects of pedagogy (Bransford et al. 1999). It is the integration of these components that requires science teachers to be highly professional adopters, adapters, and advocators of learning science for future citizens. We need research that builds on the commitments and talents of teachers, as they integrate these components, as well as research that takes account of the constraints and opportunities of classroom contexts. Teachers are more likely to embrace new ways of teaching and facilitating learning if they have been included in the development of these processes (Darling Hammond 2000). We also know that more effective outcomes for students can be achieved when teachers take account of evidence to inform their practice (Gunstone 2000; Rao 2011). Large-scale surveys can provide indications of how well education systems are meeting outcomes and what effective education systems are doing that makes a difference (Barber and Mourshed 2009; OECD 2011; Mourshed et al. 2010). An analysis of the intersections and correlations between components surveyed may provide insights for further development. In contrast, specific case studies can provide rich information about the details of pedagogical practices and how learning occurs when integration of content knowledge, practical and learning skills, thinking, and informed citizenry are promoted as part of the learning, simultaneously. In the next section, I briefly consider policy and systems analysis research. The sections following that consider more in-depth approaches to research and the emphases these might take in future research agendas.

16.1.2 Systems Evaluation

High-performing education systems ensure all students obtain quality education, regardless of their abilities. They attempt to reduce gaps between high- and low-achieving students (Aho et al. 2006; Valijarvi 2003) and are developing policies to put structures and processes in place to support ongoing teacher professional learning (Barber and Mourshed 2009; Mourshed et al. 2010). Ministries of Education can make a difference to science education reform. Therefore, there is a need to base any changes to policy (and funding) on evidence of what has worked and what shows promise for large-scale agendas.

Systems-wide national analyses of achievement (TIMMS and PISA) highlight some aspects related to what systems are doing well and how well some countries are addressing key aspects of science teaching and learning. The evidence indicates that high-performing systems nurture well-educated, well-rounded students that go beyond conceptual procedural knowledge to include learning processes and twenty-first-century skills, such as critical thinking (Facione 2007), adaptability, and leadership. For example, the Ministry of Education, Singapore (2012), announced a new phase of educational development – a student-centered, values-driven education – where education is grounded in ethics, character development, and dispositions, such as adaptability and resilience. This is not to imply that academic abilities have been de-emphasized, but that the Ministry of Education in Singapore is signaling a shift toward twenty-first-century needs. In this model, students are not just consumers of information. They are knowledge creators too. Students are encouraged to develop deeper understandings by building, adapting, and applying knowledge in real-life scenarios (Tan 2006) where they apply critical thinking as recommended by Norris and Ennis (1989). The Finnish education system has also built itself on the values inherent in Finnish society (Aho et al. 2006; Korpela 2008). Here children and adolescents grow up in an environment where education is highly valued by society and where there is a high level of preparedness and willingness to work. The value base and importance of education in Finland are rarely questioned. Teachers in Finland also have high levels of qualifications (typically master's degrees) and a high level of ongoing professional support.

Many governments are currently considering how they can add value to the data gathered from international surveys such as TIMMS and PISA to inform policy and practice about teacher professional learning (Lan and Gonzalez 2012). What are teachers doing that makes a difference? What sort of professional development have the teachers participated in and how can this participation be linked to student achievement (Conner 2014b)? What are students telling us about what makes a difference? Secondary analyses of TIMMS and PISA data will help us answer these questions and warrant attention of researchers and policy makers, to inform initial and ongoing teacher professional learning.

16.1.3 Research on Aspects of Learning Science

There has been an explosion of cognitive research within science learning contexts over the past 30 years, including how concept development can be enhanced through a wide range of pedagogical approaches (Ornek and Saleh 2012). Much research has focused on students' understanding of science as being "laden with misconceptions" and evaluating the effectiveness of teaching and learning strategies for how to redress these misconceptions. Roth et al. (2008) suggest that a more worthwhile approach might be to consider what experiences science education contexts provide for students to engage with science in order to extend their current thinking and develop new ways of talking about natural phenomena. This is not to say we should ignore misconceptions, but rather that we should move away from pathologizing about them to finding more holistic, expansive ways of building students' capacities to develop thinking skills, for progressing their understanding. The rationale for this is that students will be more motivated to participate in science if they see the development of thinking skills as a way to expand their action possibilities (Roth et al. 2008).

The explosion of research on cognitive load theory (Sweller et al. 2011) indicates that there is much interest in expanding the ideas about how to design learning environments that take account of cognitive demands. Alongside this has been an underlying tension that science education, and physics in particular, is laden with complex ideas that, in the main, still tend to be taught in transmissive ways without taking account of the effects that noncognitive factors have on students' learning. Pintrich et al. (1993) emphasized that instructional models that focus only on cognition tend to neglect the inclusion of constructs such as an individual's goals, intentions, purposes, expectations, needs, and other contextual factors of the learning environment. Some recent theoretical developments in *Evaluative Constructivism* (Conner 2014a, b), when expanded and researched further, could potentially elaborate how students' approaches to learning can be married with cognitive load theory. That is, how can students be guided to develop learning intentions, be accurate in their evaluations about their needs, and appropriately use learning strategies to help them identify cognitive load, and what can they do to reduce it? What do learning environments and guidance frameworks look like when they accommodate variations in students' backgrounds and experiences of learning, together with the use of vanishing scaffolds or "guidance fading effects" in instruction, and how does this influence understanding and achievement? Students have the capacity to become aware of and make use of more effective ways to learn, if we support them to do so.

The development of scientific literacies and, indeed, multiple literacies has been strongly advocated (Rodrigues 2010). Fortuna et al. (2010) emphasize the importance of developing these within science education that include the ability to: identify, understand, interpret, create, apply, communicate, compute, and use technology-based resources for learning about science. While mobile technologies provide promise for greater access in space and time for student learning, Roschelle (2003) warns us that we tend to complexify views of the use of technology and simplify social practices, but what we need is research that tells the story of the rich pedagogical practices that integrate both technologies and social practices for enhancing learning.

Research is needed on how teachers integrate resource materials (from a variety of sources) and the processes or conversations that help them to "reimagine" the science curriculum they design and teach (Tytler 2007). Not only is this important as science content knowledge changes over time, but it is also extremely important if teachers are to support students in their development of *knowledge-ability*: the ability to find, evaluate, and apply knowledge and to be able to communicate the connections between disparate knowledge effectively (Gilbert 2005; Anderson 2009). Assessment reform may target *knowledge-ability*, or in Anderson's (2009) words "richer knowledge networks" of relational connections between concepts and their applications. How effective are these assessments and what changes might need to be made to target students' development of knowledge-seeking capabilities and resourcefulness, so that these skills are valued by students and teachers? In general, students who are willing to learn as they discover, adapt, and evaluate their own thinking and learning are more likely to achieve better in assessments (Conner 2014a, b).

In complex times such as we live in, citizens need to be able to link ideas, question the validity and reliability of data, and apply this to engage with others and to make decisions for themselves, their families, and the society at large as they become more successful and contributors to the scientific enterprise. In science education, this is reflected in the specific skills of fallacy detection, constructing an argument, using data to develop opinions, criticizing findings and implications in terms of validity and reliability, and challenging claims based on evidence (Evagorou and Dillon 2009). Research on how students cross knowledge boundaries (between the science knowledge currently taught in schools and what they use for everyday living or to connect multiple sources of knowledge to create new innovations and discoveries) are relatively scant.

The challenge for curriculum design is to target experiences and actions so that the consequent students' outcomes, including any artifacts they produce, might be useful to their local community more generally (Roth et al. 2008). Modern knowledge-based economies focus on the utilization of knowledge (Gilbert 2005) but this does not necessarily mean that the same knowledge is relevant everywhere. Therefore, knowing multiple ways to find, create through scientific investigations, evaluate, and communicate scientific information in relation to personal and local everyday life becomes extremely important (Roth et al. 2008). For example, knowledge about the Krebs cycle is not needed for everyday living. However, people do need to know about chemical or biological pollutants and contaminants in their water, how the contaminants change over time and with the seasons, so they might take appropriate actions to improve the quality of water. We cannot live without clean water! These ideas have implications for what content is focused on and how students are encouraged to apply concepts to their lives.

The importance of information processing skills is being embraced globally. The Commission of European Communities has deemed that lifelong learning must become the guiding principle for provision and participation across all leaning contexts. Around the world, there is now a drive to embed and integrate twenty-first-century learning skills in all school subjects (Schleicher 2012). The accessibility of knowledge and its collective generation and communication through digital media enable people to be connected to knowledge and each other as part of generating new knowledge, in ways that did not exist prior to the advent of the Internet. Therefore collaborative knowledge generation can be accelerated because we now can use technology to do this. The Internet enables and tends to increase the rate of interactions between people. As well, what is learned in school (and other formal learning settings) needs to prepare people for learning within employment contexts and for everyday living. With increasingly fluid employment prospects, where people may have multiple ways of being employed (concurrently in multiple roles and throughout their lives), there needs to be more research about how people adopt, adapt, and develop their skills to learn new knowledge that is needed for new employment contexts. How do people use knowledge in its various forms and transfer it between science content contexts and from science learning contexts to employment situations? We simply don't know much about these transfers. Similarly the role of assessments in indicating the value of developing thinking skills, where learners are required to transfer, apply, and adapt science knowledge to novel situations, will be important. Such research on the effectiveness of targeted assessment could provide feedback for future curriculum and assessment developments.

16.1.4 Integrating Aspects of the Nature of Science (NoS)

The adage that we learn to do by doing is very applicable to what we want students to learn in science. As Roth (2006) has indicated, teaching and learning in science should be adjusted to address what we want students to learn. While the promise was that hands-on science would support conceptual learning and entice and enthuse more students into pursuing science careers, research on the inclusion of practical work in science in the 1990s indicated that teachers did not believe students learned what they needed to know. That is, practical work did not lead to learning as assessed in written tests (Tobin 1990). Roth suggests that if we want students to learn content in depth, there is little point in providing opportunities for practical work. If on the other hand we want them to have opportunities to learn practical skills, to experience real problem-solving, to ask real questions that need solutions, and to be involved in making decisions about how to find out the answers, we need to provide them with opportunities to learn the skills and participate in open science inquiry (Roth 2006). What is needed is research on the approaches that allow students to investigate what we do *not* know, i.e., to enable school students to generate new scientific knowledge. This presupposes they have the skills to find out what is already known and have access to resources and people who might supply the information they need. However, by being immersed in real-life purposeful inquiry projects, students would need to know what is already known (*knowing* science concepts), and they would need to learn ways to identify what is not known (*doing* science). They would also need to know how to make use of aspects of NoS to evaluate what they found out (*being* like a scientist). Then, the purpose of learning about identifying variables, interpreting data, hypothesizing, and experimenting with different techniques, connecting ideas, and integrating *knowing*, *doing*, *and being* becomes highly relevant and authentic. Research within this model could investigate how students can become adept at adapting creatively in situations of uncertainty, where they might have to transfer knowledge from another context or when the directions for them to proceed are not explicit and they have to figure things out for themselves. Transfer of thinking and learning skills to science contexts can occur from other subject domains (Conner and Gunstone 2004).

How does this shift the teachers' role from telling students what to do to using vanishing scaffolds, so that students gradually become effective independent problem-solvers, which is characteristic of good scientists?

Arguably, the *nature of science* (NoS) has always been inherent in science and science teaching. In the last decade or so, there has been a greater emphasis on explicit approaches for teaching the inclusion of the NoS (Hipkins 2012; Jules and Conner 2009). Aikenhead et al. (2011) have suggested how the incorporation of teaching about NoS more explicitly can make science more interesting, relevant, and engaging, but we need more examples of this with evaluations of specific students' outcomes. Since teachers need to develop effective pedagogical skills necessary to effectively teach about inquiry and NoS (Abd-El-Khalick et al. 2004; Conner 2010), more case studies and examples of teaching might inform teacher development and the multitude of ways teachers can use questioning to make the components of NoS more explicit.

Hipkins (2012) suggested that the emphasis of NoS in the New Zealand curriculum was also to make science more equitable, by enabling success for students from diverse backgrounds. What evidence do we have that explicit teaching about NoS can do this or make a difference to equity of access to science and science careers? Further, does an emphasis on NoS support students to understand the big ideas of science and its relevance and application for informed citizenship (decision-making)? The role teachers take in emphasizing NoS and the resources they use to help students make connections within and between content areas of science and the processes involved in scientific endeavors would also be a fruitful area for research. Alongside this is another area of potential research related to science teachers' use of metacognition (Thomas 2012). Such research could evaluate the development and effectiveness of teachers' use of metacognitive processes linked to teaching about NoS and how they model thinking for students.

16.1.5 Teaching Science and Teacher Development

The quality of teaching has a huge influence on the quality of learning (Darling Hammond 2000; Hattie 2012). A key characteristic of high-performing education systems is the high level of qualifications and professional dispositions of teachers in those systems (Barber and Mourshed 2009; Mourshed et al. 2010). We now know more about what constitutes and supports teachers to transfer and combine content knowledge and professional learning to meet a range of student learning needs (Loucks-Horsley et al. 2010) and how working in communities of learners, or professional learning groups, supports ongoing teacher development (Conner et al. 2008). Building teacher's capability and capacity is accelerated when teachers work in communities of practice (learners and thinkers), where they interrogate their own learning or lesson studies and adapt and change what they do based on evidence of students' outcomes and students' feedback (Loucks-Horsley et al. 2010; Rao 2011).

For science teachers, knowledge of science concepts is fundamental to good science teaching. Therefore, a continuing science education research agenda must be to investigate effective ways of building science teachers' core science knowledge so that teachers have enough knowledge and know-how to find out and incorporate new knowledge, as it arises. However, given the exponential rate at which new knowledge is being generated, it will also be crucial to build teachers' capacity to be ongoing learners so they can model lifelong learning for students. The structures, policy, and processes (programs) to support teachers to do this are important. One-off professional development courses for individuals rarely lead to large-scale, sustainable teacher learning. Potentially, promising models of ongoing professional learning involve teachers setting up communities of practice, supported by experts, where they are provided with time and facilitation to meet regularly to discuss their areas of concern and can learn much from each other (Timperley 2011; Kaser and Halbert 2012).

There is a wider role that technology can play to support improvements in teaching and ongoing professional learning. For example, teachers can independently develop and share lessons and learning resources, including ideas about how to enrich and enliven structured delivery through the use of a range of technologies in science. Multiple sources of information are available for updating science content knowledge and pedagogical content knowledge through websites, online journals and e-books, social media, videos, and scientists who have made themselves available through online portals such as "ask a scientist." Research on informal ways teachers learn might help to expand the possibilities for self-driven, self-paced teacher professional learning as part of the suite of ways teachers' capability is built. Policy makers might like to consider how teachers could be encouraged and supported to do this. For example, they require time to participate in and share their learning from such activities.

Research on how teachers select teaching approaches and strategies have built on concepts such as pedagogical content knowledge (PCK) (Shulman 1986) to include how teachers synthesize multiple types of knowledge to help learners make sense of particular science content (Loughran et al. 2001). Appropriating pedagogy to content and desired student outcomes is quite complex and merits more investigation. We are slowly beginning to understand the ways teachers' PCK has been developed further to incorporate technological aspects (TPACK) (Mishra and Koehler 2006; Neiss 2012), where teachers make choices about using appropriate e-tasks or other approaches, based on a wide range of contextual factors, and adapt their approaches accordingly. The research on how contexts influence teachers' pedagogical decision-making when using ICT in science learning is in its infancy (Owusu et al. 2015).

Teachers want specific examples of practice, case studies of teaching in action, "lesson studies," that are enacted with real classes and how these relate to the outcomes that were achieved. Interrogating the nuances of learning contexts is complex but can provide rich information, despite the limits of transference from one context to another. There is a continuing paradigm shift occurring about what it means to be a professional teacher, where teachers contextualize, interrogate, and reflect on their work to consider conditional aspects of "when and why" as they appropriate pedagogy to student needs, based on student feedback (Loughran et al. 2007). Treagust (2007, p. 11) states that:

> *Without research to determine whether or not innovative research-based curricula or topics within a curriculum are perceived by students as being interesting, and that successful learning outcomes ensue, there is little likelihood of arresting the decline in enrolments in science.*

Teachers, therefore, need to take notice of what students are saying. I take this even further. If teachers included students as co-collaborators of research on their teaching and on the students' outcomes, they may generate more interest in science teaching as a career as well as find out what their students think, in more depth.

16.1.6 Multimodal Approaches to Teaching and Learning

In new curriculum developments around the world, we are seeing a shift in emphasis away from predetermined content knowledge and toward information processing and thinking skills, where different forms of knowledge are integrated and ICT is used as a tool for this integration (Webb 2005, 2008). Marshall (2002) found strong evidence that educational technology "complements what a great teacher does naturally." There is a large body of evidence that supports the positive association between the use of technology and learning outcomes (So and Ching 2012). Research has not only shown that e-learning can improve students' conceptual understanding in science (Becta 2009), but that e-tools and scaffolds can be useful for developing inquiry and thinking skills within science learning contexts (e.g., White and Frederiksen 1998). Lee et al. (2011) in their review on inquiry using e-resources concluded that Internet student learning environments, in general, improved student learning outcomes including attitude, motivation, conceptual understanding, and conceptual change. Moreover, student motivation and their engagement in learning generally increased when technology was used in the teaching and learning process in many instances (Knezek et al. 2011; Osborne and Hennessy 2003).

More importantly for science education, though, is that information technology plays a central role in scientific research and in scientific communication. Therefore, it is very important that as part of learning science, students develop literacy skills that include rich experiences in using technological tools that support and develop information literacy. The use, evaluation, and application of IT-based research tools will increasingly contribute to scientific practice, and therefore research on how students and teachers use ICT as tools for learning science literacy is crucial.

With increased access to information through ICTs, there has been a burgeoning of digital or e-learning resources, including gaming and the use of virtual reality for learning (Grimley et al. 2010). This area of research is becoming increasingly important as students embrace and use new technologies quickly. Many resource-based e-learning environments (RBeLEs) have been developed and evaluated (Linn et al. 2004; Slykhuis and Krall 2012; So 2012) to help teachers create learning environments with online resources more effectively. These build on the discussion of resource-based learning environments (Hill and Hannafin 2001) and learning science-based learning environments (Blumenfeld et al. 2006) that include scaffolds for learning and processing information. While there have been system-wide analyses of how the use of technology can improve student outcomes (e.g., in the UK, Harrison et al. 2002), there is little research as yet, about how teachers might deliberately use e-resources and consider the appropriateness and multiple ways to integrate them with other components of learning environments to support students' learning. Some promising examples (So and Ching 2012; Owusu et al. 2015) indicate how teachers can match the online resources with appropriate tools for information or data manipulation and how they can include the scaffolds needed for enabling students to access the information provided by the resources. The adoption of the TPACK approach recognizes that the application of e-learning (T, for technology) builds on Shulman's pedagogical content knowledge (PCK) (Mishra and Koehler 2006). How teachers appropriate and adapt pedagogy as they integrate and appreciate the affordances of digital technologies is still an emerging endeavor (Neiss 2012). We need to know more about models of best and appropriate practice in the application of e-learning. There have been calls for longer-term studies on the sustained and embedded use of e-technologies (e.g., Becta 2009). For example, many questions arise about the teaching and learning nexus:

1. How does the content context make a difference to teachers' choices for approaches? What role do students have in determining the content contexts?
2. How can spatial and dynamic images, which portray intricate relationships among complex concepts, support student learning (Barak et al. 2011)?
3. How do simulations and visualizations support teaching and learning processes (e.g., see PhET, www. http://phet.colorado.edu/)?
4. What changes to pedagogy are required when using different modes of teaching and learning? What are the implications for teacher development?

5. How do teachers relinquish power and control over the directions students take, yet ensure that students engage in higher-order thinking activities?
6. What guidance and scaffolding (or vanishing scaffolding) support learning when the outcomes in terms of content are not predetermined?
7. What forms of assessment are appropriate when there are a range of desired learning outcomes?

Webb (2005, 2008) provides in-depth reviews of research in the field of e-in-science in which she considers that multiple modes and approaches to teaching and learning science can provide a vehicle for embedding NoS. Some typologies of e-learning have been evolving as indicated in website designs, such as that of the *Science Learning Hub* and the NASA website and many others. The research on e-learning, m-learning (mobile), and u-learning (ubiquitous) is still emerging in terms of the technological attributes of each of these and the pedagogical affordances of their use for individual and social learning (Park 2011). Other research shows exciting possibilities in the development of digital objects to illustrate abstract concepts and make objects that are invisible to the naked eye visible, as well as the use of modeling and simulations to assist learning and theory development (Slykhuis and Krall 2012). The research on the use of virtual environments for both student and teacher learning is still in its infancy. How teachers integrate the use of technologies for learning about, in and to advance learning in science education and science, are also areas to be developed and researched.

16.1.7 The Promise of Science Education Research

If science and science education were to have meaning and contribute to solving the problems of the world, provide the platform for improving the quality of life, as well as contribute to deeper understandings about how our world works, then the purpose of any science education research needs to be clear. We need to question how the research informs or will inform us about how to educate students to have the dispositions necessary to gain *knowledge-ability* about science, how they can integrate multiple sources and techniques to acquire the appropriate knowledge needed for a particular situation, and how this can be applied to novel situations. How can science education models contribute to the generation of new scientific knowledge? This is associated with a related question: How does science education develop thinking individuals who question everything in a quest for understanding our world and beyond? Students ought to be able to identify what their learning needs are, to be given opportunities and experiences of success in monitoring and being in control of their learning (Conner 2006), and to consider how they could achieve multiple intentions simultaneously. These skills and dispositions help learners transfer learning they gain in schools to practical knowledge for working and learning throughout life (Hipkins 2009). Developing such dispositions may be culturally challenging (Mok 2006). This is not an easy agenda and therefore one which research can support.

Research on how students, teachers, and education systems provide experiences and opportunities for developing themselves as individuals and collectives in the pursuit and generation of scientific knowledge could be a future focus for research, especially, if one of the end gains is to develop people who actually use science to create new knowledge and innovations for the benefit of society. What information can systems-wide analyses provide to inform this? How do we build capability and capacity for teachers to contribute to the research on practice in specific contexts, where students may also want to be part of the design of such research on science learning? For it is clear that sustainable curricular innovations require extensive opportunities for customization and flexibly adaptive designs (Linn et al. 2004) that require teachers to adapt and apply new ideas about teaching and learning. Teachers who are valued as highly professional individuals, capable of adapting to students' needs, are more likely to be empowered to adjust the learning environments appropriately to local contexts.

While much research has already been conducted on how we can enthuse students to see the possibilities that science might promise to potentially contribute to humanity (Aikenhead 2006), we need to find ways to make this both explicit and implicit in what we value and therefore what is assessed. Changes in assessment need to mirror changes in the outcomes we deem important. What is valued by teachers and students, and therefore what is taught or learned, is what is assessed. Researching the shifts in assessment practices in school science must be high on the agenda to help reform systems more quickly.

Similarly there has been considerable discussion about professional development models for teachers of science (Loucks-Horsley et al. 2010). These require the blending of research, practitioner wisdom, and a repertoire of strategies from which teachers can choose what to do. Therefore, I suggest that future research on science teacher education (in the broadest sense that encompasses both initial and ongoing professional learning) should be considered for how it might transform practice, so that future citizens can use their acquired skills to evaluate information and make decisions for their own health, society, and our precious environment.

Acknowledgment The author wishes to thank Professor Richard Gunstone very sincerely for providing valuable peer review on the content and construction of this chapter.

References

Abd-El-Khalick, F., BouJaoude, S., Duschl, R., Lederman, N. G., Mamlok-Naaman, R., Hofstein, A., Niaz, Treagust, D., & Tuan, H.-L. (2004). Inquiry in science education: International perspectives. *Science Education, 88*(3), 397–419. doi:10.1002/sce.10118.

Abell, S., & Lederman, N. (Eds.). (2009). *Handbook of research on science education*. New York: Routledge.

Aho, E., Pitkänen, K., & Sahlberg, P. (2006). *Policy development and reform principles of basic and secondary education in Finland since 1986* (Education Working Paper Series No. 2). Retreived April 21, 2016, from http://siteresources.worldbank.org/EDUCATION/Resources/278200-1099079877269/547664-1099079967208/Education_in_Finland_May06.pdf

Aikenhead, G. S. (2006). *Science education for everyday life: Evidence-based practice*. New York: Teachers College Press.

Aikenhead, G., Orpwood, G., & Fensham, P. (2011). Scientific literacy for a knowledge society. In C. Linder, L. Ostman, D. Roberts, P. Wickman, G. Erickson, & A. MacKinnon (Eds.), *Exploring the landscape of scientific literacy* (pp. 28–44). New York: Routledge.

Anderson, O. R. (2009). The role of knowledge network structures in learning scientific habits of mind: Higher order thinking and inquiry skills. In I. M. Saleh & M. S. Khine (Eds.), *Fostering scientific habits of mind: Pedagogical knowledge and best practices in science education* (pp. 59–82). Rotterdam: Sense.

Barak, M., Ashkar, T., & Dori, Y. J. (2011). Learning science via animated movies: Its effect on students' thinking and motivation. *Computers & Education, 56*(3), 839–846.

Barber, M., & Mourshed, M. (2009). *Shaping the future: How good education systems can become great in the decade ahead*. London: McKinsey & Company.

Becta. (2009). *Evidence on the impact of technology on learning and educational outcomes*. Retrieved March 19, 2013, from http://webarchive.nationalarchives.gov.uk/20110130111510/http://partners.becta.org.uk/upload-dir/downloads/page_documents/research/impact_of_technology_on_outcomes_jul09.pdf

Bielaczyc, K. (2011). When kids' ideas come first. *ReEd (Research in Education)*, 2(5). Retrieved March 28, 2013, from http://www.nie.edu.sg/files/oer/OER-NIE-ReEd2_Final%20for%20Web.pdf.

Blumenfeld, P. C., Kempler, T. M., & Krajcik, J. S. (2006). Motivation and cognitive engagement in learning environments. In R. K. Sawyer (Ed.), *The Cambridge handbook of the learning sciences* (pp. 475–488). Cambridge: Cambridge University Press.

Bransford, J. D., Brown, A. L., & Cocking, R. R. (1999). *How people learn: Brain, mind, experience, and school*. Washington: National Academy Press.

Chinn, C., & Malhorra, B. (2002). Epistemologically authentic inquiry in schools: A theoretical framework for evaluating inquiry tasks. *Science Education, 86*(20), 175–218.

Conner, L. (2006). Strategy use, self-regulation and achievement. *Academic Exchange Quarterly, 10*(2), 276–281.

Conner, L. (2010). Raising awareness and understanding of the nature of science. *New Zealand Science Teacher, 120*, 4.

Conner, L. (2014a). Students' use of evaluative constructivism: comparative degrees of intentional learning. *International Journal of Qualitative Studies in Education*, *27*(4), 472–489. doi:10.1080/09518398.2013.771228.

Conner, L. (2014b). TIMSS 2011: School principals' actions and preparedness of grade 8 science teachers in four East Asian school systems. In Ong Saw Lan, E. Gonzalez, & S. Kanageswari Suppiah Shanmugam (Eds.), *TIMSS 2011: What can we learn together?* (pp. 21–37). Penang: South East Asian Ministries of Education Organisation.

Conner, L. N., & Gunstone, R. F. (2004). Conscious knowledge of learning: Accessing learning strategies in a final year high school biology class. *International Journal of Science Education, 26*(12), 1427–1443.

Conner, L., Cowie, B., Buyers, P., Glynn, T., & Otrel-Cass, K. (2008). Cascading collaborative research on science teaching in diverse classrooms. *Education-Line*. http://www.leeds.ac.uk/educol/documents/175428.pdf.

Darling Hammond, L. (2000). Teacher quality and student achievement: A review of state policy evidence. *Education Policy Analysis Archives, 8*(1), 1–50.

Evagorou, M., & Dillon, J. (2009). Infusing thinking skills in the science classroom: System thinking and argumentation as a means to engage students in the process reasoning. In I. M. Saleh & M. S. Khine (Eds.), *Fostering scientific habits of mind: Pedagogical knowledge and best practices in science education* (pp. 107–124). Rotterdam: Sense.

Facione, P. A. (2007). *Critical thinking: What it is and why it counts*. Millbrae: California Academic Press.

Ford, M., & Forman, E. (2006). Redefining disciplinary learning in classroom contexts. In J. Green & A. Luke (Eds.), *Rethinking learning: What counts as learning and what learning counts* (30th ed., pp. 1–32). Washington: American Educational Research Association.

Fortuna, C., Henderson, S., McLuckie, J., Rodrigues, S., Syme-Smith, L., Taylor, N., & Williamson, G. (2010). A conversation between colleagues: Defining multiple literacy in science education. In S. Rodrigues (Ed.), *Multiple literacy and science education: ICTs in formal and informal learning environments* (pp. 1–10). Hershey: IGI Global.

Gilbert, J. (2005). *The knowledge wave: The knowledge society and the future of education.* Wellington: New Zealand Council of Educational Research.

Gough, N. (1998). "If this were played upon a stage": School laboratory work as a theatre of representation. In J. Wellington (Ed.), *Practical work in school science: Which way now?* London: Routledge.

Grimley, M., Nilsen, T., Kerr, R., Green, R., & Thompson, D. (2010). Virtual worlds for science learning. In S. Rodrigues (Ed.), *Multiple literacy and science education: ICTs in formal and informal learning environments* (pp. 263–279). Hershey: IGI-Global.

Gunstone, R. F. (Ed.). (2000). *Science teachers as researchers in Australia.* Brisbane: Queensland University of Technology.

Harrison, C., Comber, C., Fisher, T., Hawe, K., Lewin, C., Lunzer, E., McFarland, A., Mavers, D., Scrimshaw, P., Somekh, B., & Watling, R. (2002). *ImpaCT2: The impact of information and communication technologies on pupils learning and attainment. ICT in school.* Research and Evaluation Series No.7. DfES/Becta.

Hattie, J. (2012). *Visible learning for teachers: Maximizing impact on learning.* London: Routledge.

Hill, J. R., & Hannafin, M. J. (2001). Teaching and learning in digital environments: The resurgence of resource-based learning. *Educational Technology Research and Development, 49*(3), 37–52.

Hipkins, R. (2009). Introduction: Where to now for school science? *SET: Research information for teachers.* Set Reprints: Science 2009.

Hipkins, R. (2012). *Building a science curriculum with an effective nature of science component* (Working paper, NZCER). Retrieved March 18, 2013, from http://www.nzcer.org.nz/system/files/NOS%20role%20in%20curriculum.pdf.

Jones, A., Buntting, C., Hipkins, R., McKim, A., Conner, L., & Saunders, K. (2011). Developing students' futures thinking in science education. *Research in Science Education.* doi:10.1007/s11165-011-9214-9.

Jules, R., & Conner, L. (2009). Nature of science in 'Science in the National Curriculum of Seychelles': Recommended policy and practice changes. *Pacific-Asian Education Journal, 21*(1), 19–34.

Kaser, L., & Halbert, J. (2012). *Spirals of inquiry.* Victoria: BCPVPA.

Knezek, G., Lai, K. W., Khaddage, F., & Baker, R. (2011). TWG 2: Student technology experiences in formal and informal learning. *International summit on ICT in education* (*EDUSUMMIT 2011*).

Korpela, S. (2008). *This is Finland: Success of Finnish school children indicates society's values.* Retrieved April 2, 2013, from http://finland.fi/public/default.aspx?contentid=160131.

Krajcik, J., & Mamlock-Naaman, R. (2006). Using driving questions to motivate and sustain student interest in learning science. In K. Tobin (Ed.), *Teaching and learning science: A handbook* (pp. 317–327). Lanham: Rowman and Littlefield Education.

Lan, O. S., & Gonzalez, E. J. (Eds.). (2012). *TIMMS 2007: What can we learn? A collection of educational experiences.* Penang: Southeast Asian Ministers of Education Organisation Regional Centre for Education in Science and Mathematics.

Lederman, N. (2009). Nature of science: Past, present, and future. In S. Abell & N. Lederman (Eds.), *Handbook of research on science education* (pp. 831–879). Mahwah: Lawrence Erlbaum.

Lee, S. W. Y., Tsai, C. C., Wu, Y. T., Tsai, M. J., Liu, T. C., & Hwang, F. K. (2011). Internet-based science learning: A review of journal publications. *International Journal of Science Education, 33*(14), 1893–1925.

Lemke, J. (1992). *The missing context in science education: Science.* American Educational Research Association Annual Meeting. Atlanta, April, 1992.

Linn, M. C., Clark, D., & Slotta, S. (2004). WISE design for knowledge integration. *Science Education, 87*(4), 517–538.

Loucks-Horsley, S., Stiles, K. E., Mundry, S., Love, N., & Hewson, P. (2010). *Designing professional development for teachers of science and mathematics* (3rd ed.). Thousand Oaks: Corwin/Sage.

Loughran, J. J., Milroy, P., Berry, A., Gunstone, R., & Mulhall, P. (2001). Documenting science teacher pedagogical content knowledge through PaP –eRs. *Research in Science Education, 31*, 289–307.

Loughran, J. J., Berry, A., & Mulhall, P. (2007). Pedagogical content knowledge: What does it mean to science teachers? In R. Pinto & D. Couso (Eds.), *Contributions from science education research* (pp. 93–105). Dordrecht: Springer.

Marshall, J. M. (2002). *Learning with technology: Evidence that technology can, and does, support learning* (A white paper prepared for Cable in the Classroom). Retrieved March 26, 2013, from http://www.dcmp.org/caai/NADH176.pdf.

Ministry of Education, Singapore. (2012). *Education in Singapore*. Retrieved March 26, 2013, from http://www.moe.gov.sg/about/files/moe-corporate-brochure.pdf.

Mishra, P., & Koehler, M. J. (2006). Technological pedagogical content knowledge: A framework for teacher knowledge. *Teachers College Record, 108*(6), 1017–1054.

Mok, I. A. C. (2006). Shedding light on the East Asian learner paradox: Reconstructing student-centeredness in a Shanghai classroom. *Asia Pacific Journal of Education, 26*(2), 131–142.

Mourshed, M., Chijioke, C., & Barber, M. (2010). *Education: How the world's most improved school systems keep getting better*. London: McKinsey & Company.

Neiss, M. (2012). Teacher knowledge for teaching with technology: A TPACK lens. In R. N. Ronau, C. R. Rakes, & M. L. Neiss (Eds.), *Educational technology, teacher knowledge and classroom impact: A research handbook on frameworks and approaches* (pp. 1–15). Hershey: IGI Global.

Norris, S. P., & Ennis, R. H. (1989). *Evaluating critical thinking: The practitioners' guide to teaching thinking series*. Pacific Grove: Critical Thinking Press.

OECD. (2011). *Lessons from PISA for the United States: Strong performers and successful reformers in education*. Paris: OECD Publishing.

Ornek, F., & Saleh, I. M. (Eds.). (2012). *Contemporary science teaching approaches: Promoting conceptual understanding in science*. Charlotte: Information Age.

Osborne, J., & Hennessy, S. (2003). *Literature review in science education and the role of ICT: Promise, problems and future directions* (FUTURELAB SERIES Report 6).

Owusu, K., Conner, L., & Astall, C. (2015). Contextual influences on science teachers' TPACK levels. In M. Neiss & H. Gilloe-Wiles (Eds.), *Handbook of research on teacher education in the digital age*. Hershey: IGI Global.

Park, Y. (2011). A pedagogical framework for mobile learning: Categorizing educational applications of mobile technologies into four types. *The International Review of Research in Open and Distance Learning, 12*(2), 78–102.

Pintrich, R., Marx, R. W., & Boyle, R. A. (1993). Beyond cold conceptual change: The role of motivational beliefs, and classroom contextual factors in the process of conceptual change. *Review of Educational Research, 63*(2), 167–199.

Rao, S. M. (2011). When teachers become classroom researchers. *ReEd (Research in Education)*, 3(9). Retrieved from http://www.nie.edu.sg/files/oer/OER-NIE-ReEd3_Final%20for%20Web.pdf

Rodrigues, S. (Ed.). (2010). *Multiple literacy and science education: ICTs in formal and informal learning environments*. Hershey: IGI-Global.

Roschelle, J. (2003). Unlocking the learning value of wireless mobile devices. *Journal of Computer Assisted Learning, 19*(3), 260–272.

Roth, W.-M. (1995). *Authentic school science: Knowing and learning in open-inquiry science laboratories*. Dordrecht: Kluwer Academic Publishers.

Roth, W.-M. (2006). Learning from laboratory activities. In K. Tobin (Ed.), *Teaching and learning science: A handbook* (pp. 51–60). Lanham: Rowman and Littlefield Education.

Roth, W.-M., van Eijck, M., Reiss, G., & Hsu, P.-L. (2008). *Authentic science revisited: In praise of diversity, heterogeneity, hybridity*. Rotterdam: Sense.

Schleicher, A. (Ed.). (2012). *Preparing teachers and developing school leaders for the 21st century: Lessons from around the world.* Paris: OECD Publishing. http://www.oecd.org/site/eduistp2012/49850576.pdf

Shulman, L. S. (1986). Those who understand: Knowledge growth in teaching. *Educational Researcher, 15*, 4–14.

Slykhuis, D. A., & Krall, R. M. (2012). Successful implementation of technology into science: Research implications. In R. N. Ronau, C. R. Rakes, & M. L. Neiss (Eds.), *Educational technology, teacher knowledge and classroom impact: A research handbook on frameworks and approaches* (pp. 271–294). Hershey: IGI Global.

So, W. W. M. (2012). Creating a framework of a resource-based e-learning environment for science learning in primary classrooms. *Technology, Pedagogy and Education, 21*(3), 317–335.

So, M. W. W., & Ching, F. N. Y. (2012). Online resource-based learning environment: case studies in primary classrooms. *Asia-Pacific Forum on Science Learning and Teaching, 13*(2), Article 10.

Sweller, J., Ayres, P., & Kalyuga, S. (2011). *Cognitive load theory*. Dordrecht: Springer.

Tan, C. (2006). Creating thinking schools through 'knowledge inquiry': The curriculum challenges for Singapore. *Curriculum Journal, 17*(1), 89–105.

Tasker, R., & Freyberg, P. (1985). Facing mismatches in the classroom. In R. Osborne & P. Freyberg (Eds.), *Learning in science: The implications of children's science*. Auckland: Heinemann.

Thomas, G. P. (2012). The metacognitive science teacher: A statement for enhanced teacher cognition and pedagogy. In F. Ornek & I. M. Saleh (Eds.), *Contemporary science teaching approaches: Promoting conceptual understanding in science* (pp. 29–53). Charlotte: Information Age.

Timperley, H. (2011). *Realizing the power of professional learning*. Maidenhead: Open University Press.

Tobin, K. (1990). Research on science laboratory activities: In pursuit of better questions and answers to improve learning. *School Science and Mathematics, 90*, 403–418.

Tobin, K. (Ed.). (2006). *Teaching and learning science: A handbook.* Lanham: Rowman and Littlefield Education.

Treagust, D. F. (2007). Research-based innovation units for enhancing student cognitive outcomes and interest in science. In R. Pinto & D. Couso (Eds.), *Contributions from science education research* (pp. 11–26). Dordrecht: Springer.

Tytler, R. (2007). *Re-imagining science education: Engaging students in science for Australia's future*. Camberwell, Australia: Melbourne Australian Council for Educational Research.

Valijarvi, J. (2003). The system and how does it work: Some curricular and pedagogical characteristics of the Finnish comprehensive school. *Education Journal, 31*(1), 31–55.

Webb, M. E. (2005). Affordances of ICT in science learning: Implications for an integrated pedagogy. *International Journal of Science Education, 27*(6), 705–735.

Webb, M. E. (2008). Impact of IT on science education. In J. Voogt & G. Knezek (Eds.), *International handbook of information technology in primary and secondary education* (pp. 133–148). London: Springer.

White, B., & Frederiksen, J. (1998). Inquiry, modeling, and metacognition: Making science accessible to all students. *Cognition and Instruction, 16*(1), 3–118.

Chapter 17
Science Education in a Future World

Ben Akpan

17.1 Introduction

In January of 1998, I attended the Presidential Address of the Association for Science Education (ASE) at its Annual Meeting in the University of Liverpool in the UK. It was given by Mr David Brown, then Chairman of Motorola Ltd. Mr Brown chose to speak on the topic 'Expanding Horizons by Shrinking Worlds' (Brown 1998), in which he looked at developments in information and communications technologies (ICT), at what could be expected to happen as the world moved into the new millennium, at the implications of these changes for the society and at what, as a consequence, needed to be done to prepare young people for a life that could be different in important ways from the one people had already experienced. With respect to the future of ICT, he identified four trends: ICT will get cheaper and more powerful, networks will increasingly carry information other than speech, the user's physical location will matter less and less and the relationship between systems and users will greatly improve. According to him, the implications of the changes were breathtaking as he envisaged a world where the disadvantages of distance and time would disappear as it would not matter whether one was abled or disabled, black or white, male or female, rich or poor and working in a company or at home. Information, Mr Brown posited, would become a universal resource. Just 17 years have gone past since the lecture and we all can gauge where we are now vis-a-vis the predictions. I will return to Mr Brown's address later.

In this closing chapter, I take a closer look at the place of science education in a future world. What will (or could) science education be like, say in 2065, 50 years hence? In doing this, I will first look at the concept of the future, examine some insights into the future of science education from some veteran science educators

B. Akpan (✉)
Science Teachers Association of Nigeria, Abuja, Nigeria
e-mail: ben.b.akpan@gmail.com

B. Akpan (ed.), *Science Education: A Global Perspective*,
DOI 10.1007/978-3-319-32351-0_17

and discuss the emerging themes from the various insights. This chapter will end with a consideration of some initiatives for implementation looking ahead for 50 years.

17.2 Futurology

According to the *Oxford Advanced Learner's Dictionary*, the future refers to the time that will come after the present or the events that will happen then. The entire human race is indeed preoccupied with the future. From simple daily chores of an individual to complex setups in large establishments, the future is always the target. According to Braun (1969, p 34):

> *If our intention had been merely to bring back a handful of soil and rocks from the lunar gravel pit and then forget the whole thing, we would certainly be history's biggest fools. But that is not our intention now—it never will be. What we are seeking in tomorrow's [Apollo 11] trip is indeed that key to our future on earth. We are expanding the mind of man. We are extending this God-given brain and these God-given hands to their outermost limits and in so doing all mankind will benefit. All mankind will reap the harvest.... What we will have attained when Neil Armstrong steps down upon the moon is a completely new step in the evolution of man.*

The future has become so important (Akpan 2008) to the extent that an entire academic area, futurology, is devoted to it. Futurology is the science, art and practice of postulating possible, probable and preferable futures and the world-views and myths that underlie them, according to *Wikipedia*. It is sometimes referred to as futures, futures studies or foresight. Futurology seeks to understand what is likely to continue, what is likely to change and what is novel. Part of the discipline thus seeks a systematic and pattern-based understanding of the past and present and a basis to determine the likelihood of future events and trends.

Wikipedia contends that an assumption in futures studies is that the future is plural not singular: there are alternative futures of varying likelihood although it is impossible in principle to say with certainty which one will occur or whether they are mutually exclusive. Thus the focus in futurology is on alternative futures, and this may involve collecting various data about the possibility, probability and desirability of change. However, among the various alternatives, there may be preferred futures. These are also called normative futures. Interestingly, futurology is frequently applied in various areas of science as shown in the example in Box 17.1.

And as Thomas Henry Huxley (in Newman 2000; p. 239) has said:

> *There are some men who are counted great because they represent the actuality of their own age, and mirror it as it is. Such a one was Voltaire, of whom it was epigrammatically said: 'he expressed everybody's thoughts better than anyone.' But there are other men who attain greatness because they embody the potentiality of their own day and magically reflect the future. They express the thoughts which will be everybody's two or three centuries after them. Such a one was Descartes.*

This chapter is about the 'Descartes' in science education.

Box 17.1: The End of the Universe
Some astronomers believe that there may be enough matter in the Universe to cause its current expansion to slow down and stop. If they are correct, we are living in a Universe destined to collapse in the far future. Towards the end of this collapse, galaxies will collide and tear apart and later on, even individual stars will collide. The Universe will become a hot dense gas at temperatures of over a billion degrees. Cosmologists (scientists who study the events of the early Universe) call this the 'Big Crunch' to distinguish it from the Big Bang.

The Big Crunch will only happen if there is enough mass in the Universe for the gravity of all its parts to overcome the current expansion speeds of all the galaxies.

It doesn't seem that this is very likely since astronomers can only find about 10 % of the mass needed to cause the Universe to collapse. But don't worry! Cosmologists confirm that this would not happen for another 30 billion years or so.

Source: Jones (2002 in Akpan 2008 p. 6)

17.3 Science Education in 2065

In seeking to look at the place of science education 50 years from now, I asked some of the contributors to this book to offer their personal views on the matter. In what follows, I present their views:

Comments by Gilbert Onwu
I am stumped for words about what the place of science education is likely to be in the future. Given the pervasive secularization of virtually every aspect of our human condition, I hesitate to hold, subscribe or anticipate a cosy picture for the future of science education. I say this because if I recall correctly from what I read sometime ago, leading 19th century secular humanists had what could be described then as a rosy picture for the future. Through the irresistible advance of science (the industrial revolution) universal education, and free commercial trade, they prophesied, all major human problems would be solved in the twentieth century. They were glad to be living in that dawn of a new era, and hoped that their successors in the next century would live in the full light of the day.

But in spite of the many accomplishments, nineteenth century woes, nightmares and worse persisted into our scientifically, educationally and commercially advanced (twenty first century) times. One is not merely expressing a self-serving sentiment here. One only needs to read what was said in the Executive Summary to the 2011 State of the Future document by Glen, Gordon and Florescu (2011), in order to get a sense of what it takes to perhaps be a far better prophet than the secular establishment has so far demonstrated.

Glen, Gordon and Florescu (2011) state:

> The world is getting richer, healthier, better educated, more peaceful and better connected and people are living longer, yet half the world is potentially unstable. Food prices are rising, water tables are falling, corruption and organized crimes are increasing, environmental viability for our life support is diminishing, debt and economic insecurity are increasing, climate change continues, and the gap between the rich and poor countries continues to widen dangerously (p.1)

They further contend: “There is no question the world can be better than it is -IF we make the right decisions (p.1)”.

Clearly the ability to make those decisions and moral choices will largely be influenced by the nature and type of science education – education through science or science through education – experienced by the young ones – the learners of today and in the future.

We envisage science education programmes – education in science projects – that support partnership-based academic cooperation between researchers and institutions, within and across developing countries and their partners in the more developed world focusing on (i) research for the service of humanity (in what could be construed as a socially-responsible science education which is relevant to national, and personal development and poverty reduction), (ii) education, (iii) capacity building and sustainability, and (iv) institutional development and consolidation.

As a matter of interest, when we interviewed students involved in a collaborative South-South and South–North science education research project to give us a sense of what a socially-responsible science education meant to them, one of the doctoral students responded: “Learners shouldn’t go through science education and come out as if unaffected, they should be able to transform their thinking, they should be able to transform their lives, their attitude and their approach to life”.

Let us hope then that the place or role of science education in the future is one that will of necessity inspire initiatives oriented toward morally- and socially-responsible science education.

Gilbert Onwu, University of Pretoria, RSA June, 2013

Comments by Jack Holbrook

It is reasonable to assume that the future pace of developments in the field of Science and Technology is unlikely to slow. And assuming the desire and the will exists for systems within countries to develop, thereby raising the standard of living, or quality of life of their citizens, the need for new technologies will become increasingly important. These technologies are likely to be geared to facilitating manufacturing processes, to support better health, increased or more diverse food supplies, or enabling us to better handle the environment. All can be expected to play an increasing role. And alongside this, the competition between countries to be at the forefront of such developments can be expected to be an increasingly valuable asset.

The above surely points to the need for a future where science education plays a role in promoting scientific literacy, or to be more specific, scientific, technological and engineering literacy, where science related to the way of thinking geared to the natural and technologically-enhanced world, technology relates to the products meeting human needs and desires, while engineering refers to the processes involved in utilising the science to create the technologically-desired products. Alas, this scientific literacy is predicted to have little to do with acquiring science content. While such content will continue to increase exponentially, acquisition of smaller and smaller components will have little meaning and greater acquisition will become untenable through limited 'time on task' by students. Thus, as wide coverage of content within a school science course becomes increasingly unattractive, more and more attention can be expected to be geared to acquiring the ability, or better still, the capability, to transfer knowledge, skills and values to new situations. And here, the need for creativity, for adaptability and the need to strive for self-development, self-actualisation and self-efficacy (the so-called 21st century skills) can be expected to play an increasingly important role.

With such a vision for the future, it is uncertain how far subject-oriented sub-divisions of education can be expected to be meaningful. Scientific literacy (expressed as also encompassing technology/ engineering interactions and viewed in terms of 21st country skills), can be expected to increasingly interrelate with the needs of employers, the needs of an increasingly sophisticated society and for meeting the desires of future citizens to be more aware of the world. And as such, the importance of 'scientist-focused' learning is inevitably reduced and a wider education thrust, based on meaningful, student-relevant science ideas, takes its place. And with such a vision, subjects at the school level can be predicted to be more interdisciplinary and even merge so that conceptual knowledge is explored across subject boundaries and solicited as and when needed.

Nevertheless, within such a vision, the need for scientific literacy can be predicted to be high and in this sense, science education is predicted to remain an important education component, enabling students to interact with the unknown and cope with decision-making in socio-scientific/ technological areas. The goal can be predicted to be, to enable citizens to gain a range of attributes so as to play their role in society in such aspects as sustainable development, poverty alleviation and through the development of entrepreneurial skills and, especially including, creativity to deal with new situations and challenges.

The danger in such a vision is likely to be the pace of teacher change in recognising the diminishing role of science content acquisition and the increasing role of transfer skills, problem-solving skills and attributions such as personal interactions, and informed decision making.

Jack Holbrook, University of Tartu, Estonia, June, 2013

Comments by John Oversby

Does this mean what will science education be like or what could it be like? Since the first could be rather depressing, I will choose the latter. Just before the end of the 20th century, I gave an interview to the Lagos Times in Nigeria on what would 21st century science education be like. It was radical and quite different from the traditional science we have today. This paper is an interpreted and developed version of that interview.

We are presently in a tug-of-war between science education for the future scientists, and science education for all. The major focus is on content, high level for future scientists but still about stuff to be remembered, and if possible, understood. Science for all appears to be a subset of science for the future scientists, stripped of quantitative content since it is assumed that only future scientists either require this, or understand it. Pedagogy remains, on the whole, traditional and transmissive, concerned with the teacher as the expert scientist ('sage on the stage'), with the learners dependent on the speed of teaching the whole class, a lock-step approach reminiscent of military marching.

The scientific content could shift dramatically towards that science that will give more power to learners to have more control over their own lives and become active citizens. My vision is a content curriculum that will focus more significantly on life sciences. A major part of this would focus on the human body, exploring a variety of functions such as reproduction (within a framework of positive and loving relationships), and becoming and staying healthy by recognising the value of food and exercise. Our part in our environment, human ecology, is essential to sensitise ourselves to the value of a sustainable community, including travel. Human ecology is strongly related to ecology in general, the importance of ecosystems with their energy and feeding relationships. Humans have a substantial impact on their environment as individuals and communities, which need to be explored. Food and water supplies are central but subject to territorial disagreements, sometimes violent, especially when these supplies are not readily available. I see a role for life sciences as an integrating area, one in which both rights and responsibilities are played out. Perhaps, life sciences could constitute half of the content curriculum for all.

Physical sciences could make up around a quarter of the content. It is likely, given changes in fuel supply in 50 years' time that energy could make up a large fraction of the content, providing understanding of human energy demands as well as underpinning issues of societal mobility. To provide space for such an increase in time given to energy, I see a decrease in time given to traditional themes such as magnetism, and light and other waves. Chemistry and energy, from burning fuels to electricity production would form an application of energy and relate back to human food input.

Earth sciences and astronomy fascinate and engage adults almost more than any other area of sciences, so must form part of the content curriculum.

We will know much more about climate science, which in 50 years' time will have a more obviously impact on human existence. Its contribution to scientific understanding facilitating knowledge about uncertainty, risk and modelling will be one of the planks of the theme of modelling that is part of how science works. Not all scientific exploration will consist of traditional investigations in laboratories, with tightly controlled variables and accurate measurements. Computing capacity will enable individuals and groups of learners to construct micro-worlds to try out virtual experiments. A dramatic improvement in inter-connection will permit access to live data such as weather information collected all over the world, atmospheric composition especially pollutants, pollutants in rivers and oceans monitored by accurate sensors and stored in various 'clouds' available to all. Global collaboration by learners working in schools, homes and the field, will promote broader perspectives on data analysis. This is essential if science education is to promote citizenship.

The role of history of science in science education today is very limited. Access to source information through digitising such information could allow some to create their own view of historical scientific events. For others, dramatic productions, using the best of realistic animations, may reveal insights into how ideas changed in the past.

I have paid a lot of attention to content and to issues of the nature of science, such as modelling, and how this may change over the next 50 years. The image of a classroom is the next idea for consideration. It is likely that some learning will take place in what are presently traditional schools. Whether that involves learners of the same age learning together, or of classes drawn from the community with similar motivation and stage of learning, is questionable. Life-long learning will be well-embedded, in my view, with young learners learning alongside adults of different ages. Their groups may be small, or large, seminar-type sessions from experts, perhaps available in recorded form, and symposia for discussion in synchronous or asynchronous mode. Today's virtual conference and communication software such as Skype will seem archaic as learning takes place in regions and internationally as well as in local communities. Language difficulties may well reduce as on-line translation becomes available, with voice recognition as we cannot imagine today to make records available instantly.

Disabled learners are often excluded from science education. The transformation of aids should mean that a greater range of the community will be involved in the learning process. Of course, cameras and sensors will be available for data collection but the development of physical aids that can be manipulated by small movements of eyes or hands will mean the integration of many more in the learning process.

Visualisation, perhaps with much more powerful software tools, broadens the range of communication methods, especially between groups whose different languages can be a barrier.

Now, we should note that science education development also needs research. Perhaps freeing teachers from the tedium of 'delivering' relatively straightforward material will release more to take part in the research process as high level professionals. Not only will we see an explosion of research evidence, but the development of new research methods. From this distance it is almost impossible to tell what these new methods might be.

I have left until last the consideration of the science elite, those who need science for future employment. Many of the points I have made earlier will also apply to their education. Perhaps there will be a transformation of universities, with learners of different ages and stages, benefitting from the flexibilities I have mentioned. We should see future scientists as only one kind of career that they may take up, using science education to flip and change direction. They can decide whether to develop within a chosen career or to change direction completely.

John Oversby, University of Reading, UK, June, 2013

Comments by Keith S Taber

It is either brave or foolish to predict 50 years in the future. 50 years ago I was about to start school. At that time many would have suggested that schooling would change beyond recognition in 50 years – would we still have classrooms of children sitting at tables directed by a single adult teacher – or would the teacher have been largely replaced by the technology of programmed learning? Yet, in many ways schools now are as they were then – at least in the most fundamental sense. Technology has certainly changed, and in many parts of the world has changed how we go about teaching and learning – but at the same time our understanding of the processes of teaching and learning have developed and the importance of the dialogue between teacher and students, and among students, has made us realise the affordances of the standard classroom organisation. Perhaps in 50 years, students will sit at home at computer terminals to attend science classes – but only if the technology allows the virtual classroom to have the interactivity of a real classroom. Dialogue is not only central to effective learning, but at the core of the scientific process itself, where the arbiter of the worth of a scientific idea is the community and any new idea must win its place through reasoned, evidenced argument that persuades the community.

The 'technology' we should expect to change concerns pedagogy itself. We increasingly understand more about the nature of learning processes; more about the way learners come to think about particular science concepts; and more about the way that students' existing ideas and ways of thinking interact with instruction to shape their learning. In fifty years' time, teachers should be much better supported to know how to present key scientific ideas to help learners better understand scientific models and concepts. This will, however, depend upon the quality of teachers. There is a danger of a great rift between those countries where teachers are highly qualified and well-prepared for their classroom work and those where available public resources only allow school teachers to have minimal post-school qualifications and limited pre-service preparation.

Will science appear on the school curriculum in 50 years' time? One general trend in the last 50 years has been an increased recognition of the importance of science education – firstly as a means of supporting national economic growth by providing scientists and engineers, and more recently in terms of its role in education for citizenship, and in particular in supporting the viability of democracies that come to make sensible choices in a world facing potential crises. If there will be a structured, civilised, largely peaceful world to go to school in fifty years hence, then this will be largely because of decisions made over the coming decades about such key scientific and technological issues as environmental protection, sustainable production, and renewable energy sources.

There are arguments about how science is best presented in the curriculum – as an integrated subject, as separate disciplines, combined with technology – or even more widely integrated with other curricular areas. There will always be arguments about the most essential topics and concepts to include – and it is in the nature of science that quite possibly new concepts or applications that may come to be seen as central to a science curriculum that we do not know anything about yet. That which is core to science education today, and will continue to be in the future, is to be found in the nature of what science is, and in the particular contribution to human experience and culture that science can offer as part of a broad 'liberal' education.

Science is about enquiry into aspects of the world, to develop better ways of understanding our (physical, natural, and social) environment – to support our action in the world from a firm base of evidence-supported knowledge. When science is understood this way, there is no sharp distinction between the natural sciences, and other areas of enquiry that utilise the scientific approach to the interplay of evidence and theory – as long as they also adopt the underlying commitment to the possibility of being able to build viable objective models of aspects of the world. What is essential is the disciplined and coordinated application of two complementary ways of thinking – creativity and logical analysis – and the adoption of scientific values of curiosity; subjecting all claims and arguments to critique; respect for evidence; and a skepticism that treats all knowledge claims as in principle provisional and open to review in the light of new evidence. Whatever changes occur in the world over the next 50 years, whatever technology is adopted to support teachers, and whatever scientific topics may appear in the curriculum, a successful science education will work to develop the values and thinking skills that support the scientific attitude.

So, how do I see the science classroom in 50 years' time? I do not know what form it will take or what will be included in its curriculum; but I very much hope and expect that it will remain a highly interactive forum, which celebrates the educational value of dialogue, with teachers who prioritise instilling the scientific values and key thinking skills as they approach the teaching of particular concept areas well informed by the outcomes of a further half-century of science education research.

Keith S Taber, University of Cambridge, UK June, 2013

Comments by Sudhakar C Agarkar

In the recent past, there has been a noticeable change in our understanding of how children acquire knowledge. Instead of collecting, it is now believed that, students construct their knowledge. At the same time there have been changes in social structure. The industrial society has slowly changed to knowledge society. The demands of this new society are drastically different from the earlier one. Coupled with these changes, there has been an unprecedented growth in Information and Communications Technology (ICT). These changes have influenced the way science is taught in schools and colleges.

Due to the changes mentioned above, all the three areas of science education namely, the curriculum aims, curriculum characteristics and pedagogy have been influenced substantially. Earlier science curricula were framed to equip students with the necessary knowledge and skills to survive in a local community or meet the manpower needs of an industrial society. These aims have now changed to developing students as problem-solvers and citizens of a modern society capable of using their intelligence in specific contexts. In order to fulfil these objectives, there is a shift in curriculum characteristics too. Earlier we were happy to follow subject-focused curriculum where subject-specific characteristics were highlighted. Now, however, the characteristics need to bring out multiple intelligence skills among the students and to enable them to become global citizens. Along with curriculum aims and curriculum characteristics one witnesses drastic changes in pedagogy – the way the curriculum is taught in classrooms. Earlier the interaction was time-bound and space-bound. Now, with the advent of technology, these restrictions have gone. Earlier, a student had to depend on a teacher for getting relevant information and knowledge. Now, however, there are boundless opportunities available for getting both information and knowledge.

From the foregoing discussion, it would be clear that science education is at the crossroads and will experience significant changes in the near future. The present picture of science education is that a teacher is information provider and a student is information receiver. This approach will no longer be true. The student will have to get knowledge through different means making use of the fact that the present society is a connected society. There is a Sanskrit saying that gets translated as "We get knowledge from four sources: about one-fourth from teachers, one-fourth from experiences, one-fourth from peers and one-fourth from our own efforts". Students will have to explore all the four sources to become knowledgeable citizens. For that, they will have to become life-long learners and acquire technological skills as and when they get developed. In order to survive in the competitive society, they will have to be competent in choosing appropriate information by filtering the unnecessary background and learning the skill of sharing information with friends, colleagues, superiors, and assistants. It will be the responsibility of schools and colleges to fulfil these demands.

Sudhakar C Agarkar, Homi Bhabha Centre for Science Education
Tata Institute of Fundamental Research
Mumbai, India July, 2013

Comments by Lindsey N Conner

I think it would be great if science education could promote the following:

- Science for social responsibility
- Science for enhancing human experience
- Science for technological innovation

These are not mutually exclusive and the challenge is to consider how the emphasis in science education integrates these concepts. The format and form that learning takes may be similar to or quite different from what we advocate in today's schools.

I think we will be talking more about learning environments as potential sites and places of learning, linked to what Gruenwald talks about as place-based education. The virtual space will be developed and explored as a safe experimental space. We are already seeing this with game-based technologies that enable learners to try out scenarios and experimental designs. Informatics and contingency logistics will be built into these digital experimental spaces. Students are likely to be able to collaborate in this space (as indicated in science fiction movies such as the Matrix or Inception where learners can co-create in virtual learning spaces). The challenge for science education in the future is mainly related to assessment. How do we assess the outcomes here? That is, how do you assess whether they have taken responsibility for actions and decisions related to the development and use of science and technology? Learners are likely to be co-creators of science knowledge and skills in the future.

Lindsey N Conner, University of Canterbury
New Zealand July 2013

17.4 Emerging Themes

Arising from the above views by eminent science educators are some themes that point to the place of science education by 2065. In this section, an attempt is being made to review those themes:

(a) *Global economic competitiveness*: Several reports including one by OECD have predicted a realignment of the global economy in the next 50 years. For more detailed report on the OECD report, please visit http://www.oecd.org/economy/outlook/lookingto2060.htm.

In its report, the PricewaterhouseCoopers (PwC 2013) made the following findings:

- The world economy is projected to grow at an average rate of just over 3 % per annum from 2011 to 2050, doubling in size by 2032 and nearly doubling again by 2050.
- China is projected to overtake the USA as the largest economy by 2017 in purchasing power parity (PPP) terms and by 2027 in market exchange rate terms. India should become the third 'global economic giant' by 2050, a long way ahead of Brazil, which is expected to move up to fourth place ahead of Japan.
- Russia could overtake Germany to become the largest European economy before 2020 in PPP terms and by around 2035 at market exchange rates. Emerging economies such as Mexico and Indonesia could be larger than the UK and France by 2050 and Turkey larger than Italy.
- Outside the G20, Vietnam, Malaysia and Nigeria all have strong long-term growth potential, while Poland should comfortably outpace the large Western European economies for the next couple of decades.

Table 17.1 Actual and projected top 20 economies ranked based on GDP in PPP terms

2011			2030		2050	
PPP rank	Country	GDP at PPP (2011 US $bn)	Country	Projected GDP at PPP (2011 US $bn)	Country	Projected GDP at PPP (2011 US $bn)
1	USA	15,094	China	30,634	China	53,856
2	China	11,347	USA	23,376	USA	37,998
3	India	4531	India	13,716	India	34,704
4	Japan	4381	Japan	5842	Brazil	8825
5	Germany	3221	Russia	5308	Japan	8065
6	Russia	3031	Brazil	4685	Russia	8013
7	Brazil	2305	Germany	4118	Mexico	7409
8	France	2303	Mexico	3662	Indonesia	6346
9	UK	2287	UK	3499	Germany	5822
10	Italy	1979	France	3427	France	5714
11	Mexico	1761	Indonesia	2912	UK	5598
12	Spain	1512	Turkey	2760	Turkey	5032
13	South Korea	1504	Italy	2629	Nigeria	3964
14	Canada	1398	South Korea	2454	Italy	3867
15	Turkey	1243	Spain	2327	Spain	3612
16	Indonesia	1131	Canada	2148	Canada	3549
17	Australia	893	Saudi Arabia	1582	South Korea	3545
18	Poland	813	Australia	1535	Saudi Arabia	3090
19	Argentina	720	Poland	1415	Vietnam	2715
20	Saudi Arabia	686	Argentina	1407	Argentina	2620

Source: World Bank estimates for 2011, PwC estimates for 2050

- The projected top 10 economies ranked on the basis of gross domestic product (GDP) in PPP terms in 2050 are China, the USA, India, Brazil, Japan, Russia, Mexico, Indonesia, Germany and France (Table 17.1).
- However, even in 2050, average income per capita will still be significantly higher in the advanced economies than in the emerging economies – the current income gap being just too large to bridge fully over this period.

Overall, the predicted economic outlook for the world in 2065 is an improvement on the present. The corollary is that science education will play a pivotal role in the realignment of the economies just as economic conditions of the various countries will affect the level of investment in science education.

(b) *Technology*: At current pace, the level of technology in 50 years' time will be very high.

Technology will change the way we live, the things we do and the ways of doing things, and it is to be expected that science education will progress rapidly to be able to provide capability for the envisaged 'high-tech'. A future science education programme must provide the platform for meeting technological and engineering needs of the time.

(c) *The curriculum*: With changes in other areas such as economy and technology, there will certainly be changes in curriculum to assure relevance. It is very likely that science topics will have to earn their place in the curriculum due to the need to reduce sheer content overload. In the process, much of the traditional topics as we have them at the moment will give way to new topics, thus giving currency and dynamism to school offerings. There will be increased emphasis on topics that have the potential for adding value for human welfare at the time such that in the face of dwindling petroleum energy resources, a higher premium may be placed on capabilities on other sources of energy. Similarly, if robots are deployed in homes, offices and factories, then we should expect some attention on robotics. In addition to earth, life and physical sciences, there will be emphasis on astronomy and space technology. As knowledge moves more towards applications, the need for interdisciplinarity will increase making way for more creativity and critical thinking.

(d) *Pedagogy/teacher preparation*: With changes in content and technology available for teaching and learning, the role of the teacher as we currently know it will be redefined.

Many classrooms of the 2060s will have virtual libraries for easy access by students. Due to advancement in society, applications of science and mathematics will be commonplace, and this will greatly aid teaching. With the right type of technology, the implementation of inquiry teaching will be greatly facilitated and creativity and critical thinking will be promoted. The use of 'high-tech' in classrooms will aid cooperative learning which in turn will facilitate the development of interpersonal skills. As students acquire interpersonal skills, they will become more confident and, ultimately, independent in thinking and action. Classrooms will have gadgets that could enable real-time access to countryside and distant objects. Some classrooms may have direct links to museums and space observatories. There will be provision and more enhanced remedial measures for the physically and academically challenged. Overall, the future world classroom will be a cocktail of alternatives, providing ample opportunity for students and teachers to make choices on what, how and when to learn. The result will be a grand change in the role of the teacher and the responsibility of the learner as we currently know it.

(e) *Careers in and through science*: Availability of cutting-edge technology and a change in global technological and economic outlook will mean great emphasis on science, technology, engineering and mathematics (STEM) literacy. Science education will be seen as essential as all persons will require STEM literacy to be able to lead useful lives. It is likely that the curriculum will take adequate care of this unavoidable requirement. In conjunction with this, however, will be the need to cater for those seeking careers in science and related fields. New areas of emphasis in STEM will appear, and it will be the place of science education to ensure adequate and relevant curricula provision are made, both in terms of content and quality of delivery of instruction.

(f) *Science education research*: Much of the ideas for implementation in a science education programme of the 2060s will be drawn from research in science education (see Chap. 16). A vibrant research programme, enabled by advancement in technology with more skills by researchers in a more economically dynamic world, will provide the much-needed basis for decision-making in science education.

(g) *Promoting citizenship/sustainability of communities*: Without doubt, science education in a future world will seek to promote good citizenship through socially responsible programmes. Coupled with this also will be efforts to ensure sustainability of the environment as people will get increasingly conscious of their responsibilities to future generations. Indeed, Mathews (2007) maintains that one area of science education for the future is a transformative programme that links the social and emotional. He believes science education will continue to be integrated with society while maintaining the importance attached to cognitive learning.

17.5 Implications and Conclusion

In this section, I refer again to the address by Brown (1998). In reference to anticipated changes in ICT, he had this to say:

> *Who will make the decisions about the future? Anyone and everyone, of course, but above all, those who will live their adult lives in the 21st century – those for whose education we are responsible. We must ask ourselves, are they prepared for the challenges these developments may bring? And, if not, what should we do about it?* (Brown 1998, p.14)

Then, as now, Brown's statement is apt. In 2065, 50 years from now, the children currently in primary and secondary schools (elementary and high schools) will be spending their adult lives. It is absolutely important that we take responsibility in preparing them towards the future world of our dream. In doing so, we must be persuaded to realise that it is the right thing to do because like a former Governor of Wisconsin and founder of Earth Day, Gaylord Nelson (n.d.) has said, the ultimate test of people's conscience may be their willingness to sacrifice something today for future generations whose words of thanks will not be heard. I also agree with Brown that things happen because we engineer them into being, not of their own accord, and so young people should control the future and not be controlled by it. This is precisely why Dixon (n.d.) urged everyone to take care of their future or the future will take hold of them.

In sum, a look to a future world means we need to take the following initiatives echoed above by the eminent science educators (Gilbert Onwu, Jack Holbrook, John Oversby, Keith S Taber, Sudhakar C Agarkar and Lindsey N Conner):

- Implementation of socially responsible science education programmes – leading to transformation of lives and attitudes, promotion of citizenship and viability of democracies so as to position science, as we know it today, for its immense contribution to human experience and culture as part of the heritage of a broad liberal education

- Promotion of creativity through capability to transfer skills and values to new situations and linking with logical analysis of scientific endeavours
- Movement towards interdisciplinarity such that conceptual knowledge is explored across subject boundaries
- Re-articulation of life sciences with emphasis on healthy living and human ecology thereby making life sciences an integrating hub – guarding rights and responsibilities of individuals and groups and linking to citizenship education as earlier indicated above
- A focus on application of energy from burning fuels to electricity production and relating back to human food input and linking to environmental protection, sustainable production and renewable energy sources
- A focus on emerging technologies for research, learning, teaching and assessment
- Emphasis on the overarching role of dialogue in teaching and learning of science and a demonstrable recognition of science teachers as pivots of any science education programme

It is critical that these steps be taken by all stakeholders. Fifty years will take us far into the future if humanity is prepared. However, we need to realise that the problem with the future is that it keeps turning into the present (Watterson n.d.). And yet, according to Gore (2009), 'the choice is awesome and potentially eternal. It is in the hands of the present generation' (p. 405).

References

Akpan, B. B. (1994). *Implementing science and technology education for all: Guide to better policy and practice for teachers*. London: Commonwealth Secretariat.

Akpan, B. B. (2004). The status of science education in commonwealth Africa. *School Science Review, 86*(314), 71–80.

Akpan, B. B. (2008). *Nigeria and the future of science education*. Ibadan: Science Teachers Association of Nigeria.

Bell, D. (2008). Implementing curriculum change: What does it feel like to you? *Education in Science, 3*(228).

Braun, W. (1969). Of a fire on the moon. *Life, 67*(No.9), 34.

Brown, D. (1998). Expanding horizons by shrinking worlds. *School Science Review, 79*(288), 11–18.

Dixon, P. (n.d.). http://www.globalchange.com/take-hold-of-your-future-or-the-future-will-take-hold-of-you-by-dr-patrick-dixon-futurist.htm. Retrieved November 3, 2015.

Glen, J. C., Gordon, T. J., & Florescu, E. (2011). *2011 State of the Future*. Washington, DC: The Millennium Project. http://www.millennium-project.org/millennium/2011SOF.html. Retrieved 3 Nov 2015.

Gore, A. (2009). *Our choice: A plan to solve climate crisis*. Emmaus: Rodale.

Jones, B. W. (2002). *Life in the universe*. London: Royal Astronomical Society.

Mathews, B. (2007). Science education: Preparing pupils for the future? *School Science Review, 89*(326), 77–84.

Millar, R., & Osborne, J. (Eds.). (1998). *Beyond 2000: Science education for the future*. London: Kings College London, School of Education.

Nelson, G. (n.d.). https://www.goodreads.com/quotes/251497-the-ultimate-test-of-man-s-conscience-may-be-his-willingness?page=2. Retrieved November 3, 2015.

Newman, J. R. (2000). *The world of mathematics* (Vol. 1, p. 239). https://www.google.com.ng/webhp?sourceid=chrome-instant&ion=1&espv=2&ie=UTF-8#q=Newman%2C+J.R.+(2000).+The+World+of+Mathematics%2C+Vol.1%2C+239. Retrieved November 3, 2015.

PwC. (2013). *World in 2050: The BRICs and beyond – Prospects, challenges and opportunities*. London: PwC. https://www.pwc.com/im/en/publications/assets/pwc_world_in_2050_report_january_2013.pdf. Retrieved November 3, 2015.

Rescher, N. (1988). *Predicting the future*. Albany/New York: State University of New York Press.

Silk, J. (2005). *On the shores of the unknown: A short history of the universe*. Cambridge: Cambridge University Press.

UNESCO. (2003, April 23–24). *Future directions for national reviews of science, technology, and innovation in developing countries*. Workshop report.

Watterson, B. (n.d.). https://www.google.com.ng/webhp?sourceid=chrome-instant&ion=1&espv=2&ie=UTF-8#q=the+problem+with+the+future+is+that+it+keeps+turning+into+the+present+(Watterson. Retrieved November 3, 2015.

About the Authors

John Monday Ubada Abah was the Head, Financial Management Unit of STEP-B Project. He holds a BSc Degree in Business Administration from the University of Lagos (1984) and MBA from Ahmadu Bello University, Zaria (1998). John is a Fellow of the Institute of Chartered Accountants of Nigeria (ICAN). He is an associate member of Chartered Institute of Taxation of Nigeria (CITN) and Nigeria Institute of Management (NIM).

Michael Umale Adikwu graduated in 1986 from the Faculty of Pharmaceutical Sciences, University of Nigeria, Nsukka, where he later took up an appointment in 1990 as Lecturer II. He rose to the rank of Professor in 1998. Michael was the National Coordinator of the Science and Technology Education Post-Basic – a World Bank-assisted education project in Nigeria. He is a fellow of the Nigerian Academy of Science as well as an Honorary Fellow of the Science Teachers Association of Nigeria. He is also a Fellow of the Pharmaceutical Society of Nigeria and the Vice Chancellor of the University of Abuja.

Sudhakar C. Agarkar is a Professor and Dean VPM's Academy of International Education and Research, Thane, Maharashtra State, India. He has a long experience of conducting research in science and mathematics education. His basic interest lies in the education of the first generation learners. In this context, he has investigated learning hurdles faced by these students and has designed appropriate remedial measures to overcome them. The Remedial Instructional Strategy developed by Professor Agarkar is found to boost the performance of socially deprived students significantly. For the past decade, he has been engaged in designing open educational resources for students, teachers and parents to improve the teaching of science and mathematics in Indian schools.

Mercy Oluwaseyi Agunbiade was a Monitoring & Evaluation and Education Specialist and as consultant to the World Bank/IDA-funded STEP-B project. She holds a first degree in Economics and advanced degree in Educational Administration.

B. Akpan (ed.), *Science Education: A Global Perspective*,
DOI 10.1007/978-3-319-32351-0

She has 30 years experience at the Federal Ministry of Education in Nigeria where she rose from the position of Education Officer to a Deputy Director of Education.

Ben Akpan is the Executive Director of the Science Teachers Association of Nigeria (STAN). He served as President of the International Council of Associations for Science Education (ICASE) for 2011–2013 and currently serves on the Executive Committee of ICASE as the Chair of World Conferences Standing Committee. Ben's areas of interest include chemistry, science education, environmental education, and support for science teacher associations.

Mónica Baptista is a Professor of the Institute of Education of the University of Lisbon. She earned her PhD in Science Education. She has been involved in research projects related with science education. Presently, she participates in a national and in two European projects focused on IBSE and assessment and on responsible research and innovation issues. Mónica has published and been involved in research about science education in different contexts and teachers' professional development.

Professor Dr Lindsey N. Conner is the Director of the Science and Technology Education Research Group at the University of Canterbury, Christchurch, New Zealand. Before becoming a teacher educator in 1997, she was a scientist for the NZ government and a secondary science and biology teacher. She has since been a national examiner in NZ, a developer of NZ national standards assessments and assessment tasks, President of the New Zealand Association of Science Educators and a consultant on developing science curriculums and participatory approaches to teaching and learning, internationally. She is currently on the executive of the International Council of Associations for Science Education (ICASE). Lindsey's research interests include innovative science pedagogies, futures education and the greater role education can play in enhancing the quality of people's lives.

Paul Adewole Eniayeju (MEd, PhD, Fstan, Dsse) is a Professor of Science Education. He has held, in the past 30 years, both academic and administrative leadership positions in many establishments. Paul is currently Dean of Education, Nasarawa State University, Keffi, Nigeria, where he had also served as Dean of Postgraduate School. His research interests include meta-cognition and spatial relationships in molecular structures. Paul has several published works in books and reputable journals. Between 1985 and 2008, he served the Science Teachers Association of Nigeria (STAN) as Secretary of the Chemistry Panel, Editor-in-Chief and President.

Cláudia Faria is a Professor at the Institute of Education of the University of Lisbon. She has a PhD in Biology and a PhD in Education. She has published and been involved in research about marine biology and science education. Presently, she coordinates a national project aimed at developing and implementing multidisciplinary activities using an integrated approach of different literacies within the classroom and non-formal institutions, such as science centres and museums. Cláudia is also member of a European project centred on science teachers' education in assessment and IBSE.

Sofia Freire is a researcher at the Unidade de Investigação e Desenvolvimento em Educação e Formação of the Institute of Education of the University of Lisbon. She is a psychologist with a PhD in Education. Presently, she is investigating in the area of science education and, particularly, on the use of inquiry activities and its association to learning and competences, attitudes development and motivation. She has participated in several national and international projects focused. Presently, Sofia is involved with two projects focused on inquiry and inquiry assessment.

Cecília Galvão is Full Professor of the Institute of Education of the University of Lisbon and the coordinator of the Science Education research group. She coordinated the team responsible for the documentation of Specific Competences in Physical and Natural Sciences, which integrated the National Curriculum. She was also the coordinator responsible for the document: Curricular Guidelines for Physical and Natural Sciences for the 3rd cycle of Basic Education (12–15 age), implemented in 2002/2003. Cecília coordinated a national project aimed at evaluating the national science curriculum (financed by the FCT national agency) and she integrated the team responsible for the coordination of the Portuguese Learning Outcomes for the national curriculum. She was the Portuguese coordinator of the 2006/2009 European Project PARSEL (Popularity and Relevance of Science Education for scientific Literacy). Currently, Cecília is the Portuguese coordinator of the EU FP7 Project SAILS (Strategies for Assessment of Inquiry Learning in Science) coordinated by Dublin City University (http://sails-project.eu).

Christiane Gioppo holds a PhD in Science Education from the North Carolina State University. She is Faculty Member at the Federal University of Paraná, South Brazil, and the Coordinator of Environmental Education Programme at the Technological Institute of Transportation and Infrastructure (ITTI). Christiane is also the Latin American and Caribbean representative on the Executive Committee of the International Council of Associations for Science Education (ICASE).

Michelle Bocchi Gonçalves holds a MSc in Education from the Federal University of Paraná (UFPR), Brazil. She is a Faculty Member Countryside Teacher Education Programme at the Federal University of Paraná (UFPR). Michelle is a researcher at the group of Language, Science and Teacher Education at the same University. Michelle has experience in Science Education, French Framed Discourse Analysis and Countryside Education.

Jack B. Holbrook is a visiting professor, Centre for Science Education, University of Tartu, Estonia. He trained as a chemistry/mathematics teacher in the University of London and spent 5 years as a secondary school teacher (grades 7–13) before moving into teacher training. This was first in the UK followed by Tanzania, Hong

Kong and Estonia. Currently, Jack is involved in guiding PhD students in the field of science education and European science education projects and acting as an International Consultant in the fields of Curriculum, Teacher Education and Assessment. Jack's qualifications include PhD in Chemistry (University of London), FRSC from the Royal Society of Chemistry (UK) and Past President and Distinguished Award Holder for ICASE (International Council of Associations for Science Education). He is the ICASE representative in the PROFILES project initiated to promote motivation in science education. Jack has written a number of articles in international journals and books and as a co-editor for a book entitled *The Need for a Paradigm Shift in Science Education in Post-Soviet Countries*.

Teresa J. Kennedy is the Executive Director of the Office of International Programs and holds a joint appointment as Professor of Bilingual/ELL/STEM (Science, Technology, Engineering and Mathematics) Education in the College of Education and Psychology and as Professor of Engineering in the College of Engineering at the University of Texas at Tyler, USA. Her research interests include gender equity in education; comparative education; online teaching and learning; content-based second language teaching and learning focused on science, mathematics and social studies; brain research in relation to second language acquisition and bilingualism; and Earth science, space science and engineering education.

Cristiane Vertelino Marques holds a Baccalaureate in Elementary School Licensure. She is a teacher at Marcolino Lili Elementary School at the Aquidauana city School System. She is also a Terena and coordinates all the hardwork done by teachers from Lagoinha Indigenous Village.

Dr Christine V. McDonald is a Lecturer in Science Education at Griffith University, Australia. She received her PhD focused on nature of science and argumentation in 2008, and since that time has published her doctoral research in *Journal of Research in Science Teaching* (2010) and *Second International Handbook of Science Education* (2012). She is a member of the editorial board for *Science Education International* and is the Australia/Asia Pacific Representative for the International Council of Associations for Science Education (ICASE).

Gilbert Onwu is Emeritus Professor at the Faculty of Education of University of Pretoria, South Africa. With a background in Chemistry and Chemical/Science Education, Gilbert Onwu holds MSc and PhD degrees from the School of Chemical Sciences, University of East Anglia, UK. He is an elected Member of the Academy of Science of South Africa and a Fellow of Science Teachers Association of Nigeria. Gilbert is a recipient of the 2005 Chemical Education Gold Medal Award by the South African Chemical Institute and Assayists for outstanding contribution to chemical education based on his published work.

John Oversby is a science educator at the University of Reading, UK. He has had three phases to his career. The first was in chemistry, in the chemical industry as a technician and then at Leeds University. His doctorate was on the effect of radiation on organic chemicals in solids. Following this, after further study, a major part of his career was in school teaching, in schools in Ghana and the UK. He exercised leadership in chemistry, the sciences and ICT, and taught mathematics and politics at secondary level. The second major part of his career focused on teacher education at the University of Reading where, for many years, he led the secondary science pre-service course, in addition to primary science. He developed an interest in modelling in association with Carol Boulter and John Gilbert, extending this to the wider field of visualisation. His other research interests include science teacher education, history and philosophy in science teaching, and climate change education with innovative pedagogy. He has been involved and led European projects, recently the Changing with the Climate Comenius Network. His writing includes various chapters in science education books, academic research papers, and, recently, editing *The ASE Guide to Research in Science Education.* He is an officer for The Royal Society of Chemistry and for the Association for Science Education (as Chair of its Research Committee and other posts).

Robert Peter Devadason, PhD, was seconded from the Ministry of Education Malaysia to SEAMEO RECSAM from 2003 to 2013. While in RECSAM, he served in various capacities; as a senior specialist, Deputy Director and as acting Centre Director for over a year. He was a secondary school science and mathematics teacher as well as teacher educator for 19 years before joining the Centre. Currently, he serves the Ministry of Education, Malaysia, as a senior lecturer at the Teacher Education Institution Ipoh Campus (IPGK Ipoh). While his expertise is in science education with particular emphasis on technology, his managerial experience has, in recent years, provided him with opportunities to look at the centre from a regional perspective.

Miia Rannikmae is a Professor and Head of the Centre for Science Education, University of Tartu. She was a chemistry teacher for more than 20 years but also taught at the University of Tartu in the chemistry and philosophy departments before joining the Centre for Science Education. Miia has a PhD in science education, a Master's in Education and a BSc in Chemistry and has recently been awarded an honorary doctorate at the University of Joensuu, Finland. She is involved in teaching Master's and PhD students undertaking degrees in science education. Miia has written a number of articles in international journals associated with scientific literacy, the nature of science and relevance in science education and is a co-author for the book entitled *The Need for a Paradigm Shift in Science Education in Post-Soviet Countries*.

Steven S. Sexton is a Senior Lecturer at the University of Otago's College of Education. He is currently the Master of Teaching and Learning (MTchgLn) programme coordinator. His research areas are in perceptions of teachers and teaching, education through a science context and masculinity in schools.

Cheryl W. Sundberg, Ronin Institute (United States) Research Scholar, is an independent scholar with a substantial portion of her current scholarship centred on the use of Web 2.0 to facilitate science learning at all levels. Her research interests include interactive online teaching, teaching with emerging technology and science teaching.

Dr Keith S. Taber is a Professor in Science Education at the University of Cambridge. He is Chair of the Science, Technology and Mathematics Academic Group in the Faculty of Education. Keith is the editor of the journal entitled *Chemistry Education Research and Practice*, published by the Royal Society of Chemistry. He taught science (mainly physics and chemistry) in secondary schools and further education. Keith undertakes research into aspects of student learning and thinking in science including understanding of science concepts and aspects of the nature of science, as well as conceptual development and integration in science.

Index

B. Akpan (ed.), *Science Education: A Global Perspective*,
DOI 10.1007/978-3-319-32351-0

MIX
Papier aus verantwortungsvollen Quellen
Paper from responsible sources
FSC® C105338

If you have any concerns about our products,
you can contact us on
ProductSafety@springernature.com

In case Publisher is established outside the EU,
the EU authorized representative is:
Springer Nature Customer Service Center GmbH
Europaplatz 3, 69115 Heidelberg, Germany

Printed by Libri Plureos GmbH
in Hamburg, Germany